Methods in Molecular Biology

For further volumes:
http://www.springer.com/series/7651

For over 35 years, biological scientists have come to rely on the research protocols and methodologies in the critically acclaimed *Methods in Molecular Biology* series. The series was the first to introduce the step-by-step protocols approach that has become the standard in all biomedical protocol publishing. Each protocol is provided in readily-reproducible step-by-step fashion, opening with an introductory overview, a list of the materials and reagents needed to complete the experiment, and followed by a detailed procedure that is supported with a helpful notes section offering tips and tricks of the trade as well as troubleshooting advice. These hallmark features were introduced by series editor Dr. John Walker and constitute the key ingredient in each and every volume of the *Methods in Molecular Biology* series. Tested and trusted, comprehensive and reliable, all protocols from the series are indexed in PubMed.

Zebrafish

Methods and Protocols

Third Edition

Edited by

James F. Amatruda

Department of Pediatrics Children's Hospital Los Angeles; Keck School of Medicine, University of Southern California, Los Angeles, CA, USA

Corinne Houart

Centre for Developmental Neurobiology, MRC Centre for NeuroDevelopmental Disorders, King's College London, London, UK

Koichi Kawakami

Laboratory of Molecular and Developmental Biology, National Institute of Genetics, Mishima, Shizuoka, Japan

Kenneth D. Poss

Duke Regeneration Center, Department of Cell Biology, Duke University Medical Center, Durham, NC, USA

Editors
James F. Amatruda
Department of Pediatrics Children's
Hospital Los Angeles
Keck School of Medicine
University of Southern California
Los Angeles, CA, USA

Corinne Houart
Centre for Developmental Neurobiology
MRC Centre for NeuroDevelopmental Disorders
King's College London
London, UK

Koichi Kawakami
Laboratory of Molecular and
Developmental Biology
National Institute of Genetics
Mishima, Shizuoka, Japan

Kenneth D. Poss
Duke Regeneration Center
Department of Cell Biology
Duke University Medical Center
Durham, NC, USA

ISSN 1064-3745 ISSN 1940-6029 (electronic)
Methods in Molecular Biology
ISBN 978-1-0716-3400-4 ISBN 978-1-0716-3401-1 (eBook)
https://doi.org/10.1007/978-1-0716-3401-1

Cover Illustration Caption: Image created by Richard Taylor, King's College London.

This Humana imprint is published by the registered company Springer Science+Business Media, LLC, part of Springer Nature.
The registered company address is: 1 New York Plaza, New York, NY 10004, U.S.A.

Paper in this product is recyclable.

Preface

2021 marked two important milestones in the zebrafish field. The first was the 25-year anniversary of the seminal 1996 publication of a special issue of *Development*, describing the fruits of the first large-scale zebrafish mutagenesis screens in Tübingen and Boston, led by Wolfgang Driever, Christiane Nüsslein-Volhard, and Mark Fishman. The second marked 40 years since the pioneering work of George Streisinger, reporting the generation of homozygous clones of zebrafish and launching the use of zebrafish as a versatile and powerful tool for vertebrate genetics.

2021 also marked five years since the publication of the previous volume in this Methods in Molecular Biology series: *Zebrafish: Methods and Protocols, Second Edition*, edited by Koichi Kawakami, E. Elizabeth Patton, and Michael Orger. Following the 2009 first edition (edited by Graham J. Lieschke, Andrew C. Oates, and Koichi Kawakami), the 2016 2nd edition was organized into three sections: Genetics and Genomics, Disease Models and Mechanism, and Neuroscience, reflecting the rapid growth within subject areas as groups of researchers around the world applied emerging technologies to further broaden the scope of research using zebrafish as a model system.

As the editors of the current 3rd edition of this series began in 2021 to organize the new edition, the scientific community, just as the rest of the world, was feeling the effects of the Covid-19 global pandemic beginning to take hold. For scientists, the pandemic introduced severe challenges including laboratory and institute closures, supply-chain problems that caused significant research delays, the cancellation or postponement of countless seminars and conferences, and of course the human toll on individuals and families dealing with Covid-related illnesses. The chapters in this current volume serve, therefore, as a concrete testament to the resiliency, creativity, and commitment of the scientists who found ways to move the field forward, under the most difficult circumstances, introducing new tools, models, and analytic insights that position the zebrafish even more strongly as an engine of discovery for developmental and disease biology.

The current volume is organized into four sections. Part I, Disease Models, provides detailed methods for use of zebrafish to model a variety of human diseases. Chapter 1 describes a standardized, high-throughput system to create traumatic brain injury in larvae, providing a preclinical model to study the consequences of this prevalent injury type. Chapter 2 addresses the need for rapid functional genomic methods to interrogate the activity of fusion oncogenes, important emerging drivers in a wide range of childhood cancers. Chapter 3 provides detailed protocols for the study of liver disease using larval zebrafish, including injury, fibrosis, and pre-neoplastic changes. Chapter 4 is concerned with toxicology and methods for automated zebrafish developmental high-throughput screening, with emphasis on exposure methods, morphological and behavioral readouts, and quality control. Chapter 5 discusses the use of electroporation to introduce transgenes into adult tissues, allowing expression of oncogenes or genome editing via CRISPR/Cas9 or similar approaches. Chapter 6 describes elegant lineage tracing approaches in medaka to study the biology of fin regeneration and the repair of osteoporosis-like vertebral bone lesions.

Part II, Neuroscience, reflects the key ongoing role of the fish model in studies of the vertebrate nervous system. Chapter 7 provides an enhanced protocol for the generation of primary cell cultures from zebrafish embryos, enabling trans-well cell compartment separation and relatively long-term culture of neurons. The methods described in Chap. 8 combine optogenetics and holographic illumination for spatial selective control of neuronal activity, coupled with high-speed behavioral monitoring to dissect circuit function. Chapter 9 describes methods for staining and mounting larval zebrafish to facilitate whole-brain fluorescence imaging, as well as a template for aligning brain images to a reference atlas. Chapter 10 describes the use of tracking microscopy employing a motion cancellation system along with structured illumination microscopy for imaging the brain of a freely swimming larval zebrafish. Chapter 11 describes a setup for active avoidance fear conditioning of the adult fish and a method for inhibiting a specific neuronal circuit via the Gal4-UAS system.

Part III, Regeneration, provides tools and approaches using zebrafish to study stem cell and regenerative biology, increasingly an important aspect of human health. Chapter 12 focuses on the scale as a model for tissue regeneration, describing methods for live, tissue-wide imaging of scale regeneration along with methods to quantify cell, tissue, and signal dynamics. Chapter 13 provides a detailed protocol for the generation of conditional knockout (cKO) zebrafish using a Cre-dependent genetic switch and the use of this technique for heart regeneration research. Chapter 14 addresses the critical problem of spinal cord injury, describing tools for the study of spinal cord regeneration, including injury, assessment of swim endurance, and quantitative histologic techniques to measure axonal regeneration. Chapter 15 details improvements in nitroreductase-mediated tissue ablation, including a NTR 2.0 variant that enables the ablation of resistant cell types and novel cell ablation paradigms. Chapter 16 provides optimized methods for section immunostaining of regenerating fin using paraffin and frozen fin sections.

Part IV is in the subject area of Genetics and Genomics. Chapter 17 concerns the innovative use of optogenetics to drive liquid-liquid phase separation mediated by intrinsically disordered protein regions, in neurons in vivo, thus modeling key potential drivers of neurodegenerative disease. Chapter 18 describes the use of colorimetric barcoding for in vivo tracking of clonal diversity in hematopoietic stem cell (HSC) development, including techniques for quantifying the number and size of HSC clones during development. Chapter 19 discusses technologies used for generating precise base pair-specific mutations in the zebrafish genome, with an emphasis on CRISPR/Cas9 knock-in procedures using single-stranded DNA as the donor template. Chapter 20 provides methods to improve the efficiency of CRISPR/Cas9-mediated targeted mutagenesis. By delivering multiple sgRNAs flanked by tRNAs, the authors describe a strategy to overcome functional redundancy of genes and genetic compensation that can limit effectiveness of gene modification in vivo. Chapter 21 describes the development of Multiplexed Intermixed CRISPR Droplets (MIC-Drop), and, by injecting the droplet into embryos, the authors provide a platform that makes large-scale reverse genetic screens possible in zebrafish. Finally, Chapter 22 describes the recent studies analyzing the genome and mutants of the goldfish, a relative of zebrafish, and how the goldfish can contribute to advancing the zebrafish research.

It is our hope that the methods described in this volume will prove useful to a wide range of researchers in the community, and will help to propel advances in developmental biology, disease modeling, and regeneration research that use zebrafish and medaka as model systems. We are indebted to the authors for their contributions, and for their patience with us as editors in gathering these chapters together. We thank Amy E. Jackson for her help with communications and organization.

Los Angeles, CA, USA *James F. Amatruda*
London, UK *Corinne Houart*
Mishima, Japan *Koichi Kawakami*
Durham, NC, USA *Kenneth D. Poss*

Contents

Contributors

W. TED ALLISON • *Centre for Prions & Protein Folding Disease, University of Alberta, Edmonton, AB, Canada; Department of Biological Sciences, University of Alberta, Edmonton, AB, Canada; Department of Medical Genetics, University of Alberta, Edmonton, AB, Canada*

HADEEL ALYENBAAWI • *Department of Medical Laboratories, Majmaah University, Majmaah, Saudi Arabia*

KAZUHIDE ASAKAWA • *Laboratory of Molecular and Developmental Biology, National Institute of Genetics, Mishima, Shizuoka, Japan*

JASON N. BERMAN • *Children's Hospital of Eastern Ontario Research Institute, Ottawa, ON, Canada; Departments of Pediatrics and Cellular and Molecular Medicine, University of Ottawa, Ottawa, ON, Canada*

ASHWIN A. BHANDIWAD • *Division of Developmental Biology, Eunice Kennedy Shriver National Institute of Child Health and Human Development, Bethesda, MD, USA*

DOROTHEE BORNHORST • *Stem Cell Program and Division of Hematology/Oncology, Boston Children's Hospital, Boston, MA, USA; Department of Stem Cell and Regenerative Biology, Harvard University, Boston, MA, USA*

HAROLD A. BURGESS • *Division of Developmental Biology, Eunice Kennedy Shriver National Institute of Child Health and Human Development, Bethesda, MD, USA*

SHAWN M. BURGESS • *Translational and Functional Genomics Branch, National Human Genome Research Institute, Bethesda, MD, USA*

BROOKE BURRIS • *Department of Developmental Biology, Washington University School of Medicine in St. Louis, St. Louis, MO, USA*

ALEXANDER H. BURTON • *Departments of Chemical and Biomedical Engineering, College of Engineering, Carnegie Mellon University, Pittsburgh, PA, USA*

EDWARD A. BURTON • *Department of Neurology, University of Pittsburgh, Pittsburgh, PA, USA; Geriatric Research, Education and Clinical Center, Pittsburgh VA Healthcare System, Pittsburgh, PA, USA*

DELIA CALDERON • *Center for Childhood Cancer & Blood Diseases, The Abigail Wexner Research Institute, Nationwide Children's Hospital, Columbus, OH, USA; Molecular, Cellular, and Developmental Biology Ph.D. Program, The Ohio State University, Columbus, OH, USA*

JAIME CHU • *Department of Pediatrics, Icahn School of Medicine at Mount Sinai, New York, NY, USA*

SUBHAM DASGUPTA • *Sinnhuber Aquatic Research Laboratory, Department of Environmental and Molecular Toxicology, Oregon State University, Corvallis, OR, USA; Department of Biological Sciences, Clemson University, Clemson, SC, USA*

PATRICE DELANEY • *Program in Biology, New York University Abu Dhabi, Abu Dhabi, United Arab Emirates*

ALESSANDRO DE SIMONE • *Department of Cell Biology, Duke University Medical Center, Durham, NC, USA; Duke Regeneration Center, Duke University, Durham, NC, USA; Department of Genetics and Evolution, University of Geneva, Geneva, Switzerland*

BRANDON GHELLER • *Stem Cell Program and Division of Hematology/Oncology, Boston Children's Hospital, Boston, MA, USA; Harvard Medical School, Boston, MA, USA*
TAYLOR GILL • *Centre for Prions & Protein Folding Disease, University of Alberta, Edmonton, AB, Canada; Department of Biological Sciences, University of Alberta, Edmonton, AB, Canada*
TRIPTI GUPTA • *Division of Developmental Biology, Eunice Kennedy Shriver National Institute of Child Health and Human Development, Bethesda, MD, USA*
HIROSHI HANDA • *Department of Molecular Pharmacology, Center for Future Medical Research, Tokyo Medical University, Tokyo, Japan*
VICTORIA HEIGH • *Division of Developmental Biology, Eunice Kennedy Shriver National Institute of Child Health and Human Development, Bethesda, MD, USA*
GEORGE A. HOLMES • *Division of Developmental Biology, Eunice Kennedy Shriver National Institute of Child Health and Human Development, Bethesda, MD, USA*
CORINNE HOUART • *Centre for Developmental Neurobiology, MRC Centre for NeuroDevelopmental Disorders, King's College London, London, UK*
XINYU JIA • *Sorbonne Université, Institut du Cerveau (ICM), Paris, France*
KOICHI KAWAKAMI • *Laboratory of Molecular and Developmental Biology, National Institute of Genetics, Mishima, Shizuoka, Japan; Department of Genetics, Graduate University for Advanced Studies (SOKENDAI), Mishima, Shizuoka, Japan*
GENEVIEVE C. KENDALL • *Center for Childhood Cancer & Blood Diseases, The Abigail Wexner Research Institute, Nationwide Children's Hospital, Columbus, OH, USA; Department of Pediatrics, The Ohio State University College of Medicine, Columbus, OH, USA*
MATTHEW R. KENT • *Center for Childhood Cancer & Blood Diseases, The Abigail Wexner Research Institute, Nationwide Children's Hospital, Columbus, OH, USA*
KAZU KIKUCHI • *Department of Cardiac Regeneration Biology, National Cerebral and Cardiovascular Center Research Institute, Osaka, Japan*
JACK KUCINSKI • *Center for Childhood Cancer & Blood Diseases, The Abigail Wexner Research Institute, Nationwide Children's Hospital, Columbus, OH, USA; Molecular, Cellular, and Developmental Biology Ph.D. Program, The Ohio State University, Columbus, OH, USA*
PRADEEP LAL • *Fish Biology and Aquaculture Group, Climate & Environment Department, NORCE Norwegian Research Centre, Bergen, Norway*
JENNIFER M. LI • *Max Planck Institute for Biological Cybernetics, Tuebingen, Germany*
LASZLO F. LOCSKAI • *Centre for Prions & Protein Folding Disease, University of Alberta, Edmonton, AB, Canada; Department of Biological Sciences, University of Alberta, Edmonton, AB, Canada*
YILUN MA • *Cancer Biology and Genetics, Memorial Sloan Kettering Cancer Center, New York, NY, USA*
ELENA MAGNANI • *Program in Biology, New York University Abu Dhabi, Abu Dhabi, United Arab Emirates*
IAN MCBAIN • *Program in Biology, New York University Abu Dhabi, Abu Dhabi, United Arab Emirates*
MAYSSA H. MOKALLED • *Center of Regenerative Medicine, Washington University School of Medicine in St. Louis, St. Louis, MO, USA*
EMILY MONTAL • *Cancer Biology and Genetics, Memorial Sloan Kettering Cancer Center, New York, NY, USA*

TIMOTHY S. MULLIGAN • *Department of Ophthalmology, Wilmer Eye Institute, Johns Hopkins University, Baltimore, MD, USA*
JEFF S. MUMM • *Department of Ophthalmology, Wilmer Eye Institute, Johns Hopkins University, Baltimore, MD, USA*
ANJANA RAMDAS NAIR • *Program in Biology, New York University Abu Dhabi, Abu Dhabi, United Arab Emirates*
MASAHITO OGAWA • *Department of Cardiac Regeneration Biology , National Cerebral and Cardiovascular Center Research Institute, Osaka, Japan*
YOSHIHIRO OMORI • *Laboratory of Functional Genomics, Graduate School of Bioscience, Nagahama Institute of Bioscience and Technology, Nagahama, Japan*
SABA PARVEZ • *Department of Pharmacology & Toxicology, University of Utah, Salt Lake City, UT, USA*
RANDALL T. PETERSON • *Department of Pharmacology & Toxicology, University of Utah, Salt Lake City, UT, USA*
SERGEY V. PRYKHOZHIJ • *Children's Hospital of Eastern Ontario Research Institute, Ottawa, ON, Canada*
DREW N. ROBSON • *Max Planck Institute for Biological Cybernetics, Tuebingen, Germany*
KIRSTEN C. SADLER • *Program in Biology, New York University Abu Dhabi, Abu Dhabi, United Arab Emirates*
TOMOYA SHIRAKI • *Laboratory of Molecular and Developmental Biology, National Institute of Genetics, Mishima, Shizuoka, Japan*
KATHERINE SILVIUS • *Center for Childhood Cancer & Blood Diseases, The Abigail Wexner Research Institute, Nationwide Children's Hospital, Columbus, OH, USA*
MICHAEL T. SIMONICH • *Sinnhuber Aquatic Research Laboratory, Department of Environmental and Molecular Toxicology, Oregon State University, Corvallis, OR, USA*
KRYN STANKUNAS • *Institute of Molecular Biology, University of Oregon, Eugene, OR, USA; Department of Biology, University of Oregon, Eugene, OR, USA*
SCOTT STEWART • *Institute of Molecular Biology, University of Oregon, Eugene, OR, USA*
ABHIGNYA SUBEDI • *Division of Developmental Biology, Eunice Kennedy Shriver National Institute of Child Health and Human Development, Bethesda, MD, USA*
SHRUTHY SURESH • *Cancer Biology and Genetics, Memorial Sloan Kettering Cancer Center, New York, NY, USA*
MOHITA M. TAGORE • *Cancer Biology and Genetics, Memorial Sloan Kettering Cancer Center, New York, NY, USA*
WEN HUI TAN • *Department of Developmental Genetics, Max Planck Institute for Heart and Lung Research, Bad Nauheim, Germany*
HIDEYUKI TANABE • *Laboratory of Molecular and Developmental Biology, National Institute of Genetics, Mishima, Shizuoka, Japan*
ROBYN L. TANGUAY • *Sinnhuber Aquatic Research Laboratory, Department of Environmental and Molecular Toxicology, Oregon State University, Corvallis, OR, USA*
RICHARD TAYLOR • *Centre for Developmental Neurobiology and Medical Research Council Centre for Neurodevelopmental Disorders, Institute of Psychiatry, Psychology & Neuroscience, Guy's Campus, King's College London, London, UK*
TRAVIS WHEELER • *Department of Cell Biology, University of Pittsburgh, Pittsburgh, PA, USA*
RICHARD M. WHITE • *Cancer Biology and Genetics, Memorial Sloan Kettering Cancer Center, New York, NY, USA*

CHRISTOPH WINKLER • *Department of Biological Sciences and Centre for Bioimaging Sciences, National University of Singapore, Singapore, Singapore*
CLAIRE WYART • *Sorbonne Université, Institut du Cerveau (ICM), Paris, France*
TEJIA ZHANG • *Department of Pharmacology & Toxicology, University of Utah, Salt Lake City, UT, USA*
LEONARD I. ZON • *Stem Cell Program and Division of Hematology/Oncology, Boston Children's Hospital, Boston, MA, USA; Department of Stem Cell and Regenerative Biology, Harvard University, Boston, MA, USA; Harvard Medical School, Boston, MA, USA; Howard Hughes Medical Institute, Boston, MA, USA*

Part I

Disease Models

Chapter 1

Delivering Traumatic Brain Injury to Larval Zebrafish

Taylor Gill, Laszlo F. Locskai, Alexander H. Burton, Hadeel Alyenbaawi, Travis Wheeler, Edward A. Burton, and W. Ted Allison

Abstract

We describe a straightforward, scalable method for administering traumatic brain injury (TBI) to zebrafish larvae. The pathological outcomes appear generalizable for all TBI types, but perhaps most closely model closed-skull, diffuse lesion (blast injury) neurotrauma. The injury is delivered by dropping a weight onto the plunger of a fluid-filled syringe containing zebrafish larvae. This model is easy to implement, cost-effective, and provides a high-throughput system that induces brain injury in many larvae at once. Unique to vertebrate TBI models, this method can be used to deliver TBI without anesthetic or other metabolic agents. The methods simulate the main aspects of traumatic brain injury in humans, providing a preclinical model to study the consequences of this prevalent injury type and a way to explore early interventions that may ameliorate subsequent neurodegeneration. We also describe a convenient method for executing pressure measurements to calibrate and validate this method. When used in concert with the genetic tools readily available in zebrafish, this model of traumatic brain injury offers opportunities to examine many mechanisms and outcomes induced by traumatic brain injury. For example, genetically encoded fluorescent reporters have been implemented with this system to measure protein misfolding and neural activity via optogenetics.

Key words Neurotrauma, Animal model, Blast injury, Closed head injury, Concussion, Pressure transducer

1 Introduction

The global incidence of traumatic brain injury (TBI) is estimated to be 69 million per year [1]. TBI contributes to a substantial burden of disability worldwide as approximately 15–30% of TBIs lead to post-concussion syndrome with long-term consequences such as depression, anxiety, seizures, and sleep disorders [2, 3]. Moreover, TBI is a leading risk factor for debilitating neurodegenerative

Joint first authors: Taylor Gill and Laszlo Locskai contributed equally to this work

Supplementary Information The online version contains supplementary material available at https://doi.org/10.1007/978-1-0716-3401-1_1.

James F. Amatruda et al. (eds.), *Zebrafish: Methods and Protocols*, Methods in Molecular Biology, vol. 2707, https://doi.org/10.1007/978-1-0716-3401-1_1,

diseases like Chronic Traumatic Encephalopathy and Alzheimer's disease [4–6].

Despite growing understanding, significant knowledge gaps still exist around the cellular mechanisms involved in the complex outcomes of TBI and what therapeutic interventions may act as protective measures against potentially debilitating consequences [7, 8]. The molecular study of TBI and its long-term effects in humans is primarily limited to postmortem examinations, whereas cellular or other in vitro models fail to represent the complexity of the injury or tissue responses. Thus, animal models are essential to studying the multifaceted pathophysiological mechanisms of TBI and treatment thereof.

Rodent models of TBI are commonly employed for these studies, and diverse biomechanical methods of producing injury are available. Other mammalian models, including cats, dogs, pigs, lambs, and macaque monkeys are also used. These animals' large size facilitates ease in measuring intracranial pressure, blood oxygen, and cerebral blood flow following injury [9, 10]. However, large animals incur high maintenance costs and require lengthy study periods, precluding high-throughput or discovery-driven approaches to elucidate the molecular mechanisms mediating outcomes from TBI.

Zebrafish (*Danio rerio*) offer many advantages as a preclinical model of TBI, and several of these advantages are amplified when studying larval stages. Zebrafish models are accessible to experimentation since they are economical in time, space, and cost. They are also relevant and applicable to human TBI. Zebrafish share substantial genetic homology with humans, and many pharmacological compounds have similar effects in zebrafish as they do in humans [11–13]. Zebrafish larvae also offer ease of drug delivery, as many small molecules are absorbed systemically following their addition to the medium containing the fish [14]. Importantly, zebrafish are amenable to genetic engineering. For example, CRISPR-induced null mutations and knock-ins, coupled with transgene expression support experimental elucidation of injury and treatment mechanisms. Additionally, transparent larvae and transgenic fluorescent reporters enable potent imaging modalities, including both live and longitudinal imaging of the injury and the subsequent healing or neurodegenerative response. For example, the model of TBI described in this chapter (recently presented by Alyenbaawi et al. 2021) was assessed with optogenetics and fluorescent reporters of tauopathy [15]. This work has provided insights into seizures and tau aggregation following TBI, and the potential prevention/treatment of tau aggregation using antiepileptic pharmacology [15].

TBI models have also been developed in fruit fly (*Drosophila melanogaster*) and roundworm (*Caenorhabditis elegans*). Although these are excellent genetic models, they lack the similarity of CNS tissue architecture shared amongst the vertebrates [15–37]. These considerations together suggest that zebrafish models provide an optimal balance of experimental accessibility with biological and translational relevance.

The TBI method presented here uses larval zebrafish at 3 days post-fertilization (dpf). As described in detail below, a group of larvae in their typical bath media is loaded into a syringe. The syringe is subsequently sealed with a stopper valve. A weight of known mass is dropped on the syringe plunger from a consistent height, depressing the plunger and producing a reproducible pressure wave that causes TBI in the larvae contained in the syringe (*see* Figs. 1 and 2). The pressure generated in the syringe, and the severity of the resulting injury, can be readily modulated by adjusting the mass of the weight or the height of the drop and/or the number of times it is dropped on the syringe plunger (*see* Fig. 3). Herein, the weight drop is repeated three times to mitigate technical variability that might arise from variation in the position or orientation of each larva in the syringe.

Congruent with patients and other animal model symptoms of mild closed-head TBI, this larval zebrafish model of TBI induces cardinal signs of diffuse neurotrauma. These include increased cell

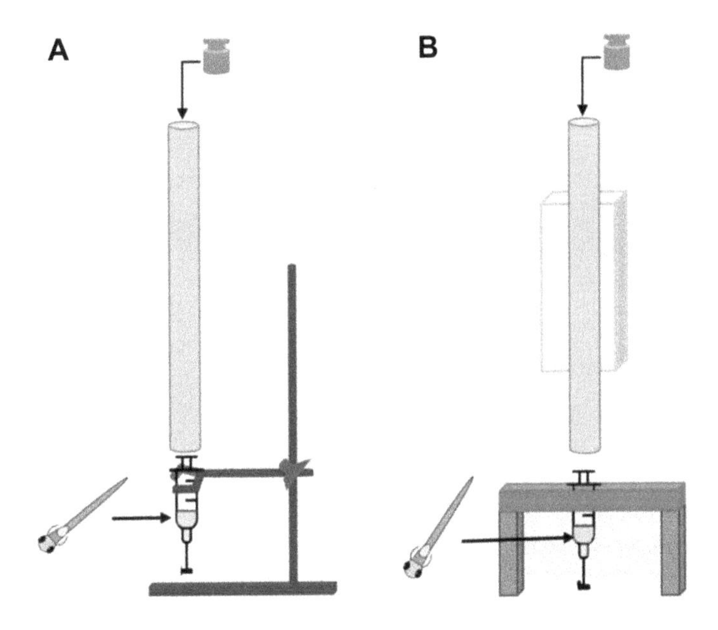

Fig. 1 Equipment configuration for inducing TBI in larval zebrafish. Various weights are dropped through a 121.92 cm (48″) tube onto the plunger of a syringe; The syringe is loaded with 10–20 zebrafish larvae in 1 mL of E3 medium and sealed with a stopper valve. The syringe can be held by a three-pronged clamp (**a**) or a foam block (**b**). The pressure waves generated when the weight depresses the syringe plunger simulates a blast injury, causing traumatic brain injury in the larvae

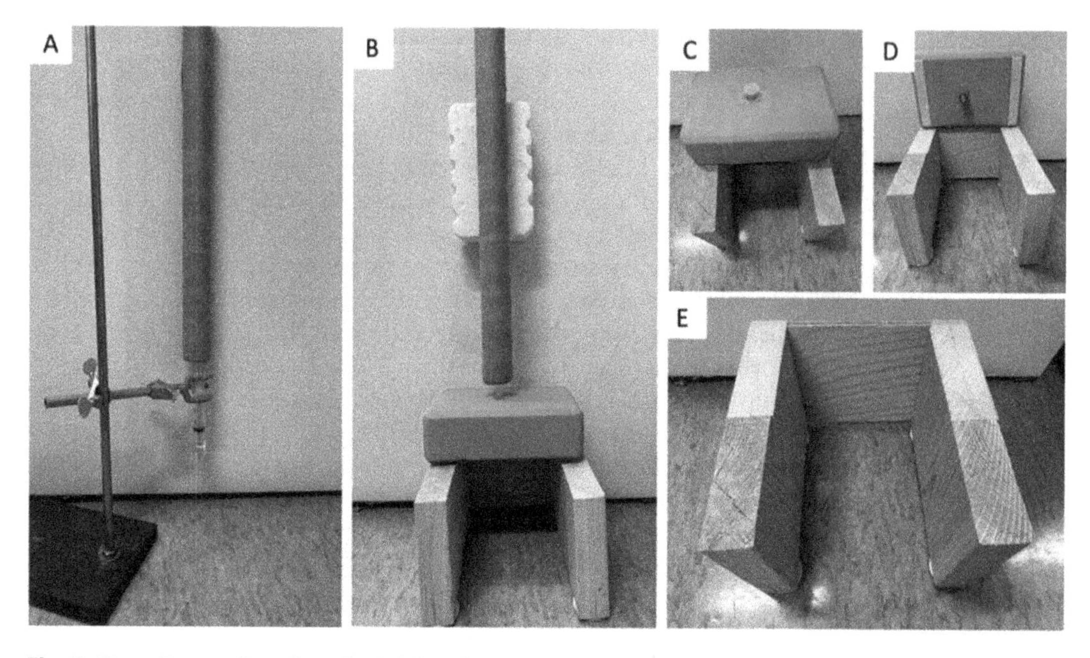

Fig. 2 Alternative configurations for holding the syringe: the clamp or the foam block. (**a**) The three-prong clamp attached to a support stand holds the syringe below the cardboard tube, which acts as a guide for the dropped weight. (**b**) The syringe can be held in position by a foam block. The cardboard tube is held out from the wall by spacer to ensure the syringe is centered under the tube. (**c**) A wooden support stand acts as a base for the foam block. (**d**) The foam block attaches to the wooden base via Velcro tape. (**e**) The syringe is securely fit into a hole in the foam block, with the syringe arms resting against the foam

death, hemorrhage, blood flow abnormalities, tauopathy, and post-traumatic seizures [15]. Investigating additional TBI outcomes in this model is warranted.

The methods described here align most closely with a "closed head" animal model of TBI. This contrasts widely used models of TBI where windows are surgically removed from the skull to give direct access to injure the neural tissue. Open- and closed-head models each have advantages. Arguably, closed-head animal models of TBI are underrepresented in the field and this model may assist in filling that gap.

Most animal models of TBI rely on the use of anesthetic, whereas the model presented here can avoid anesthetic and any of its associated confounds, assuming of course that appropriate ethics approval is granted. For example, some biomechanical methods use anesthetic for surgical preparation when craniotomy is necessary, while others use it to ensure immobilization during injury or for pain management [20, 38–40]. Unfortunately, anesthetics can induce artefacts such as neuroprotective effects, and interact with the secondary outcomes of TBI in intricate ways [41–44]. For example, isoflurane use increases hippocampal neuronal survival, possibly due to increased blood flow and decreased excitotoxicity [45, 46]. At least one mouse model of TBI abstained from

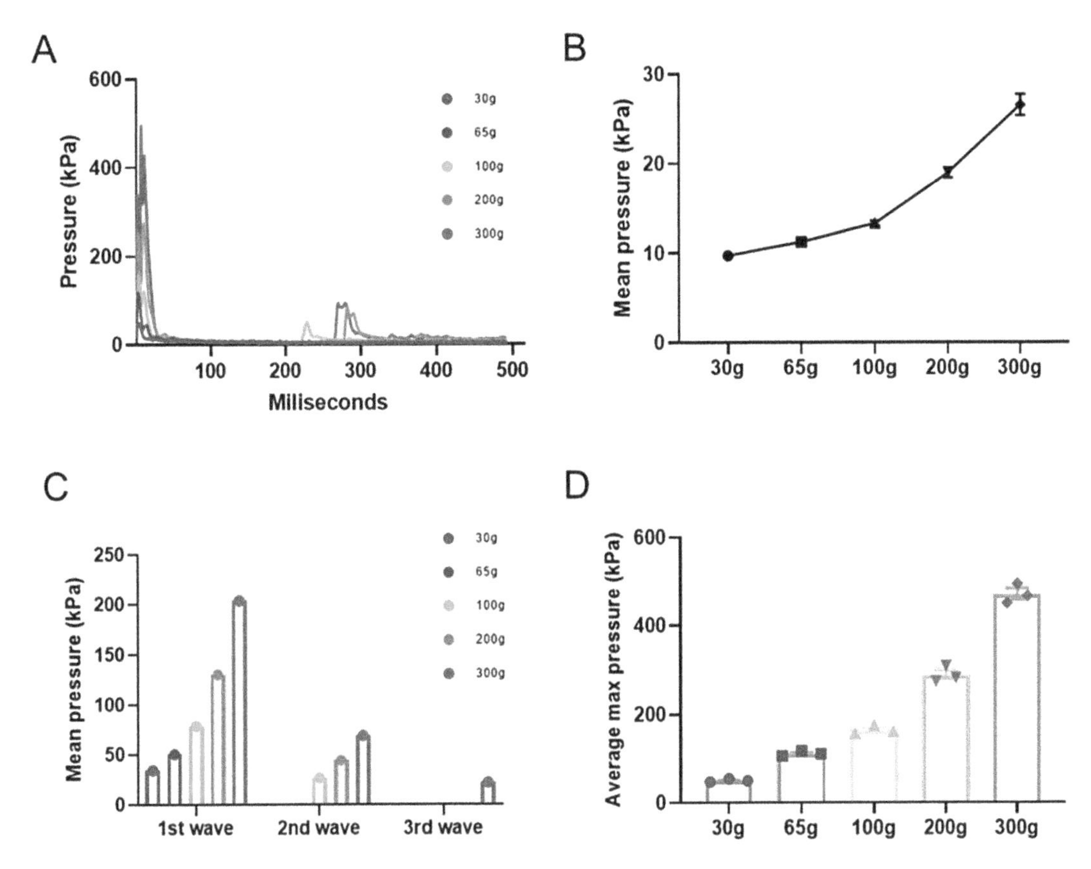

Fig. 3 Pressure kinetics and measurements from the larval zebrafish TBI model. (**a**) Representative examples of the pressure dynamics following weight drops. Heavier weights produce pressure change "waves" not only at first impact but also 200–400 ms later, depending on the weight's mass, that we interpret to be the weight bouncing on the plunger. (**b**) Mean pressure measured from 3 repeated drops of each weight over a 450 ms timeframe. (**c**) The average pressure of each pressure wave created after weight drop. Pressure waves were measured from the first recorded data point of increased pressure to the last recorded data point before pressure returned to baseline. (**d**) The max pressure averaged from 3 repeated drops of each weight averaged. Graphs present variance as ±SEM

anesthetic use due to its potentially confounding effects [47]. While some TBI techniques used in fruit flies and roundworms do not require anesthetic, the method described here is unique among zebrafish models of TBI in not requiring anesthetic or other metabolic agents like Quinolinic acid, which can cause excitotoxic brain injury [20, 34]. The ability to forgo anesthetic improves the applicability of this model to TBI in humans and is particularly important when studying factors like the sleep pathway and the glymphatic clearance of waste (such as amyloid-beta or tau proteins), that are impacted by anesthetic [48, 49]. Notably, while the methods presented here do not require anesthetic or other pharmacology, they could readily be included in the embryo media within the syringe as needed.

While there are many advantages of the TBI methods described in this chapter, some caveats should be considered. The pressure wave generated by the weight drop impacts the entirety of the larvae rather than being targeted at the head specifically [40, 50]. Amongst the familiar forms of TBI experienced by patients, the model presented herein is perhaps most akin to being in the blast radius of an explosion. Injuries to the spinal cord and other organs are likely and should be considered when comparing this method to animal models where the neurotrauma is more localized to the head. Additionally, the cranium of zebrafish is not fully developed in larvae, so this TBI model may not fully reproduce the intracranial pressure and its downstream effects that are commonly associated with cerebral edema in humans [25]. This method successfully models primary blast injury, which is the direct injury caused by the pressure wave of a blast, but it does not include the other hazards of blast injury including projectile objects, heat, and smoke [51]. Additional considerations include that the animal is young and developing rapidly during the injury response, which may in some regards lead to different outcomes compared to injury responses in an adult brain.

This chapter also describes a method to measure the pressure wave inside the syringe that houses the larvae, which can be useful for calibrating the method, assessing sources of technical variability, and/or reducing variability between users or labs. The method presented here is broadly equivalent in concept to the methods presented previously (Alyenbaawi et al. 2021) but utilizes components that are less expensive and more readily acquired [15]. A pressure to voltage transducer (designed for automotive fuel lines) is attached to the syringe in place of the stopper valve (*see* Fig. 4). The signal from the pressure transducer during the weight drop event is read by an inexpensive microcontroller board (Arduino Uno) using open source software (*see* Fig. 3).

While reporting pressure kinetics during injury is a mainstay in the TBI modelling literature, caution is warranted when comparing the pressures reported in this method with the pressures reported in blast injury victims or larger animal models; for example, it remains to be determined if the values generated from large animal models (e.g., injured with ballistic pistons) are directly comparable to TBI in tiny larval zebrafish (that have yet to develop skeletal calcification, etc.). Moreover, investing in these pressure measurements may not be a top priority depending on the researcher's familiarity with assembling electronics, or if the researchers are instead able to calibrate the method empirically in their own hands towards consistently inducing biological TBI outcomes. That is, if the method consistently produces a TBI-related defect that can be investigated robustly (e.g., varies rationally with the injury severity/dose), then knowing the exact pressure generated may be of secondary importance (assuming other researchers can also empirically optimize the

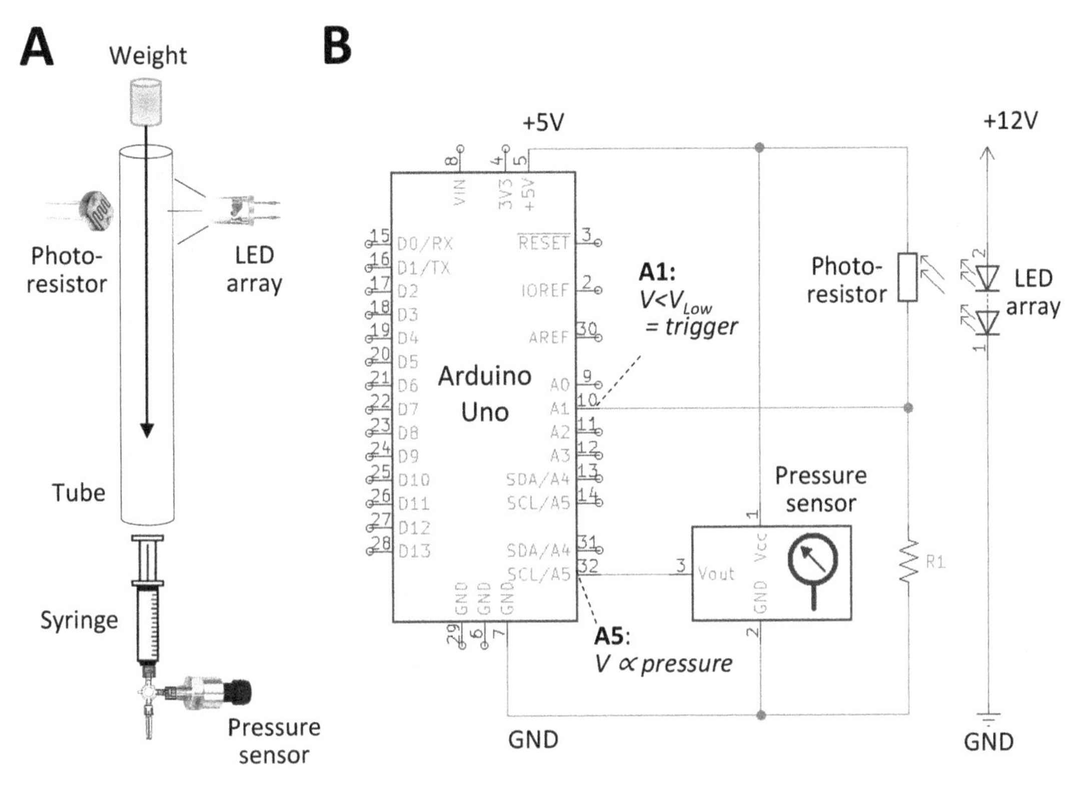

Fig. 4 Measuring the pressure waves inside the syringe. (**a**) Experimental arrangement similar to Figs. 1 and 2, but with a pressure transducer (pressure sensor) attached to the syringe to record events within the syringe. Recording of data from the pressure sensor is triggered when the falling weight breaks a light beam detected by a photoresistor. (**b**) Circuit diagram showing the connections between the Arduino Uno microcontroller, the pressure transducer, and the photoresistor, and the separate LED circuit

method to find similar outcomes). In sum, when establishing this larval TBI method some laboratories may benefit greatly from including a pressure transducer if it is easy to implement in those users' hands; for other labs, it may instead be more practical to first explore biological outcomes following TBI and then decide if those outcomes are interesting enough to warrant quantifying the pressures. Regardless, measuring pressures generated during TBI is immensely useful for assessing sources of technical variation and optimizing methodological parameters while importantly reducing the use of animals (and the inter-individual variation they inevitably add to such measures).

Overall the model of TBI described in this chapter successfully models a closed-skull blast injury. Similar to other blast-like injury models, it is highly reproducible and allows researchers to study the outcomes of TBI, including inflammation, protein folding stress, and axonal injury [13, 21]. The genetic and pharmacological accessibility of larval zebrafish powerfully enables experimental intervention to probe TBI mechanisms. The affordability, flexibility in injury severity, and ease of use promote this method as a valuable approach to elucidating the cellular pathophysiological responses to TBI.

2 Materials

2.1 Inducing TBI

The setup of the experimental equipment is shown in Figs. 1 and 2.

1. Zebrafish larvae 3 dpf.
2. E3 medium for embryos and larvae, or any other suitable husbandry media. For 2 L of 60× E3 stock, dissolve 34.4 g NaCl, 1.52 g KCl, 5.8 g CaCl2.2H2O, and 9.8 g MgSO4.7H2O, add double distilled water up to 2000 mL.
3. Disposable transfer pipettes.
4. Petri dishes.
5. 20 mL plastic syringe with Luer-Lok tip (e.g., Becton Dickinson #302830) (*see* **Notes 1 and 2**).
6. Stop valve with Luer-Lok attachment (e.g., Cole-Parmer # UZ-30600-00).
7. Cardboard guide tube. The tube can be made with thin cardboard poster material and tape. The length of the tube should be 121.92 cm (48"). The tube diameter must accommodate the circumference of the weight without excess space. A standard 300 g calibration weight has a diameter of 38 mm, so a tube used with this weight should be approximately 40 mm diameter) (*see* **Notes 3 and 4**).
8. Tape to attach guide tube to the wall.
9. 300 g calibration weight such as the type used to calibrate a scientific balance (e.g., Ohaus #80780137 or Fisher # 01921300) (*see* **Note 5**).

   There are two options for securing the syringe. Materials from either the clamp setup (*see* Figs. 1a, 2a; **step 10**) or the foam block setup (*see* Figs. 1b, 2b–e; **step 11**) are required (*see* **Note 6**):
10. Securing the syringe with a clamp.
    (a) Support stand with a vertical rod. A heavy base to increase stability is ideal.
    (b) Three-prong extension clamp. Must securely attach to the rod on the support stand and close tightly around the syringe barrel.
11. Securing the syringe in a foam bock.
    (a) High-density foam block. Approximately 3″ (7.6 cm) thick. A standard foam yoga block works well (9″ x 6″ × 3″(23 x 15 x 7.6 cm)).
    (b) Drill bit (a few millimeters smaller than the syringe diameter) to drill a hole in the foam block.

(c) Three wood pieces for support stand (*see* Fig. 2e). The width should match the foam block width. The height must accommodate syringe and stopper with Luer-Lok tip for TBI procedure or pressure transducer for carrying out pressure measurements.

(d) Wood screws or nails to secure the backing of wooden support to the sidewalls.

(e) Velcro with adhesive backing. To secure the foam block to the wooden support stand in order to prevent movement of the syringe when weight is dropped (*see* Fig. 2d).

(f) 2″ (5.08 cm) foam or Styrofoam piece. This will be attached to the wall and used as a spacer to keep the cardboard guide tube positioned over the syringe stopper (*see* Fig. 2b).

2.2 Setup of Pressure Transducer for Measuring Pressure Within the Syringe

The setup of the equipment is shown in Figs. 4 and 5.

1. Arduino Uno Rev3 microcontroller board (Arduino #A000066).
2. Breadboard (Haraqi ESH-PB-01). Any solderless breadboard can be used.

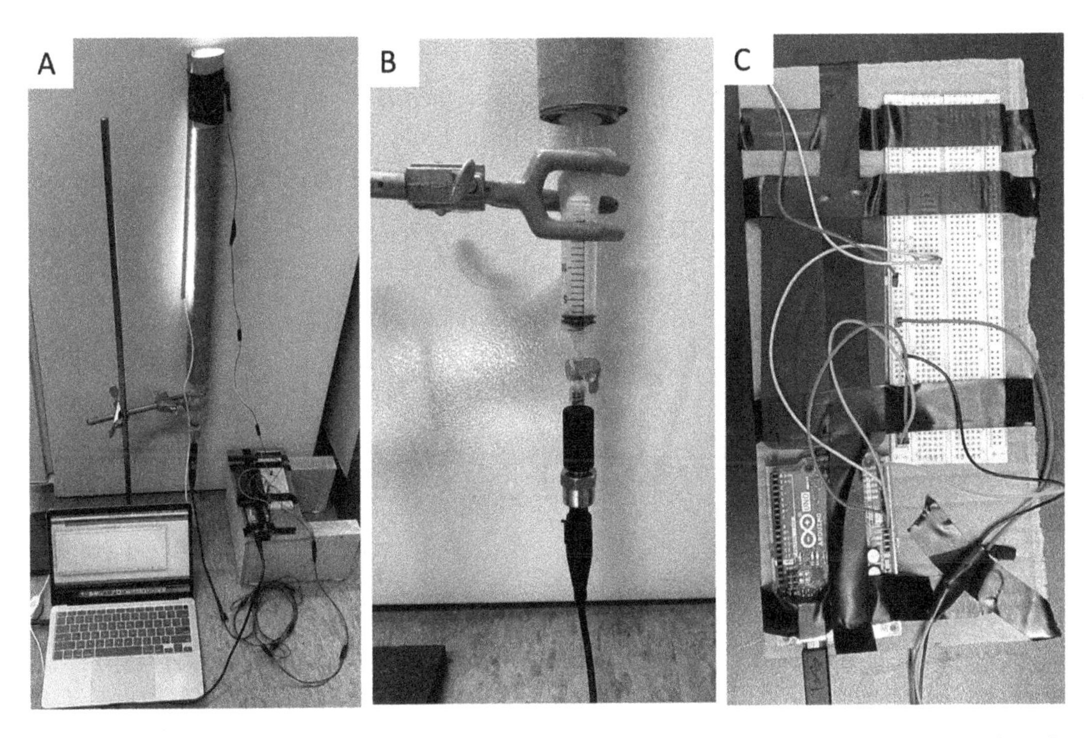

Fig. 5 Experimental arrangement to measure pressure inside the syringe. (**a**) The pressure transducer is connected to the syringe via a custom adaptor, and its outputs are sent via a breadboard to a computer running Arduino software. (**b**) The syringe is connected to the pressure transducer via a 3-way stop valve. (**c**) Picture showing connections between the Arduino board

3. Seven jumper wires (Haraqi ESH-PB-01). Any solderless jumper wires can be used.
4. 2700 Ω Resistor (Aniann An-resistor02). Any metal film resistor can be used.
5. Photoresistor (eBoot EBOOT-RESISTOR-05). Any light-sensitive resistor can be used.
6. LED light array.
7. Fuel line pressure transducer (AUTEX GSND-0556629788) (*see* **Notes 7 and 8**).
8. Arduino IDE 1.8.15 software (or newer; *see* **Note 8**) from www.arduino.cc/en/software. Computer (minimum hardware specs: 256 MB RAM, CPU with Pentium 4 or above, and proper operating system according to Arduino IDE recommendations).
9. 20 mL plastic syringe with Luer-Lok tip (e.g., Becton Dickinson #302830) (*see* **Note 2**).
10. 3-way stop valve with Luer-Lok attachment (e.g., Cole-Parmer RK-30600-02) (*see* **Note 9**).
11. 3D printed adaptor to connect the fuel line transducer to the Luer-Lok of the syringe (*see* **Note 10**).
    (a) Model specifications for 3D printing (*see* Fig. 6). The adapter can be designed and modeled using SolidWorks (see the Electronic Supplementary Material design file,

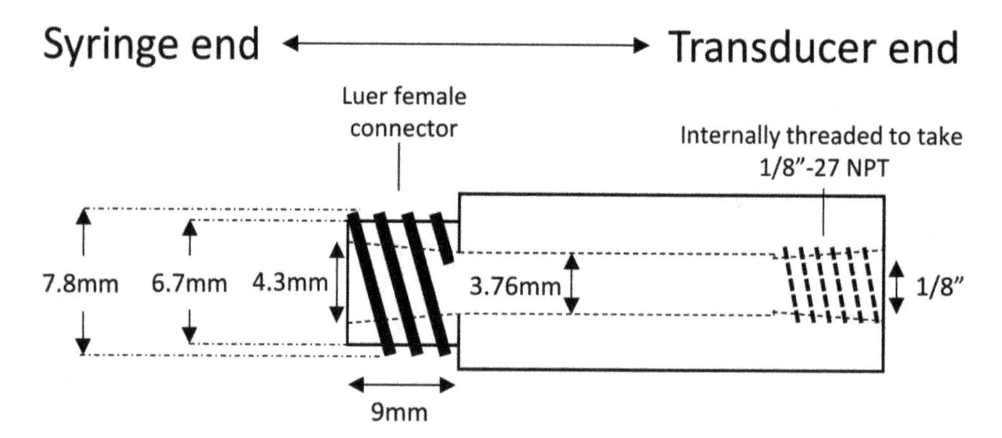

Fig. 6 Schematic for custom 3D printed adaptor that connects syringe to the pressure transducer. The female Luer-Lok to 1/8″ 27 NPT adapter was designed and modeled using SolidWorks and produced by 3D printing using a Formlabs Form2 SLA 3D printer and Formlabs Clear Resin. After 3D printing post-processing is completed, the internally tapered Luer fitting should be smoothed using a Luer reamer and gauge (Cockrill Precision Products) to ensure a proper seal with the Luer fitting. The pressure gauge end was then tapped using a 1/8–27 tapered pipe tap. All connections should be tested to ensure a snug and watertight fit prior to data collection

ADAPTER.SLDPRT. The file can be viewed with the SolidWorks viewer program available for download at https://www.solidworks.com/support/free-downloads).

(b) 3D print of adaptor. 3D printing can be completed by using a Formlabs Form2 SLA 3D printer and Formlabs Clear Resin.

(c) Postprint processing of adaptor (*see* **Note 11**).

(i) Luer reamer and gauge (Cockrill Precision Products).

(ii) 1/8–27 tapered pipe tap.

3 Methods

3.1 Inducing TBI

1. Collect embryos and raise them as per standard husbandry conditions.
2. Carry out TBI protocol at 3 dpf (*see* **Note 12**).
3. Using the 20 mL syringe with valve stopper attached, open stopper valve and pull 15–20 zebrafish larvae into the syringe along with E3 medium.
4. Expel excess medium until a final volume of 1 mL is reached (*see* **Note 13**). Remove all bubbles by inverting syringe and depressing plunger, flicking the side gently if needed.
5. Close the valve on the syringe stopper.
6. If securing the syringe in the clamp setup (*see* Fig. 2a):

(a) Secure the guide tube to the wall using tape. Considering the length of tube used, it is most practical to attach it at height such that the bottom of the guide tube is near the floor (but allowing room to insert and remove the syringe at the bottom of the guide tube). The guide tube must be exactly vertical to ensure the weight drops through it consistently.

(b) Tighten the three-pronged clamp around the syringe. Position the syringe vertically with the plunger at the top so that the falling weight will strike it. Center the plunger of the syringe in the bottom of the guide tube (*see* **Notes 1** and **14**).

7. Or, if securing the syringe in the foam block setup (*see* Fig. 2b–e):

(a) Drill a hole in the foam block a few millimeters smaller than the syringe diameter.

(b) Construct a support stand from three pieces of wood, using nails to secure the backing of wooden support to the sidewalls (*see* Fig. 2c, d).

(c) Use Velcro to attach the foam block to the support stand. Adhere Velcro to the top of the support stand and the bottom of the foam block (*see* Fig. 2e).

8. Position the guide tube relative to the syringe.
   (a) Attach the guide tube to a wall so that it is positioned to guide the weight onto the syringe plunger (but allowing room to insert and remove the syringe at the bottom of the guide tube). A spacer (e.g., 2″ (5.08 cm) foam or Styrofoam block attached with lab tape) can be used between the wall and the guide tube to accommodate the centering of the guide tube over the syringe (*see* Fig. 2b). The guide tube must be exactly vertical to ensure the weight drops through it consistently.
   (b) Attach guide tube to the top of the spacer to ensure the bottom of the guide tube will be centered over the syringe plunger.
   (c) Slide syringe into hole in foam block and secure foam block to wooden base with the installed Velcro. Position the plunger of the syringe so it is centered in the bottom of the guide tube (*see* Fig. 2b–e, **Note 14**).

   Continue for both setups:
9. Drop the weight from the top of the guide tube so it falls through the guide tube and depresses the syringe plunger upon impact. Maximizing consistency of the weight release and its fall, and minimizing its friction against the guide tube reduce intertrial technical variability.
10. Reposition the larvae between weight drops to minimize the potential artefact of interindividual variation in position/orientation during impact. Remove the syringe from the holding mechanism (if using the foam block setup, the syringe can remain in the foam block and both can be lifted off the wooden support stand). Invert syringe and open the stopper valve. Pull the plunger (air will be pulled into syringe) in order to pull the larvae from the tip back into the barrel of the syringe. Remove air bubbles before returning the plunger to the 1 mL mark and reclosing the stopper valve. Return the syringe to the holding mechanism below the guide tube.
11. Repeat the weight drop (**steps 6** and **7**) two more times (drop the weight onto the plunger three times in total) (*see* **Note 15**).
12. Open the valve stopper and draw up additional medium into the syringe. Expel the larvae into a petri dish, but stop short of fully depressing the plunger so the larvae are not crushed by the plunger as it is depressed. Draw in air or more medium before depressing the plunger again until all the larvae are ejected from the syringe.
13. Return the larvae to incubator to support their continued development or prepare them for analysis as appropriate.

3.2 Setup of Pressure Transducer for Pressure Measurement Within the Syringe

1. Connect the pressure transducer to the Arduino board as shown in Fig. 4b. Connect the Arduino +5 V and GND terminals to the positive (+) and negative (−) rails of the breadboard, respectively. Connect the positive (+) rail of the breadboard to the 5 V terminal of the transducer and the negative (−) rail of the breadboard to the GND terminal of the transducer. Connect the V_{out} terminal of the transducer to the analog input A5 pin of the Arduino board (*see* **Note 16**).
2. Connect the LED array to a power source. Cut holes on either side of the tube and position the LED array so it shines across the tube directly to the photoresistor. Connect the photoresistor in a voltage divider circuit as shown in Fig. 4b. Briefly, connect the positive (+) rail of the breadboard to one side of the photoresistor. Connect the other side of the photoresistor to a central terminal on the breadboard. Connect the central breadboard terminal to both (i) the Arduino analog input A1 using a jumper wire; and (ii) the negative (−) rail of the breadboard through a 2700 Ω resistor (R1 in Fig. 4b). When the light beam incident on the photoresistor is broken by the falling weight, the change in voltage at pin A1 is recognized by the software, so that pressure recording is started. This avoids collecting thousands of data points between trial runs and before the weight is dropped, as the sampling frequency can reach >1 kHz (Fig. 5) (*see* **Note 17**).
3. Connect the Arduino board to the computer via a USB connection.
4. Open the Arduino IDE and open the code (available in Appendix 1). Upload the code to the Arduino board. Open the serial monitor window and set the Baud rate to 2,000,000. The serial monitor will remain blank until a recording is triggered by the weight passing the photoresistor.
5. OPTIONAL: After setting up the pressure transducer apparatus, connections and pressure calculations can be verified by using static measurements of weights of a known mass resting directly on the syringe plunger. These values can be compared to a predicted pressure inside the syringe calculated as [mass × gravity]/[surface area of the plunger]. These static measurements of pressures provoked by weights with a known mass can be compared with the predicted values of the same weights to verify the experimental setup of the sensor (*see* Fig. 7). A line graph of measured versus calculated pressure should be close to the line of identity – although the dynamic pressure waves caused by dropping the weights are much higher than the pressures caused by resting the same weights on the plunger. If values measured in Arduino are inaccurate, see **Note 18**.

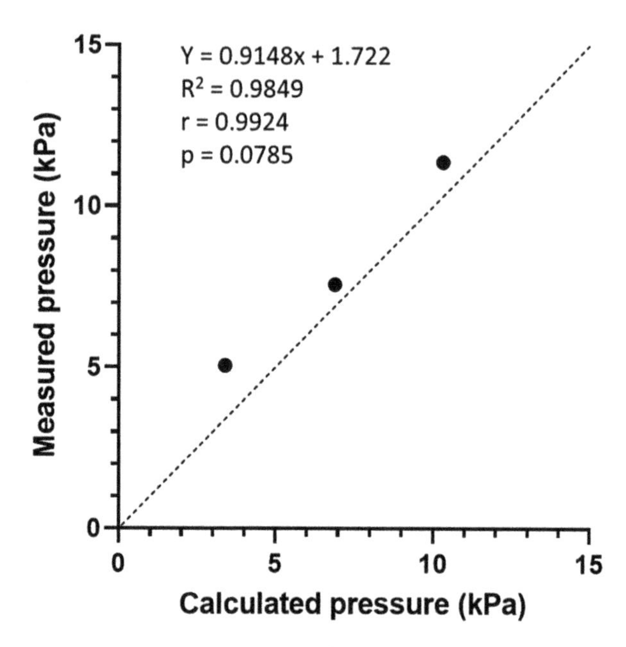

Fig. 7 Verification of fuel line pressure transducer measurements. Static pressure measurements are compared to predicted pressure calculated from elementary physics. Known weights are placed on the syringe plunger (rather than being dropped onto it) and the static pressure inside the syringe is measured using the transducer as shown above. Predicted pressure is calculated using the formula [(mass (kg) × gravity (m/s^2))/cross-sectional internal area of the syringe (m^2)]. Measured pressure was calculated by converting the 10-bit binary reading, B, from the Arduino board into a voltage reading, V, from the sensor [$V = (5 \times B)/1023$] and then converting the sensor voltage to pressure according to the manufacturer's specifications [$PSI = (V - 0.5) \times (37.5)$], [$kPa = PSI \times 6.895$]. The correlation between the measured and calculated pressure was found using a Pearson's correlation test, which gave a correlation coefficient (r) of 0.9924 and a coefficient of determination (R^2) of 0.9849

3.3 Pressure Measurement within the Syringe

1. Prepare the guide tube and syringe-holding setup in the same manner as the zebrafish TBI method (no zebrafish are used for pressure measurements) (*see* Subheading 3.1, **steps 6–8**).
2. Attach the pressure transducer at the syringe tip after connecting the transducer as above (Subheading 3.2).
3. Once the pressure transducer is connected, fill the syringe with 1 mL of water and then position the valve so it is open only to the transducer (*see* **Note 19**).
4. Drop the weight in the same manner as for an experiment with larvae. When the light breaks the LED array beam incident on the photoresistor, a pressure recording epoch of 2000 readings is triggered and the resulting data is sent to the serial monitor.

5. After each data run, copy and paste the data from the serial monitor into a spreadsheet program such as Excel or LibreOffice Calc. The data are organized as timestamps in the first column and 10-bit binary measurements from analog input pin A1 in the second column. After saving the data from each run, close and reopen the serial monitor, which resets the board so it is ready for the next run.
6. Convert the 10-bit binary readings at pin A1 to voltage, allowing the outputs from the pressure transducer to then be converted to units of pressure. First, in the spreadsheet, convert the binary reading, B, to voltage ($V = 5 \times B/1023$), where B is the 10-bit binary value written in the second column of the serial monitor. Second, convert these voltage values to PSI ($PSI = [V - 0.5] \times [37.5]$) (*see* **Note 18**). Finally, convert PSI to kPa ($kPa = PSI \times 6.895$).

4 Notes

1. To improve the grip of the clamp, paper tape can be wrapped around the top of the syringe barrel. This improved grip also seems to reduce breakage of the barrel flange upon impact of the weight. Tightening the clamp securely around the syringe helps mitigate the downward movement of the syringe in the clamp upon weight impact, thereby increasing the pressure consistency and preventing breakage of the barrel flange. For example, pliers can be used to close the three-pronged clamp around the syringe more tightly than is possible with hand strength alone.
2. Various syringe sizes can be used. For example, a 10 mL syringe can be used instead of a 20 mL syringe. If other parameters are kept constant, higher pressures are generated in a smaller syringe barrel, so this can be used to modulate injury severity. In our experience, the smaller syringes tend to be more fragile so we prefer using larger syringes.
3. If the tube diameter is too large, excess space around the weight may cause it to wobble in the tube. This wobbling can lead to slowing of the weight as it drops. Additionally, excess space in the tube at the base can lead to the weight becoming lodged between the syringe plunger and the tube. Overall, having a good match between the tube diameter and the weight will minimize technical variability between weight drops.
4. Dropping the weight from greater heights relative to the syringe generally results in higher pressure within the syringe, at least to some extent. Guide tubes of different lengths can be used to facilitate this height adjustment.

5. Weights of greater or smaller mass can be used to modulate injury severity (30–300 g have been tested). The intra-syringe pressure varies in an approximately linear fashion with increasing weight dropped (*see* Fig. 3). Markers of injury (tau aggregates in GFP+ tau reporter zebrafish larvae) increased with heavier weights dropped [13]. Alternatives to the cylindrical calibration weights suggested (used to calibrate scientific balances) could be imagined, such as the use of a dense/heavy metal ball.
6. The foam block setup more successfully immobilizes the syringe in a rigid position, restricting dissipation of pressure associated with syringe movement, and therefore higher and more consistent pressures within the syringe. Either setup can be used as long as it is consistent across experiments. If a more severe injury is desired, the foam block setup may be preferred.
7. Most automotive fuel line pressure sensors that require a +5 V input and give readouts between 0 and +5 V are suitable for this purpose. We used an AUTEX 150 psi pressure transducer.
8. Other pressure transducer systems exist with corresponding software and calibration procedures. We chose to use this method due to the easy access of the components and relatively affordable supplies needed to make it.
9. We used a three-way stop valve to attach the syringe and the pressure transducer together because it allowed us to easily reset the syringe plunger without disassembling the apparatus.
10. To make accurate pressure measurements in the range produced by the weight drop, we employed a car fuel line pressure transducer. Since an adapter for connecting a syringe to an automotive pressure sensor is not currently marketed, we fabricated a custom adapter by 3D printing. The attached design file may require modification if the thread on the pressure sensor differs from the one we used.
11. After 3D printing, post-processing needs to be completed. Smooth the internally tapered Luer fitting using a Luer reamer and gauge (Cockrill Precision Products) to ensure a proper seal with the Luer fitting. The pressure gauge end was then tapped using a 1/8–27 tapered pipe tap. Test all connections to ensure a snug and watertight fit prior to data collection.
12. Larvae can be injured at various ages. Repetitive injury over multiple days can also be implemented.
13. Tilt the syringe upward at a 45° angle and expel the excess liquid slowly into a petri dish in order to not eject the larvae.
14. When placing the syringe under the tube in preparation for dropping the weight, ensure the top of syringe plunger is exactly parallel to the floor and the bottom of the tube (i.e.,

the syringe is exactly vertical), and that it is centered in the tube opening in order to improve the consistency of pressure and injury between drops.

15. Carrying out the weight drop three times is not intended to model repetitive injury, but to ensure consistency, for example, to mitigate anticipated random differences in position and orientation of the larvae.
16. The exact supplies used, and layout of the circuit may need to be adjusted according to the transducer used.
17. If the pressure recording starts before the light plane is broken, the value of "light threshold" can be increased. If the number of consecutive recordings needs to be changed, the value of x in the coding line "for (int i = 0; i < x; i++)" can be changed, where x equals the number of consecutive pressure recording readouts.
18. When converting voltage to PSI, the pressure transducer used will have a manufacturer-recommended conversion value for zeroing the pressure. For the transducer we used, the value was 0.5 if another pressure transducer is used simply change the 0.5 value in the formula PSI = (voltage − 0.5) × (150/4). Additionally, if the converted values are not accurate the output voltage can be measured when zero pressure is applied to the syringe and the measured value used to replace 0.5 in the above equation. Similarly, if weights of known large mass are available, it is possible to derive a calibration experimentally and to substitute the manufacturer's calibration with an equation derived from experimental measurements.
19. The syringe must be filled once the pressure transducer is already connected to ensure there are no air pockets in the transducer that could affect the pressure measurement.

References

1. Dewan MC, Rattani A, Gupta S et al (2018) Estimating the global incidence of traumatic brain injury. J Neurosurg 130:1–18. https://doi.org/10.3171/2017.10.JNS17352
2. Capizzi A, Woo J, Verduzco-Gutierrez M (2020) Traumatic brain injury: an overview of epidemiology, pathophysiology, and medical management. Med Clin North Am 104:213–238. https://doi.org/10.1016/j.mcna.2019.11.001
3. Hou R, Moss-Morris R, Peveler R et al (2012) When a minor head injury results in enduring symptoms: a prospective investigation of risk factors for postconcussional syndrome after mild traumatic brain injury. J Neurol Neurosurg Psychiatry 83:217–223. https://doi.org/10.1136/jnnp-2011-300767
4. Hay J, Johnson VE, Smith DH et al (2016) Chronic traumatic encephalopathy: the neuropathological legacy of traumatic brain injury. Annu Rev Pathol 11:21–45. https://doi.org/10.1146/annurev-pathol-012615-044116
5. Gavett BE, Stern RA, McKee AC (2011) Chronic traumatic encephalopathy: a potential late effect of sport-related concussive and subconcussive head trauma. Clin Sports Med 30(179–188):xi. https://doi.org/10.1016/j.csm.2010.09.007
6. Ramzan F, Khan MUG, Rehmat A et al (2019) A deep learning approach for automated diagnosis and multi-class classification of

Alzheimer's disease stages using resting-state fMRI and residual neural networks. J Med Syst 44:37. https://doi.org/10.1007/s10916-019-1475-2
7. Bramlett HM, Dietrich WD (2015) Long-term consequences of traumatic brain injury: current status of potential mechanisms of injury and neurological outcomes. J Neurotrauma 32: 1834–1848. https://doi.org/10.1089/neu.2014.3352
8. Alyenbaawi H, Allison WT, Mok SA (2020) Prion-like propagation mechanisms in tauopathies and traumatic brain injury: challenges and prospects. Biomol Ther 10. https://doi.org/10.3390/biom10111487
9. Vink R (2018) Large animal models of traumatic brain injury. J Neurosci Res 96:527–535. https://doi.org/10.1002/jnr.24079
10. Sorby-Adams AJ, Vink R, Turner RJ (2018) Large animal models of stroke and traumatic brain injury as translational tools. Am J Physiol Regul Integr Comp Physiol 315:R165–R190. https://doi.org/10.1152/ajpregu.00163.2017
11. Stewart AM, Ullmann JF, Norton WH et al (2015) Molecular psychiatry of zebrafish. Mol Psychiatry 20:2–17. https://doi.org/10.1038/mp.2014.128
12. Rihel J, Prober DA, Arvanites A et al (2010) Zebrafish behavioral profiling links drugs to biological targets and rest/wake regulation. Science 327:348–351. https://doi.org/10.1126/science.1183090
13. Nishimura Y, Okabe S, Sasagawa S et al (2015) Pharmacological profiling of zebrafish behavior using chemical and genetic classification of sleep-wake modifiers. Front Pharmacol 6:257. https://doi.org/10.3389/fphar.2015.00257
14. Cassar S, Adatto I, Freeman JL et al (2020) Use of zebrafish in drug discovery toxicology. Chem Res Toxicol 33:95–118. https://doi.org/10.1021/acs.chemrestox.9b00335
15. Alyenbaawi H, Kanyo R, Locskai LF et al (2021) Seizures are a druggable mechanistic link between TBI and subsequent tauopathy. elife 10. https://doi.org/10.7554/eLife.58744
16. Putnam LJ, Willes AM, Kalata BE et al (2019) Expansion of a fly TBI model to four levels of injury severity reveals synergistic effects of repetitive injury for moderate injury conditions. Fly (Austin) 13:1–11. https://doi.org/10.1080/19336934.2019.1664363
17. Katzenberger RJ, Loewen CA, Wassarman DR et al (2013) A drosophila model of closed head traumatic brain injury. Proc Natl Acad Sci U S A 110:E4152–E4159. https://doi.org/10.1073/pnas.1316895110
18. Maheras AL, Dix B, Carmo OMS et al (2018) Genetic pathways of neuroregeneration in a novel mild traumatic brain injury model in adult zebrafish. eNeuro 5. https://doi.org/10.1523/ENEURO.0208-17.2017
19. Markaki M, Tavernarakis N (2010) Modeling human diseases in caenorhabditis elegans. Biotechnol J 5:1261–1276. https://doi.org/10.1002/biot.201000183
20. Zulazmi NA, Arulsamy A, Ali I et al (2021) The utilization of small non-mammals in traumatic brain injury research: a systematic review. CNS Neurosci Ther 27:381–402. https://doi.org/10.1111/cns.13590
21. McCutcheon VP, Liu E, Wang Y, Wen X, Baker AJ (2016) A model of excitotoxic brain injury in larval zebrafish: potential application for high-throughput drug evaluation to treat traumatic brain injury. Zebrafish 13:161–169. https://doi.org/10.1089/zeb.2015.1188
22. McCutcheon V, Park E, Liu E et al (2017) A novel model of traumatic brain injury in adult zebrafish demonstrates response to injury and treatment comparable with mammalian models. J Neurotrauma 34:1382–1393. https://doi.org/10.1089/neu.2016.4497
23. Gan D, Wu S, Chen B, Zhang J (2020) Application of the zebrafish traumatic brain injury model in assessing cerebral inflammation. Zebrafish 17:73–82. https://doi.org/10.1089/zeb.2019.1793
24. Herzog C, Pons Garcia L, Keatinge M et al (2019) Rapid clearance of cellular debris by microglia limits secondary neuronal cell death after brain injury in vivo. Development 146: dev174698. https://doi.org/10.1242/dev.174698
25. Crilly S, Njegic A, Laurie SE et al (2018) Using zebrafish larval models to study brain injury, locomotor and neuroinflammatory outcomes following intracerebral haemorrhage. F1000Res 7:1617. https://doi.org/10.12688/f1000research.16473.2
26. Zhang J, Ge W, Yuan Z (2015) In vivo three-dimensional characterization of the adult zebrafish brain using a 1325 nm spectral-domain optical coherence tomography system with the 27 frame/s video rate. Biomed Opt Express 6: 3932–3940. https://doi.org/10.1364/boe.6.003932
27. Diotel N, Vaillant C, Gabbero C et al (2013) Effects of estradiol in adult neurogenesis and brain repair in zebrafish. Horm Behav 63:193–

207. https://doi.org/10.1016/j.yhbeh.2012.04.003

28. Kishimoto N, Shimizu K, Sawamoto K (2012) Neuronal regeneration in a zebrafish model of adult brain injury. Dis Model Mech 5:200–209. https://doi.org/10.1242/dmm.007336
29. Wu CC, Tsai TH, Chang C et al (2014) On the crucial cerebellar wound healing-related pathways and their cross-talks after traumatic brain injury in Danio rerio. PLoS One 9:e97902. https://doi.org/10.1371/journal.pone.0097902
30. Ayari B, El Hachimi KH, Yanicostas C et al (2010) Prokineticin 2 expression is associated with neural repair of injured adult zebrafish telencephalon. J Neurotrauma 27:959–972. https://doi.org/10.1089/neu.2009.0972
31. Kroehne V, Freudenreich D, Hans S et al (2011) Regeneration of the adult zebrafish brain from neurogenic radial glia-type progenitors. Development 138:4831–4841. https://doi.org/10.1242/dev.072587
32. März M, Schmidt R, Rastegar S et al (2011) Regenerative response following stab injury in the adult zebrafish telencephalon. Dev Dyn 240:2221–2231. https://doi.org/10.1002/dvdy.22710
33. Baumgart EV, Barbosa JS, Bally-Cuif L et al (2012) Stab wound injury of the zebrafish telencephalon: a model for comparative analysis of reactive gliosis. Glia 60:343–357. https://doi.org/10.1002/glia.22269
34. Skaggs K, Goldman D, Parent JM (2014) Excitotoxic brain injury in adult zebrafish stimulates neurogenesis and long-distance neuronal integration. Glia 62:2061–2079. https://doi.org/10.1002/glia.22726
35. Lim FT, Ogawa S, Parhar IS (2016) Spred-2 expression is associated with neural repair of injured adult zebrafish brain. J Chem Neuroanat 77:176–186. https://doi.org/10.1016/j.jchemneu.2016.07.005
36. Schmidt R, Beil T, Strähle U et al (2014) Stab wound injury of the zebrafish adult telencephalon: a method to investigate vertebrate brain neurogenesis and regeneration. JVE:e51753–e51753. https://doi.org/10.3791/51753
37. Ferrier GA, Kaur R, Park E et al (2017) An image-guided focused ultrasound system for generating acoustic shock waves that induce traumatic brain injury in wild-type zebrafish. Can Acoust 45:90–91
38. Weber B, Lackner I, Haffner-Luntzer M et al (2019) Modeling trauma in rats: similarities to humans and potential pitfalls to consider. J Transl Med 17:305. https://doi.org/10.1186/s12967-019-2052-7
39. Sempere L, Rodriguez-Rodriguez A, Boyero L et al (2019) Experimental models in traumatic brain injury: from animal models to in vitro assays. Med Intensiva (Engl Ed) 43:362–372. https://doi.org/10.1016/j.medin.2018.04.012
40. Marmarou CR, Prieto R, Taya K et al (2009) Marmarou weight drop injury model. In: Xu Z, Chen J, Xu XM, Zhang JH (eds) Animal models of acute neurological injuries. Humana Press, Totowa, NJ, pp 393–407
41. Kalish BT, Whalen MJ (2016) Weight drop models in traumatic brain injury. Methods Mol Biol 1462:193–209. https://doi.org/10.1007/978-1-4939-3816-2_12
42. Archer DP, McCann SK, Walker AM et al (2018) Neuroprotection by anaesthetics in rodent models of traumatic brain injury: a systematic review and network meta-analysis. Br J Anaesth 121:1272–1281. https://doi.org/10.1016/j.bja.2018.07.024
43. Schifilliti D, Grasso G, Conti A et al (2010) Anaesthetic-related neuroprotection: intravenous or inhalational agents? CNS Drugs 24:893–907. https://doi.org/10.2165/11584760-000000000-00000
44. Rowe RK, Harrison JL, Thomas TC et al (2013) Using anesthetics and analgesics in experimental traumatic brain injury. Lab Anim (NY) 42:286–291. https://doi.org/10.1038/laban.257
45. Kline AEDCE (2009) Contemporary in vivo models of brain trauma and a comparison of injury responses. In: Miller LP (ed) Head trauma: basic, preclinical, and clinical directions. John Wiley & Sons, New York, NY, pp 65–84
46. Statler KD, Alexander H, Vagni V et al (2006) Isoflurane exerts neuroprotective actions at or near the time of severe traumatic brain injury. Brain Res 1076:216–224. https://doi.org/10.1016/j.brainres.2005.12.106
47. Petraglia AL, Plog BA, Dayawansa S et al (2014) The spectrum of neurobehavioral sequelae after repetitive mild traumatic brain injury: a novel mouse model of chronic traumatic encephalopathy. J Neurotrauma 31:1211–1224. https://doi.org/10.1089/neu.2013.3255
48. McCarroll MN, Gendelev L, Kinser R et al (2019) Zebrafish behavioural profiling identifies GABA and serotonin receptor ligands related to sedation and paradoxical excitation. Nat Commun 10:4078. https://doi.org/10.1038/s41467-019-11936-w
49. van Lessen M, Shibata-Germanos S, van Impel A et al (2017) Intracellular uptake of

macromolecules by brain lymphatic endothelial cells during zebrafish embryonic development. elife 6. https://doi.org/10.7554/eLife.25932

50. Eakin K, Rowe RK, Lifshitz J (2015) Modeling fluid percussion injury: relevance to human traumatic brain injury. In: Brain Neurotrauma: molecular, neuropsychological, and rehabilitation aspects. F. H. Kobeissy, Boca Raton

51. Cernak I, Merkle AC, Koliatsos VE et al (2011) The pathobiology of blast injuries and blast-induced neurotrauma as identified using a new experimental model of injury in mice. Neurobiol Dis 41:538–551. https://doi.org/10.1016/j.nbd.2010.10.025

Chapter 2

Functional Genomics of Novel Rhabdomyosarcoma Fusion-Oncogenes Using Zebrafish

Matthew R. Kent, Katherine Silvius, Jack Kucinski, Delia Calderon, and Genevieve C. Kendall

Abstract

Clinical sequencing efforts continue to identify novel putative oncogenes with limited strategies to perform functional validation in vivo and study their role in tumorigenesis. Here, we present a pipeline for fusion-driven rhabdomyosarcoma (RMS) in vivo modeling using transgenic zebrafish systems. This strategy originates with novel fusion-oncogenes identified from patient samples that require functional validation in vertebrate systems, integrating these genes into the zebrafish genome, and then characterizing that they indeed drive rhabdomyosarcoma tumor formation. In this scenario, the human form of the fusion-oncogene is inserted into the zebrafish genome to understand if it is an oncogene, and if so, the underlying mechanisms of tumorigenesis. This approach has been successful in our models of infantile rhabdomyosarcoma and alveolar rhabdomyosarcoma, both driven by respective fusion-oncogenes, *VGLL2-NCOA2* and *PAX3-FOXO1*. Our described zebrafish platform is a rapid method to understand the impact of fusion-oncogene activity, divergent and shared fusion-oncogene biology, and whether any analyzed pathways converge for potential clinically actionable targets.

Key words Fusion-oncogene, Functional genomics, Rhabdomyosarcoma, Pediatric sarcoma, Zebrafish cancer models

1 Introduction

Fusion-oncogenes are typically generated by chromosomal translocations that juxtapose two genes that now have gain-of-function activities as a chimera. There is a predilection for fusion-oncogenes in sarcoma, with a predicated 20–49% of sarcomas containing a fusion-oncogene [1–4]. In the case of pediatric rhabdomyosarcoma, the fusion-oncogene is typically the defining oncogenic driver and tumors contain few other cooperating mutations [5–7]. This quiet genomic landscape is suggestive of the fusion-

Matthew Kent and Katherine Silvius contributed equally to this work

James F. Amatruda et al. (eds.), *Zebrafish: Methods and Protocols*, Methods in Molecular Biology, vol. 2707, https://doi.org/10.1007/978-1-0716-3401-1_2,

oncogene's role as a tumorigenic driver; however, clinical sequencing efforts continue to rapidly identify novel gene fusions with limited methods to functionally validate whether they are indeed oncogenes, and how they are leveraging transcriptional programs for oncogenesis. Developing rapid animal models is necessary to understand the features of the disease and compare and contrast these features within the same vertebrate context across other RMS fusion-driven models. Further, this in vivo strategy could identify clinically tractable therapeutic targets for rare cancers.

Zebrafish are a powerful complementary genetic model that we can rapidly implement to understand tumor biology and identify new therapeutic opportunities [8–10]. Eighty percent of disease-causing genes in humans have a zebrafish ortholog, making them highly relevant to studying disease [11]. Zebrafish are vertebrate models that we have utilized in the development of fusion-driven rhabdomyosarcoma, namely, for the *VGLL2-NCOA2* and *PAX3-FOXO1* fusions [12, 13]. Zebrafish have also been utilized in the fusion-negative form of rhabdomyosarcoma that is predominantly driven by mutated *kras* [14], and in other fusion-driven sarcoma models such as *EWS-FLI1*-driven Ewing Sarcoma [15] and *CIC-DUX4*-driven Ewing-like sarcoma [16]. These fusion-driven sarcoma models were all generated with the human form of the fusion-oncogene, highlighting the success of this strategy that will be detailed here. Further, zebrafish have many experimental advantages such as high fecundity, rapid external development, and the ability to inject hundreds of embryos in a single morning as mosaic transgenics and generate sarcomas that recapitulate the human disease [12]. These advantages make zebrafish a powerful tool to utilize in pediatric sarcoma research and to understand rare disease.

One of the challenges for developing rapid new sarcoma and rhabdomyosarcoma models is that the cell of origin is unknown; thus, it is not clear what tissue lineage the fusion-oncogene should be expressed in. Further, an alignment of events needs to take place for tumorigenesis to occur, including the appropriate cellular context, developmental timing, level of fusion-oncogene expression, and duration of fusion-oncogene expression. We have found that utilizing the CMV promoter to drive the human form of fusion-oncogenes captures this context in the case of rhabdomyosarcoma and can generate tumors in fish that recapitulate the human disease [12, 13]. However, depending on the subtype of rhabdomyosarcoma the inclusion of secondary cooperating mutations may be necessary. We have predominantly used the $tp53^{M214K}$ missense mutation and have found that in this context the human *PAX3-FOXO1* fusion generates rhabdomyosarcoma [13, 17]. This is similar to the mouse model of the disease that requires floxing of the *Trp53* allele [18]. However, in the case of infantile rhabdomyosarcoma, the human *VGLL2-NCOA2* fusion does not require a secondary cooperating event to generate tumors in the fish and has much earlier onset than the *PAX3-FOXO1* models [12, 13]. These differences highlight the true biology of the disease and suggest that zebrafish models capture these clinical differences in presentation. Importantly, as the number of zebrafish models for fusion-driven RMS increases, this increases the power of understanding in a shared model system how fusions behave relative to each other, and whether there are shared therapeutic vulnerabilities.

Zebrafish fusion-driven rhabdomyosarcoma tumors recapitulate the human disease at the histological and molecular level and can be utilized to understand disease biology and therapeutic opportunities. This is achieved through pathological review and transcriptomic analysis of tumor samples as would be done for patient samples. This strategy has dual utility of validating the tumor type and provides data to understand tumor biology. We have implemented this approach to identify how developmental gene targets are reactivated or inappropriately persist in fusion-driven tumors. Examples include the cooperation of *PAX3-FOXO1* and *HES3* in alveolar RMS and *VGLL2-NCOA2* and *ARF6* in infantile RMS [12, 13]. This highlights how zebrafish models have contributed to understanding fusion-driven rhabdomyosarcoma disease biology, and in the case of *PAX3-FOXO1* and *HES3*, have identified prognostic markers.

Here, we outline a streamlined approach to generate fusion-driven models of rhabdomyosarcoma for novel fusion-oncogenes that have been identified from clinical sequencing efforts (*see* Fig. 1). Key points include using the Tol2 transposon-based system [19, 20], generating an expression construct that includes the CMV promoter driving the expression of the human fusion-oncogene cDNA with a GFP viral 2A tag [13, 21], and initially injecting into wild-type and $tp53^{M214K}$ mutant zebrafish embryos for tumor modeling [17]. This approach will generate a mosaic model and will provide initial information if the fusion-oncogene is tumorigenic or not. We utilize this as a first step in validating whether this particular human fusion is in fact an oncogene, and if so, whether we can study it using our zebrafish systems. As the number of fusion-driven RMS zebrafish models continues to increase, the advantages of the system and the power to identify nuances in disease biology will continue to improve (*see* **Note 1**).

2 Materials

2.1 Zebrafish Lines

1. Wild-type zebrafish lines: AB, TL, and WIK obtained from ZIRC (https://zebrafish.org) or other stock center.
2. $tp53^{M214K}$ mutant zebrafish are used as homozygotes [17].

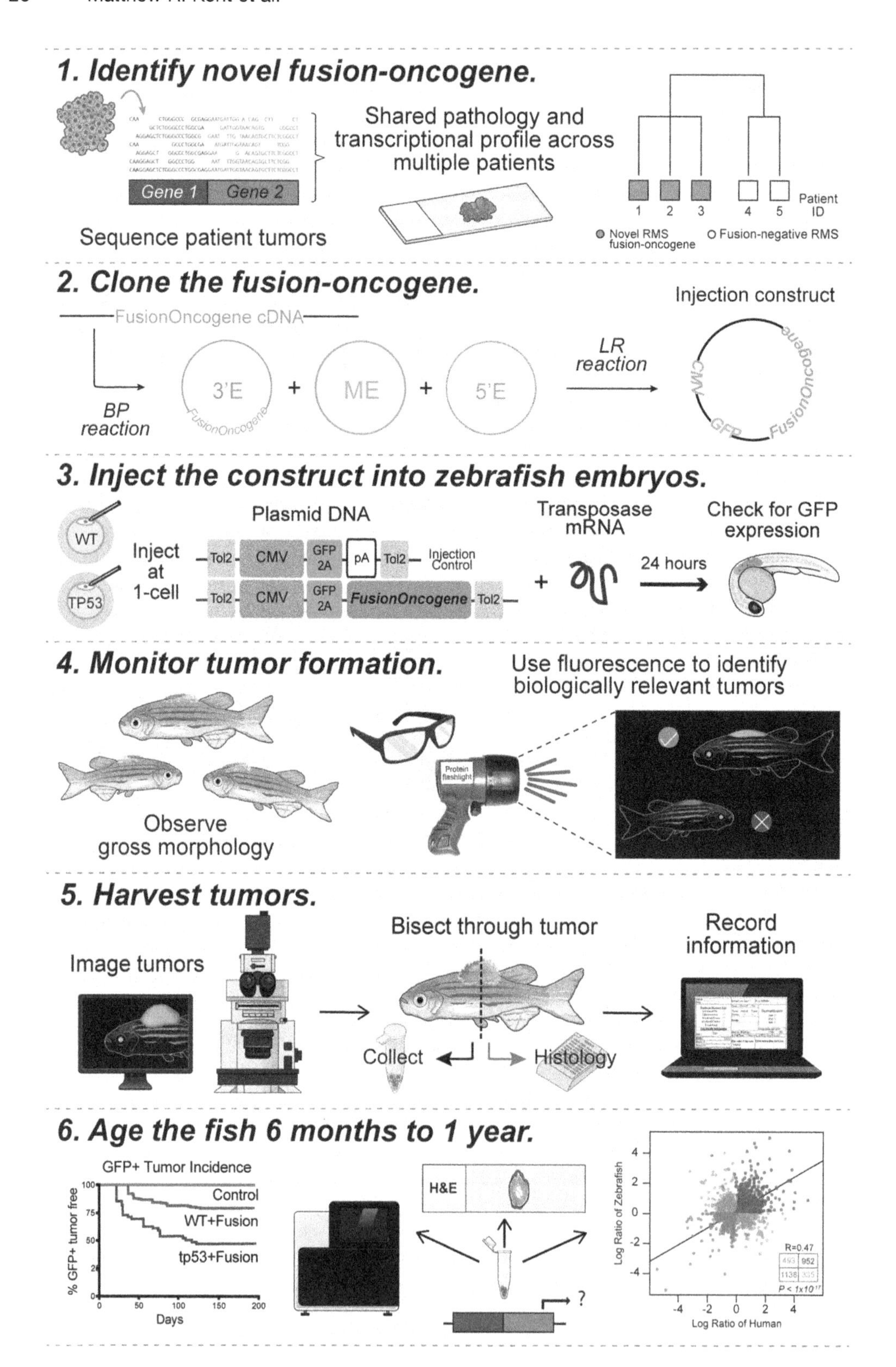

Fig. 1 Overview of the workflow for zebrafish as a functional genomics model to validate and study potential rhabdomyosarcoma (RMS) fusion-oncogenes. The zebrafish system provides a flexible platform to address the

2.2 Cloning Materials

2.2.1 Plasmids

1. 5′ entry plasmid, p5E-CMV/SP6 [21].
2. 5′ donor plasmid, pDONRP2R-P3 [21].
3. Middle entry plasmid, pmE GFP2A [13].
4. Middle entry plasmid, pmE mCherry2A [13].
5. 3′ entry plasmid, USER GENERATED.
6. 3′ entry polyA plasmid [21].
7. 3′ donor plasmid, pDONRP2R-P3 [21].
8. Destination plasmid, pDestTol2pA2 [21].
9. pCS2FA-transposase plasmid [19, 20].

2.2.2 Supplies

1. One Shot TOP10 cells.
2. BP Clonase II Enzyme Mix.
3. LR Clonase II Plus Enzyme Mix.
4. Platinum *Pfx* DNA Polymerase.
5. QIAquick Gel Extraction Kit.
6. QIAprep Spin Miniprep Kit.
7. TE Buffer pH 8.
8. Kanamycin.
9. Ampicillin.
10. Chloramphenicol.

Fig. 1 (continued) growing need for functional validation of the novel gene fusions that are being rapidly discovered through clinical sequencing efforts. In this strategy, putative fusion-oncogenes identified from patient tumors are prioritized for modeling based on the identification of shared pathology and transcriptional profiles across multiple patients with a similar disease presentation. The cDNA sequence of the selected fusion is used to generate a construct with the fusion-oncogene driven by a high-level CMV promoter and linked to a GFP Viral2A tag for transgene visualization. The fusion-oncogene is then injected into wild-type and $tp53^{M214K}$ embryos and is integrated into the zebrafish genome through Tol2-mediated transgenesis. Separate cohorts injected with a polyadenylated construct serve as experimental controls. A potent fusion-oncogene may lead to tumors by 1–6 months. During this time, fish are monitored for tumor formation by regular observation of gross morphology. GFP fluorescence detected with a royal blue protein flashlight can be used to rapidly identify biologically relevant tumors. Throughout the course of the study, fish are euthanized in accordance with the respective IACUC protocol, and each is screened and imaged using a fluorescent microscope. Tumor tissue is collected for downstream applications, and then a transverse cut (or cut that is most appropriate for the tumor presentation) is made through the tumor to send for histological analysis. Appropriate records should be kept as tumors are harvested (e.g., *see* Fig. 2). This process should be continued for 6 months to 1 year depending on the activity of the fusion-oncogene, with all fish euthanized at the end of the study. Pathology and immunohistochemistry for classical RMS diagnostic markers can be used to validate the putative fusion-oncogene as a true cancer-driving mutation. Additionally, the tumor tissue can be used to investigate the underlying disease biology and tumorigenic mechanisms, rendering this transgenic zebrafish system a powerful in vivo vertebrate model for studying rare disease

A

Initials:
Date:

Entered into logs?

TI or DFISH#:

Reason for Screening Fish:
Intentional/Old
Intentional/Sick
Intentional/Tumor
Intentional/Control
Found Dead

Collected tissue ○Yes ○No

Tissue Type(s): Normal Tumor

Details:

Fluorescent Protein(s)
GFP
RFP
BFP

Image taken and saved?

Fish Specific Information

EID: TID:

Send for Histology: ○Yes ○No

@ Cut Type: ○Transverse AND is *already* precut

Date of Birth:
Parent #1:
Parent #2:
Injection Construct: ○ N/A

Else, select 1 cut type:
○Sagittal
○Coronal
○Step Sections

Extra instructions for Core:

Labels and Miscellaneous Information:

B

Widget HTML Editor

Source

```
        <td colspan="3" style="width: 360px;"><b>@ Cut Type: </b><input name="cut_type"
type="radio" />Transverse AND is <em>already</em> precut</td>
      </tr>
      <tr>
        <td colspan="2" style="width: 227px; background-color: rgb(204, 204, 204);">Date of
Birth: <input name="birth_date" size="12" type="text" /></td>
        <td colspan="1" rowspan="5" style="width: 155px;">
          <p><b>Else, select 1 cut type:</b><br />
            <input name="cut_type" type="radio" />Sagittal<br />
            <input name="cut_type" type="radio" />Coronal<br />
            <input name="cut_type" type="radio" />Step Sections</p>
        </td>
        <td colspan="2" rowspan="5">
          <p><strong>Extra instructions for Core:</strong><br />
            <textarea cols="25" name="instructions_histology" rows="7"></textarea></p>
        </td>
      </tr>
      <tr>
        <td colspan="2" style="width: 227px;">Parent #1: <input maxlength="100"
name="parent_one" size="17" type="text" /></td>
      </tr>
      <tr>
        <td colspan="2" style="width: 227px;">Parent #2: <input maxlength="100"
name="parent_two" size="17" type="text" /></td>
```

Fig. 2 Tumor collection datasheet for streamlined documentation. (**a**) The datasheet includes examples of information to record when euthanizing fish included in your tumor study. One datasheet would be filled out

11. LB Broth.
12. LB Agar.

2.3 Injection Materials

2.3.1 Injection Apparatus

1. Microinjector.
2. Stereomicroscope.
3. Micropipette needle puller.
4. Borosilicate glass standard wall tubing with filament.

2.3.2 Injection Components

1. Petri dishes.
2. Microloader tips.
3. Microscope stage calibration slide.
4. Mineral oil.
5. Embryo injection mold.
6. Agarose.

2.3.3 Tol2 mRNA Synthesis for DNA Injections

1. SP6 Transcription Kit.
2. NotI.

2.3.4 Injection Mixes and Preparation

1. 0.5% phenol red solution.
2. 30× Danieau's buffer: 1740 mM sodium chloride, 21 mM potassium chloride, 12 mM magnesium chloride heptahydrate, 18 mM calcium nitrate tetrahydrate, 150 mM HEPES.
3. Tol2 transposase mRNA (user prepared, see above).

Fig. 2 (continued) per tumor fish or per control group. The data tracked includes the following: *euthanasia information* such as lab member initials, date and purpose of screening, verification that data is appropriately recorded in other applicable databases, and the form identifier denoting the unique fish (DFISH#) or designating a tumor incidence group (TI) in the case of euthanizing a cohort of control fish; *fish-specific information* such as unique experiment (EID) and tank (TID) identifiers attached to the respective fish, date of birth, parental cross or genotype, and injection construct if applicable; *collected tissue information* to designate if there is normal and/or tumor tissue available, along with a section for extra details; *imaging information* including verification that an image has been taken and saved, the file path, and checkboxes to indicate the fluorescent protein(s) expressed; and *histology information* including cut type and extra instructions for processing. Additionally, there is space for extra miscellaneous information at the bottom. The datasheets can be printed out and manually marked up. Alternatively, this datasheet can be utilized as an HTML-based widget on an electronic lab notebook such as LabArchives where it can be modified as desired (*see* Subheading 4). TI = Tumor Incidence, TID = Tank ID (unique tank identifier), DFISH# = Dead Fish Number (unique fish identifier). (**b**) Example of HTML editor. See Subheading 4 or URL https:/mynotebook.labarchives.com/share/Kendall%2520Lab%2520Notebook/NjM1Ljd8NTk1NDY5LzQ4OS9UcmVlTm9kZS80NDgxMTU2NTI8MTYxMy42OTk5OTk5OTk4

2.4 Zebrafish Screening Material

2.4.1 Microscope

1. Fluorescence stereomicroscope with GFP filter.

2.4.2 Embryo Preparation

1. Pronase: 50 mg/mL dissolved in E3 buffer.
2. Tricaine: 4 mg/mL dissolved in miliQ water, pH to 7.0 with Tris-HCl. Tricaine is diluted to 0.2 mg/mL for anesthetizing and 2 mg/mL for euthanizing.
3. E3 Buffer: 5 mM sodium chloride, 0.17 mM potassium chloride, 0.33 mM calcium chloride dihydrate, 0.33 mM magnesium chloride heptahydrate.

2.5 Tumor Identification and Collection

2.5.1 Tumor Identification

1. GFP protein flashlight (e.g., NIGHTSEA).

2.5.2 Fish Euthanasia and Collection of Tumor Samples

1. Tumor collection kit: razor blades, dissecting blunt-pointed forceps, 140 mm × 20 mm petri dishes, disposable transfer pipettes, E3 buffer (*see* Subheading 2.4.2), tricaine for both anesthetizing and euthanizing, aquarium fish net, Ziploc bags, 70% ethanol, kimwipes, labeling tape, pencils, tumor collection datasheet (*see* Fig. 2).

2.5.3 Liquid Nitrogen Collection

1. Benchtop liquid nitrogen containers.
2. 1.5 mL microcentrifuge tubes.

2.5.4 Histology Collection

1. Biopsy cassettes.
2. 4% Paraformaldehyde (PFA): dissolved 4% by volume in 1x phosphate buffer saline (PBS). pH to 7.4.
3. Glass wide-mouth Erlenmeyer flask.
4. 0.5 M EDTA pH 7.8.

2.6 Histological Assessment

2.6.1 Hematoxylin and Eosin Stain

1. Hematoxylin stain solution.
2. Eosin Y solution.

2.6.2 RT-PCR for myoD, myoG, Desma

1. RNA isolation kit.
2. cDNA synthesis kit.
3. Taq polymerase.

2.6.3 Primer Sequences for Gene Targets [12, 13]

1. *myod*: forward primer (ATCCAACTGCTCTGATGGC), reverse primer (AGACAATCCAAACTCGACACC).
2. *myog:* forward primer (GCTATACAGTACATCGAGAGGC), reverse primer (AGAGCCCTGATCACTAGAGG).
3. *desma:* forward primer (GCAGTGAACAAGAATAACGAGG), reverse primer (CACTCATTTGCCTCCTCAGAG).
4. *rpl13a*: forward primer (CGGTCGTCTTTCCGCTATT), reverse primer (TTCCAGAGATGTTGATACCCTCAC).

2.6.4 Immunohistochemistry

1. 60 °C hybridization oven.
2. Histological clearing agent.
3. Trilogy solution for deparaffinization and demasking (Cell Marque Corporation).
4. Pressure cooker.
5. 3% and 0.03% H_2O_2.
6. 1% Bovine Serum Albumin (BSA).
7. 1× PBST (Phosphate Buffered Saline (PBS): 0.1% Tween-20 dissolved in 1× PBS.
8. Primary antibody against fusion protein of interest.
9. Secondary antibody with HRP conjugate matching primary antibody.
10. DAB solution.
11. Hematoxylin stain solution.

2.7 RNA Isolation Materials

2.7.1 Sample/RNA Preparation

1. Electric pestle motor.
2. RNase-free disposable pellet pestles.
3. RNA isolation kit.
4. RNase AWAY™ surface decontaminant.
5. RNase-Free DNase kit.
6. Bioanalyzer instrument or strategy to determine RNA integrity.

2.7.2 RNA Library Preparation

1. Poly(A) mRNA enrichment kit.
2. RNA library preparation kit.

2.8 Statistical Analysis

1. GraphPad Prism.

3 Methods

3.1 Choosing a Rhabdomyosarcoma Fusion-Oncogene to Study

Clinical and biological significance is critical for choosing a fusion-oncogene to study. The implementation of clinical sequencing efforts is rapidly identifying new potential cancer driver genes with limited mechanisms to functionally validate their effects. This is true for rhabdomyosarcoma, in which patient tumors that do not contain the classical genetic drivers, including *RAS* mutations or *PAX3/7-FOXO1* fusions, will undergo RNA-seq. This has yielded a variety of new fusions or fusion variants that could represent variation of disease biology and unique processes and highlights the pressing need for nimble genetic models to study their biological significance. We prioritize which fusions to study in zebrafish models based on these criteria: (1) there is no existing animal model; (2) fusions are present in multiple patient tumors that have shared transcriptional signatures; and (3) patients have a significant therapeutic need and limited treatment options. Based on these criteria, we select fusion-oncogenes to develop novel transgenic zebrafish models.

3.2 Cloning Your Fusion Protein into a Gateway Expression Plasmid for Tol2 Transgenesis

3.2.1 Create 3′ Entry Fusion and Control Plasmids

1. Design primers for the coding sequence of your gene fusion for cloning into pDONRP2R-3, the 3′ donor plasmid. Add an attb2r sequence in your forward primer, and an attb3 sequence and stop codon in your reverse primer.
   (a) FWD: 5′-GGGGACAGCTTTCTTGTACAAAGTGGCC ... fusion sequence-3′.
   (b) REV: 5′-GGGGACAACTTTGTATAATAAAGTTGC ... stop codon ... fusion sequence-3′.
2. Amplify using a high-fidelity DNA polymerase.
3. Gel purify your PCR product.
4. Set up a BP reaction with 50 fmol of purified PCR product and 50 fmol of the pDONR plasmid. Add TE pH 8 buffer to 8 μL, and 2 μL of BP Clonase II enzyme.
5. Incubate overnight at room temperature.
6. Add 1 μL of proteinase K and incubate at 37 °C for 10 min.
7. Transform 1 μL of BP reaction into chemically competent cells.
8. Plate on kanamycin (50 μg/mL) LB agar plates and incubate overnight at 37 °C.
9. Miniprep colonies and confirm assembly by Sanger sequencing. We find the BP reaction is efficient enough that a colony PCR or restriction digest screen is usually not necessary.

3.2.2 Create Expression Constructs Using Either Gene Fusion 3′ Entry or Control Plasmid

1. Set up an LR reaction with 20 fmol of pDESTtol2 plasmid and 10 fmol each of 5′ entry plasmid (CMV promoter), middle entry plasmid (GFP-P2A), and 3′ entry plasmid (either gene fusion or polyA control). Add TE buffer pH 8 to 8 μL, then add 2 μL of LR Clonase II enzyme.
2. Incubate overnight at room temperature.
3. Add 2 μL of proteinase K and incubate at 37 °C for 10 min.
4. Transform 1–3 μL of LR reaction into chemically competent cells and plate on ampicillin (100 μg/mL) LB agar plates. Incubate overnight at 37 °C.
5. Miniprep positive colonies and confirm by Sanger sequencing.

3.3 Injecting Expression Construct

1. The day before injections, set up a minimum of six breeding chambers, each containing two to three males and two to three females separated by dividers. You will need enough embryos for three different groups: uninjected, GFP-control injection, and GFP-fusion-oncogene injection.
2. Injection mixes can be made either the day of injections, or the day before and stored at −80 °C. For novel fusions, it is suggested to use a variety of concentrations of the fusion expression construct, such as 25 ng/μL, 50 ng/μL, and 100 ng/μL. Combine these with 50 ng/μL of Tol2 transposase mRNA in 0.3× Danieau's buffer and 0.1% phenol red.
3. Inject at least 300 one-cell or two-cell stage wild-type embryos each into the cell body with either the control or fusion expression vector plasmid mixture.
4. Incubate embryos at 28.5 °C in 0.5 mg/L methylene blue in E3 water. Remove dead embryos at 6 h postfertilization, changing the water for fresh methylene blue in E3 water. Remove dead embryos once again at 24 hours postfertilization, and daily from then on.
5. Dechorionate embryos manually with forceps or using pronase at 24 h postfertilization.
6. Screen embryos for fluorescence at 1 day. Confirm that the majority of the embryos are GFP-positive.
7. Separate embryos into petri dishes of 20–30 embryos each.
8. Place embryos on system nursery at 5 days postfertilization, with no more than 30 per 1.8 L nursery tank. At 30 days postfertilization, split these fish into 2 × 2.8 L tanks, depending on the survival, with a density of ~5 fish/L. The target number of fish that reach juvenile stage is 100 fish per group.
9. Repeat this process the following week, injecting into at least 300 one-cell or two-cell stage *tp53*^{M214K} embryos.

3.4 Screening for Tumors

1. Beginning at ~15 days, screen fish for fluorescent cell masses using the protein flashlight. As soon as you can see the fish without a microscope, you can start screening.
2. Screen fish weekly on Mondays for fluorescent tumors. Depending on the genetics of your chosen gene fusion, tumors may begin forming anywhere from 15 days to 1 year postfertilization.
3. Tumors will present as GFP-positive masses. There may also be accompanying physical issues such as depigmentation or deformity (*see* Fig. 3a). Once a tumor is identified, monitor the fish. If the tumor size is at least 20% of the fish, or if the fish is failing

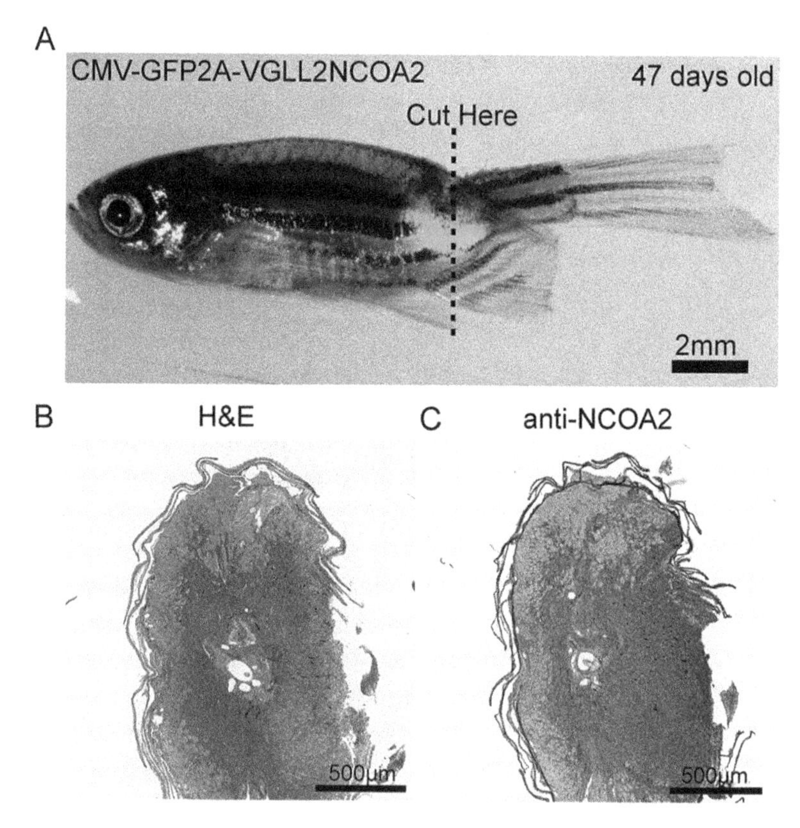

Fig. 3 Example of fusion-oncogene-driven tumor presentation in zebrafish. Wild-type zebrafish were injected with CMV-GFP2A-*VGLL2NCOA2* construct at the single-cell stage and were monitored for tumor formation. (**a**) At 47 days of age, this fish presented with a curved tail and was euthanized and imaged with DIC and GFP fluorescence. Scale bar, 2 mm. Transverse cuts were made through the GFP-positive mass and serial section obtained for (**b**) hematoxylin and eosin staining confirming that the morphology of these cells are consistent with a mesenchymal tumor and human sarcoma and (**c**) that the tumor cells express the human form of the fusion by immunohistochemistry. Here, a human anti-NCOA2 antibody was used to detect the human VGLL2-NCOA2 fusion. Scale bar, 500 microns

to thrive, mark the tank so that the fish is not fed on Tuesday and can be harvested on Wednesday. This is because certain foods are autofluorescent, potentially confounding actual tumor fluorescent signal.

4. On Wednesday, after marking a tank, sacrifice the fish according to approved institutional animal protocols.
5. Image the fish and tumor both with DIC and fluorescence on a fluorescent stereomicroscope (*see* Fig. 3a).
6. Use the tumor collection datasheet (*see* Fig. 2) to record information on the euthanized fish. See **Note 2**.
7. Make a transverse cut through the tumor.
8. Snap-freeze half of the tumor in liquid nitrogen for downstream molecular analyses (*see* Subheading 3.6.1).
9. In addition, snap-freeze nontumor tissue in liquid nitrogen for downstream molecular analysis.
10. Place the other half of the fish in a histology cassette and then in 4% paraformaldehyde (PFA) rocking for 24 h at room temperature, or 48 h at 4 °C.
11. Transfer the fish to 0.5 M EDTA, pH 7.8. Incubate rocking at room temperature for a minimum of 5 days.
12. Prepare the fixed tissue for downstream processing (*see* Subheading 3.6.3).

3.5 Ending the Experiment

1. After 1 year postfertilization, sac all remaining tumor injected and control fish.
2. Screen all tumor and control fish for GFP positivity.
3. Record all remaining tumors, save tissue, and fix and collect tissue as in Subheading 3.4.
4. Plot the overall tumor incidence in Prism using a log-rank Kaplan-Meier analysis.

3.6 Downstream Processing (See Note 3)

3.6.1 RNA-Seq

1. Isolate RNA from frozen samples using the QIAGEN RNeasy Mini kit with on-column DNase digestion.
2. Submit RNA samples for library preparation and RNA-seq. For RNA-seq, arrange for 150 bp paired end reads, with 60–80 million reads per sample. Run samples on a NovaSeq Sp (for example), which can process up to 32 samples per flow cell. This allows for a differential expression and alternative splicing analysis.

3.6.2 RT-PCR

1. Use a cDNA synthesis kit to generate cDNA from your RNA samples. As little as 100 ng of RNA can be used for the reaction. However, lower input concentrations might not identify lower-expression RNAs.

2. Using RT-PCR primers (*myod, myog, desma*, or other targets of interest) and your generated cDNA, set up and run an RT-PCR reaction according to your PCR protocol.
3. Run your product on an agarose gel to determine if diagnostic rhabdomyosarcoma markers are expressed. See Watson and LaVigne et al. [12].

3.6.3 Histology and Immunohistochemistry

1. Embed fixed and decalcified tissue/tumor in paraffin and take 5–8 μm sections.
2. Perform an H&E stain on one section to determine gross morphology of the tissue/tumor through review with a pediatric pathologist (*see* Fig. 3b, c).
3. Use additional sections for either immunohistochemistry to detect pertinent markers, such as the expression of the human form of the integrated fusion-oncogene (*see* Fig. 3b, c) [12].

3.7 Verify that Zebrafish Tumors Recapitulate the Human Disease

1. Conduct differential expression analysis comparing RNA-seq data from your zebrafish tumors to another zebrafish sarcoma such as CIC-DUX4-driven sarcoma, and then patient tumors with the same gene fusion compared to patient CIC-DUX4 sarcoma (see Watson and LaVigne et al. [12]).
2. Perform an Agreement of Differential Expression (AGDEX) analysis [22] on your differential expression results to determine the level of similarity between your zebrafish tumor to the patient tumor (see Watson and LaVigne et al. [12]).

4 Notes

1. Evaluating fusion-oncogene cooperating genes in transgenic zebrafish models.

   Mosaic fusion-oncogene zebrafish models can also be used to evaluate the effect of putative cooperating genes as these systems are especially amenable to loss and gain-of-function genetic approaches. For genes that have reduced expression in the cancer, one strategy is to use CRISPR/Cas9 systems to generate a zebrafish germline knockout of the ortholog of the cooperating gene [23]. Then, the CMV-Fusion Oncogene injection model can be implemented to determine if the knockout suppresses the tumor phenotype. Alternatively, interest may revolve around overexpression of a gene in tumors. In that case, the human form of the cooperating gene can be subcloned into the Gateway expression system. The one difference being that an mCherry viral 2A tag is included to demarcate its expression from the CMV-GFP2A-Fusion Oncogene construct. Then, the CMV constructs can be co-injected into the developing fish, and fish evaluated for tumor formation as detailed above. Tumors that develop because of both the fusion-oncogene and the cooperating gene should be monitored for fluorescence of both GFP-Fusion Oncogene and mCherry-Cooperating Gene. We previously used this strategy to understand how overexpression of human HES3 impacted PAX3FOXO1 function in the developing fish [13].

2. DNA isolation and genomic analysis.

   Whole genome analysis of zebrafish tumors is a compelling next step when verifying the model system and determining if there are additional mutations that are acquired during the tumorigenic process. If you would like to collect tumors and perform Whole Genome Sequencing, you should also collect a "normal" sample from the same fish. The normal sample should be from the same fish and represent tissue that is not fluorescent. This way you can sequence a tumor-normal pair, much like in patients. This is especially important in zebrafish models because they are not inbred; therefore, every individual could be slightly different, making SNP calling a challenge when comparing to the reference genome.

3. Tumor collection datasheet HTML code.

   The HTML code can be copied into the source code editor of an electronic lab notebook such as LabArchives to create an interactive widget and further streamline data entry via standardized electronic form. The widget can be modified and tailored as desired to suit the needs of the particular lab.

```
<tbody>
<tr>
<td colspan="2" style="width: 227px;">Initials: <input max-
length="20" name="initials_sac" size="5" type="text" /><br />
Date: <input name="sac_date" size="12" type="text" /></td>
<td style="width: 155px;">Entered into logs? <input name="-
data_logged" type="checkbox" /></td>
<td colspan="2" style="width: 201px;">TI or DFISH#: <input
maxlength="20" name="fish_identifier" size="10" type="text" /
></td>
</tr>
<tr>
<td colspan="2" rowspan="3" style="text-align: center;
width: 227px;">
<p><u>Reason for Screening Fish</u>:<br />
Intentional/Old          <input name="euthanasia_purpose"
type="radio" /><br />
Intentional/Sick         <input name="euthanasia_purpose"
type="radio" /><br />
```

```
 Intentional/Tumor        <input name="euthanasia_purpose"
type="radio" /><br />
 Intentional/Control      <input name="euthanasia_purpose"
type="radio" /> <br />
     Found Dead           <input name="euthanasia_purpose"
type="radio" />  </p>
 </td>
 <td colspan="2" rowspan="1" style="width: 175px;">Collected
tissue <input name="tissue_collected" type="radio" />Yes <in-
put name="tissue_collected" type="radio" />No</td>
 <td colspan="1" rowspan="3" style="width: 181px;">
 <p style="text-align: center;"><u>Fluorescent Protein(s)</
u></p>
 <p style="text-align: center;">GFP  <input name="pro-
tein_gfp" type="checkbox" /><br />
 RFP  <input name="protein_rfp" type="checkbox" /><br />
 BFP  <input name="protein_bfp" type="checkbox" /></p>
 </td>
 </tr>
 <tr>
 <td colspan="2" rowspan="1" style="width: 175px;">
 <p> Tissue  Normal   Tumor<br />
 Type(s):     <input name="tissue_normal" type="checkbox" />
<input name="tissue_tumor" type="checkbox" /></p>
 </td>
 </tr>
 <tr>
 <td colspan="2" rowspan="2" style="width: 175px;">
 <p><strong>Details: <br />
 <textarea cols="21" name="tissue_details" rows="3"></textar-
ea></strong></p>
 </td>
 </tr>
 <tr>
 <td colspan="2" style="width: 227px; text-align: center;"><-
strong><u>Fish Specific Information</u></strong></td>
 <td style="width: 181px;">Image taken and saved? <input
name="image_taken" type="checkbox" /></td>
 </tr>
 <tr>
 <td rowspan="2">
 <p>EID: <input maxlength="15" name="experiment_identifier"
size="5" type="text" /></p>
 </td>
 <td rowspan="2" style="width: 118px;">TID: <input max-
length="15" name="tank_identifier" size="6" type="text" /></
td>
 <td colspan="3" style="width: 360px;">Send for Histology:
```

```
<input name="histology" type="radio" />Yes    <input name="-
histology" type="radio" />No</td>
 </tr>
 <tr>
  <td colspan="3" style="width: 360px;"><b>@ Cut Type: </
b><input name="cut_type" type="radio" />Transverse AND is
<em>already</em> precut</td>
 </tr>
 <tr>
 <td colspan="2" style="width: 227px; background-color: rgb
(204, 204, 204);">Date of Birth: <input name="birth_date"
size="12" type="text" /></td>
 <td colspan="1" rowspan="5" style="width: 155px;">
 <p><b>Else, select 1 cut type:</b><br />
 <input name="cut_type" type="radio" />Sagittal<br />
 <input name="cut_type" type="radio" />Coronal<br />
 <input name="cut_type" type="radio" />Step Sections</p>
 </td>
 <td colspan="2" rowspan="5">
 <p><strong>Extra instructions for Core:</strong><br />
  <textarea cols="25" name="instructions_histology"
rows="7"></textarea></p>
 </td>
 </tr>
 <tr>
  <td colspan="2" style="width: 227px;">Parent #1: <input
maxlength="100" name="parent_one" size="17" type="text" /></
td>
 </tr>
 <tr>
  <td colspan="2" style="width: 227px;">Parent #2: <input
maxlength="100" name="parent_two" size="17" type="text" /></
td>
 </tr>
 <tr>
  <td colspan="2" style="width: 227px; background-color: rgb
(204, 204, 204);">
  <p>Injection Construct:      <input name="no_injection_con-
struct" type="radio" /> N/A<br />
 <input name="injection_construct" size="27" type="text" /></
p>
 </td>
 </tr>
 </tbody>
<p><u>Labels and Miscellaneous Information</u>:</p>
  <p><textarea cols="79" name="information_miscellaneous"
rows="4"></textarea></p>
```

Acknowledgments

J.K. is supported by a Graduate Student Research Award from the Abigail Wexner Research Institute's Trainee Association at Nationwide Children's Hospital. D.C is supported by a Graduate Enrichment Fellowship from The Ohio State University. G.C.K is supported by an Alex's Lemonade Stand Foundation "A" Award and a V Foundation for Cancer Research V Scholar Grant.

References

1. Mitelman F, Johansson B, Mertens F (2007) The impact of translocations and gene fusions on cancer causation. Nat Rev Cancer 7:233–245
2. Mitelman F, Johansson B, Mertens F (2021). Mitelman database of chromosome aberrations and gene fusions in cancer. https://mitelmandatabase.isb-cgc.org
3. Mertens F, Antonescu CR, Mitelman F (2016) Gene fusions in soft tissue tumors: recurrent and overlapping pathogenetic themes. Genes Chromosomes Cancer 55:291–310
4. Watson S, Perrin V, Guillemot D et al (2018) Transcriptomic definition of molecular subgroups of small round cell sarcomas. J Pathol 245:29–40
5. Barr FG, Galili N, Holick J et al (1993) Rearrangement of the PAX3 paired box gene in the paediatric solid tumour alveolar rhabdomyosarcoma. Nat Genet 3:113–117
6. Davis RJ, D'Cruz CM, Lovell MA et al (1994) Fusion of PAX7 to FKHR by the variant t(1;13)(p36;q14) translocation in alveolar rhabdomyosarcoma. Cancer Res 54:2869–2872
7. Shern JF, Chen L, Chmielecki J et al (2014) Comprehensive genomic analysis of rhabdomyosarcoma reveals a landscape of alterations affecting a common genetic axis in fusion-positive and fusion-negative tumors. Cancer Discov 4:216–231
8. Kashi VP, Hatley ME, Galindo RL (2015) Probing for a deeper understanding of rhabdomyosarcoma: insights from complementary model systems. Nat Rev Cancer 15:426–439
9. Amatruda JF (2021) Modeling the developmental origins of pediatric cancer to improve patient outcomes. Dis Model Mech 14
10. Patton EE, Zon LI, Langenau DM (2021) Zebrafish disease models in drug discovery: from preclinical modelling to clinical trials. Nat Rev Drug Discov 20:611–628
11. Howe K, Clark MD, Torroja CF et al (2013) The zebrafish reference genome sequence and its relationship to the human genome. Nature 496:498–503
12. Watson S, LaVigne C, Xu L et al (2023) VGLL2-NCOA2 leverages developmental programs for pediatric sarcomagenesis. Cell Rep 42
13. Kendall GC, Watson S, Xu L et al (2018) PAX3-FOXO1 transgenic zebrafish models identify HES3 as a mediator of rhabdomyosarcoma tumorigenesis. elife 7
14. Langenau DM, Keefe MD, Storer NY et al (2007) Effects of RAS on the genesis of embryonal rhabdomyosarcoma. Genes Dev 21:1382–1395
15. Leacock SW, Basse AN, Chandler GL et al (2012) A zebrafish transgenic model of Ewing's sarcoma reveals conserved mediators of EWS-FLI1 tumorigenesis. Dis Model Mech 5:95–106
16. Watson S, Kendall GC, Rakheja D et al (2019) CIC-DUX4 expression drives the development of small round cell sarcoma in transgenic zebrafish: a new model revealing a role for ETV4 in CIC-mediated sarcomagenesis. bioRxiv
17. Berghmans S, Murphey RD, Wienholds E et al (2005) tp53 mutant zebrafish develop malignant peripheral nerve sheath tumors. Proc Natl Acad Sci U S A 102:407–412
18. Keller C, Arenkiel BR, Coffin CM et al (2004) Alveolar rhabdomyosarcomas in conditional Pax3:Fkhr mice: cooperativity of Ink4a/ARF and Trp53 loss of function. Genes Dev 18:2614–2626
19. Kawakami K, Takeda H, Kawakami N et al (2004) A transposon-mediated gene trap approach identifies developmentally regulated genes in zebrafish. Dev Cell 7:133–144
20. Urasaki A, Morvan G, Kawakami K (2006) Functional dissection of the Tol2 transposable element identified the minimal cis-sequence and a highly repetitive sequence in the subterminal region essential for transposition. Genetics 174:639–649

21. Kwan KM, Fujimoto E, Grabher C et al (2007) The Tol2kit: a multisite gateway-based construction kit for Tol2 transposon transgenesis constructs. Dev Dyn 236:3088–3099

22. Pounds S, Gao CL, Johnson RA et al (2011) A procedure to statistically evaluate agreement of differential expression for cross-species genomics. Bioinformatics 27:2098–2103

23. Talbot JC, Amacher SL (2014) A streamlined CRISPR pipeline to reliably generate zebrafish frameshifting alleles. Zebrafish 11:583–585

Chapter 3

Methods to Study Liver Disease Using Zebrafish Larvae

Elena Magnani, Anjana Ramdas Nair, Ian McBain, Patrice Delaney, Jaime Chu, and Kirsten C. Sadler

Abstract

Liver disease affects millions of people worldwide, and the high morbidity and mortality is attributed in part to the paucity of treatment options. In many cases, liver injury self-resolves due to the remarkable regenerative capacity of the liver, but in cases when regeneration cannot compensate for the injury, inflammation and fibrosis occur, creating a setting for the emergence of liver cancer. Whole animal models are crucial for deciphering the basic biological underpinnings of liver biology and pathology and, importantly, for developing and testing new treatments for liver disease before it progresses to a terminal state. The cellular components and functions of the zebrafish liver are highly similar to mammals, and zebrafish develop many diseases that are observed in humans, including toxicant-induced liver injury, fatty liver, fibrosis, and cancer. Therefore, the widespread use of zebrafish larvae for studying the mechanisms of these pathologies and for developing potential treatments necessitates the optimization of experimental approaches to assess liver disease in this model. Here, we describe protocols using staining methods, imaging, and gene expression analysis to assess liver injury, fibrosis, and preneoplastic changes in the liver of larval zebrafish.

Key words Zebrafish, Liver, Cancer, Toxicology, Fibrosis, Senescence

1 Introduction

Liver diseases pose a serious global health threat as it currently accounts for over two million deaths per year worldwide, with prevalence increasing each year [1]. These terminal cases are caused by liver failure, cirrhosis, and hepatocellular carcinoma (HCC) [2], all of which can result from a cascade of events that start with hepatic injury, progresses to hepatic inflammation, and ultimately the formation of scar tissue, known as fibrosis or cirrhosis. Hepatic injury in humans is most commonly caused by viral infection, excessive alcohol intake or toxicant exposure, fatty liver, metabolic and immune-related disorders, or biliary injury. In children, fatty liver is a leading cause of liver disease [3] in addition to inborn

James F. Amatruda et al. (eds.), *Zebrafish: Methods and Protocols*, Methods in Molecular Biology, vol. 2707, https://doi.org/10.1007/978-1-0716-3401-1_3,

errors of metabolism, developmental disorders, structural abnormalities, drugs, and toxicants.

Liver damage caused by viral infection or exposure to chemicals that damage hepatocytes results in a regenerative response both the chemicals themselves and the process of metabolizing them can cause cell stress and dysfunction. In most cases, toxin-induced liver injury is remedied by sophisticated cell stress response pathways. In cases of severe acute or chronic injury, these responses are overwhelmed, causing hepatocyte death, inflammation, and formation of a fibrotic scar. Liver fibrosis is characterized by increased accumulation of extracellular matrix that disrupts the physiology of hepatocytes and biliary cells and is often accompanied by the infiltration of immune cells. Fibrosis can progress to cirrhosis, liver failure, and forms a conducive environment for development of HCC. A major barrier to the ability to treat patients with advanced liver disease is the lack of available antifibrotic treatments [4].

The rise in liver disease incidence has not been matched by advances in effective treatments; transplant remains the only curative option for most patients with advanced disease. To date, many research efforts are focused on harnessing the regenerative capacity of the liver to develop new therapies to aid in liver disease and repair. However, the basic biology of how liver cell populations can be stimulated to repair or replace damaged cells has not yet been elucidated. Studies in zebrafish have been extremely impactful in uncovering the genes and pathways required for liver development, the response to liver injury and regeneration [5–7].

Zebrafish are a widely used model to study the fundamental mechanisms that lead to liver disease [5]. Zebrafish livers share many features of the mammalian liver, including a similar cellular composition, cellular functions, and development patterns and processes [8, 9]. The major cell types found in human liver are the same as those found in zebrafish including hepatocytes, biliary cells, fenestrated endothelium, hepatic stellate cells, and resident immune cells [8, 10]. Despite some differences between mammals and teleosts in the architecture of the liver [5], the basic biology of liver cell function and the mechanisms of liver disease are similar. Zebrafish have been a powerful model to study liver cancer [11–17], metabolic and fatty liver disease [5], alcoholic liver disease [18, 19], environmental toxins [20–23], drug-induced liver injury [24], and fibrosis [25]. The availability to carry out these studies on large populations of larvae where the liver is mature by 5 days postfertilization (dpf) allows for high statistical power, with a significantly shorter time frame and lower cost compared to rodent models [8]. Therefore, there is a growing need to standardize the approaches to study liver disease in zebrafish.

The protocols described here outline commonly used approaches to assess the effects of genetic or environmental perturbations which cause precancerous lesions in the liver, toxin-induced liver injury, and fibrosis in zebrafish larvae. The protocols for generating these models have been extensively reviewed elsewhere [26–30]. Although histological analysis is a key diagnostic tool for liver disease, excellent and detailed protocols have been described previously [30, 31], and researchers are encouraged to utilize those in addition to the protocols described here which are based on imaging and gene expression analysis.

2 Materials

2.1 Generating, Raising, Imaging, and Manipulating Zebrafish Larvae

1. Healthy colony of breeding zebrafish, including wild-type strains and transgenic lines that express fluorescent markers in hepatocytes such as *Tg(fabp10a:nls-mCherry)* [17] where nuclei are labeled with mCherry, *Tg(fabp10a:CAAX-EGFP)* where the hepatocyte membrane is marked with EGFP or *Tg (fabp10a:ER-dTomato)* where endoplasmic reticulum (ER) of hepatocytes are labeled. The "LiPan line" (*Tg(fabp10a: dsRed; ela:EGFP))* [32] is also useful as this has markers of both hepatocytes (red) and exocrine pancreas (green).
2. Embryo medium (*see* **Note 1**): Add 1.6 g NaCl, 0.8 g KCl, 0.00716 g Na2HPO4, 0.012 g KH2PO4, 0.288 g CaCl2, 0.492 g MgSO4·7H2O to 1 L of Reverse Osmosed (RO) water, bring to pH 7.2–7.4 with 0.42 M NaHCO3. Add approximately 60 μL of 3% methylene blue to 1 L of the embryo medium to obtain a final concentration of 0.00018%.
3. 3% Methylene blue: Dissolve 3 g of methylene blue in 100 mL of RO water. Store at room temperature.
4. Tricaine solution: 4 g Tricaine powder (MS-222) per liter of embryo medium, pH adjusted to 7.0. Store in aliquots at 4 °C protected from light.
5. 2% Methylcellulose: 2 g methylcellulose in 100 mL reverse osmosis (RO) water (*see* **Note 2**).
6. Glass depression slide.
7. Fine gauge needles for dissecting, e.g., 27 gauge 1 cc insulin syringes.
8. Manipulating tips: Needle microloader tips manually cut to shorten the tip to 1–2 cm.
9. 0.5 mL polypropylene microcentrifuge tubes.
10. 1.5 mL polypropylene microcentrifuge tubes.
11. 2 mL polypropylene microcentrifuge tubes.
12. Glass Pasteur pipettes.
13. 3 mL plastic Pasteur pipettes.
14. 10 cm plastic petri dishes.

2.2 Apoptosis Using TUNEL Assay

1. In Situ Cell Death Detection Kit, TMR red (*see* **Note 3**).
2. 4% Paraformaldehyde (Formaldehyde) Aqueous Solution, Electron Microscopy Grade.
3. 100% methanol.
4. Phosphate-Buffered Saline (PBS) pH 7.4 without calcium and magnesium.
5. 0.1% Sodium Citrate: 0.1 g in 100 mL of RO water.
6. 100% Triton-X100.
7. 20 mg/mL Proteinase K: 20 mg of Proteinase K in 1 mL of RO water.
8. Hoechst DNA staining solution: dilute 1:1000 Hoechst 33258 solution 10 mg/mL in PBS.
9. Glass microscope slides.
10. Clear nail polish.
11. Coverslips.

2.3 Senescence Beta-Galactosidase Staining

1. Senescence β-Galactosidase Staining Kit (Cell Signaling Technology) containing 10× fixative solution, 10× staining solution, solution A, solution B, and X-galactosidase.
2. X-galactosidase solution: 20 mg X-galactosidase in 1 mL of Dimethylformamide.
3. Staining solution 1×: dilute 2 mL of 10× staining solution from the Cell Signaling Kit in RO water, bring to pH 6.2, aliquot in 930 μL (*see* **Note 4**) and freeze at −20 °C.
4. SA-β-galactosidase staining solution: thaw 930 μL of 1× staining solution previously prepared and aliquoted, add 10 μL of 100× solution A (provided by the kit), 10 μL of 100× solution B (provided by the kit), and 50 μL of 20 mg/mL X-galactosidase solution.
5. Phosphate-Buffered Saline (PBS) pH 7.4 without calcium and magnesium.

2.4 Cell Proliferation Using EdU Incorporation

1. Embryo medium without methylene blue.
2. 20 mM 5-ethynyl-2′-deoxyuridine (EdU) solution: 5 mg EdU in 1 mL of embryo medium. Aliquot and store at −20 °C in tubes for single use.
3. 100% Dimethyl Sulfoxide (DMSO).
4. 4% Paraformaldehyde (Formaldehyde) Aqueous Solution, Electron Microscopy Grade.
5. 100% methanol.
6. Phosphate Buffer Saline (PBS) pH 7.4 without calcium and magnesium.

7. 100% Tween-20.
8. Phosphate Buffer Saline Tween (PBST): PBS pH 7.4 without calcium and magnesium with 0.1% Tween-20.
9. Triton ×100.
10. Proteinase K 20 mg/mL: 20 mg of Proteinase K in 1 mL of PBS, aliquot and store at −20 °C.
11. 50 mL conical tubes.
12. Alexa Azide (Thermo Fisher), aliquot upon arrival and store at 4 °C protected from light.
13. CuSO4 100 mM: 1.6 g of CuSO4 in 100 mL of RO water, store at room temperature protected from light.
14. Ascorbic Acid 0.5 M: 0.88 g of ascorbic acid in 10 mL of RO water, aliquot and store at −20 °C.
15. Hoechst DNA staining solution: dilute 1:1000 Hoechst 33258 solution 10 mg/mL in PBS.

2.5 Assessing Hepatocyte Organelle Dynamics in Liver Disease Models by Live Imaging

1. Embryo medium without methylene blue at pH 7.
2. 1% Low melting agarose: 1 g of low melting point agarose in 100 mL embryo medium. Store as 1 mL aliquots at −20 °C.
3. Tissue culture dish with cover glass bottom ($d = 35$ mm, glass thickness 0.17 mm).
4. Superglue.
5. Eyebrow hair loop glued to a glass Pasteur pipette or thinned-out paintbrush with a single bristle (*see* **Note 5**).

2.6 Assessing Fibrosis

1. Pap pen.
2. Phosphate Buffer Saline (PBS) pH 7.4 without calcium and magnesium.
3. 30% Sucrose solution: 30 g of sucrose to 100 mL of PBS.
4. 4% Paraformaldehyde (Formaldehyde) aqueous solution, EM grade.
5. 100% Tween-20.
6. Embedding mold (15 mm × 15 mm × 5 mm; Fisher Scientific; Cat No. 22-363-553).
7. Tissue-Tek O.C.T.
8. Positively charged microscope slides.
9. Slide boxes or trays.
10. Phosphate Buffer Saline Tween (PBST): PBS pH 7.4 without calcium and magnesium with 0.1% Tween-20.
11. Triton X-100.
12. Permeabilization solution (PBS + 0.4% Triton X-100).

13. 100% acetone.
14. Blocking Solution: 5 mL of fetal bovine serum (FBS) and 2 g of bovine serum albumin (BSA) in 100 mL PBST.
15. Antifade Mounting media, Prolong Gold (Thermo Fisher) or similar.
16. Anti-Collagen 1 antibody (Abcam, ab23730) stored at −20 °C.
17. Goat Anti-Rabbit Alexa Fluor 647 stored at −20 °C.
18. 4′,6-diamidino-2-phenylindole (DAPI) stored at −20 °C.

2.7 Equipment

1. Dissecting fluorescent microscope with RFP and GFP filters, such as the Nikon SMZ25 or SMZ18.
2. 28 °C incubator with light cycle set to maintain light/dark cycles that match those in aquaculture facility (typically 14 h dark:10 h light).
3. Confocal microscope such as Leica SP8 or SP5 equipped with 5×, 40× or 63× objectives or equivalent.
4. Inverted confocal microscope such as Leica SP8 STED 3× Microscope equipped with 10× dry and 40× or 63× glycerol/water objective or equivalent.
5. Refrigerated centrifuge (Eppendorf microcentrifuge).
6. Cryostat such as the Leica CM3050 S Research Cryostat or equivalent.

2.8 Software

1. ImageJ Software (https://imagej.nih.gov/ij/download.html).
2. Imaris Software, the Imaris Essentials package as a minimum (https://imaris.oxinst.com/products/imaris-essentials).
3. LasX software to use with images from Leica (https://www.leica-microsystems.com/products/microscope-software/p/leica-las-x-ls/).

3 Methods

3.1 Liver Size Measurements

Measuring liver size is a simple but valuable method to determine whether a given experimental condition is affecting liver development or causing liver disease. For example, in our studies whereby an oncogene activates tumor-suppressive mechanisms such as senescence or apoptosis, we observe a decrease in liver size [17], while other models of liver cancer [14, 33] and fatty liver disease cause enlarged livers. This protocol describes how to quantify the size of the left lobe of the liver of zebrafish larvae.

1. Set up mating as an incross of *Tg(fabp10a:nls-mCherry)*$^{Tg/Tg}$ or any transgenic line with a hepatocyte-specific fluorescent marker to enable easy identification of the liver using fluorescent stereomicroscopy.
2. Collect embryos and transfer 50–60 healthy embryos (*see* **Note 6**) into 10 cm petri dish with 30–40 mL of the culture medium of choice (*see* **Note 1**) in each plate. Maintain embryos in an incubator set to 28 °C with a 14 h light:10 h dark cycle.
3. After 5 days, anesthetize the larvae by adding 1 mL of Tricaine Solution (4 g/L) to the petri dish and wait until they stop swimming (*see* **Note 7**).
4. Using a plastic Pasteur pipette, transfer three to five larvae into a glass depression slide containing one drop of approximately 100 μL of 2% methylcellulose (*see* **Note 8**).
5. Orient each larva using manipulating tips such that the head of the larva is towards the left side and the eyes are aligned, allowing maximum exposure of the left liver lobe as shown in Fig. 1a.
6. Use the camera attached to the stereomicroscope to take high-resolution images at the same magnification for each picture. Include a scale bar on each image.
7. Repeat until all desired images are obtained.
8. Open ImageJ software and open an image from a file (*see* Fig. 1a).
9. Select the tool "Analyze > Measure" (or Command M) to measure the length of the scale bar in pixels. Annotate this value (*see* Fig. 1b).
10. Select the tool "Analyze > Set Scale." A window will open. Fill in the appropriate scale depending on the magnification of the image taken.
    (a) Enter the following parameters:
        (i) "Distance in pixel" is the value measured on the image.
        (ii) "Known distance" is the value of the scale bar.
        (iii) "Pixel aspect ratio" is 1.
        (iv) "Unit of length" is the unit of the scale bar present in the image (typically μm).
    (b) Check the "global" box to ensure the same scale is maintained for the analysis of all images (*see* Fig. 1c).
11. Select the polygon tool and manually trace the outline of the liver in the image based on the fluorescence (*see* Fig. 1a).
12. Select "Analyze > Measure" (or Command M) (*see* **Note 9**).

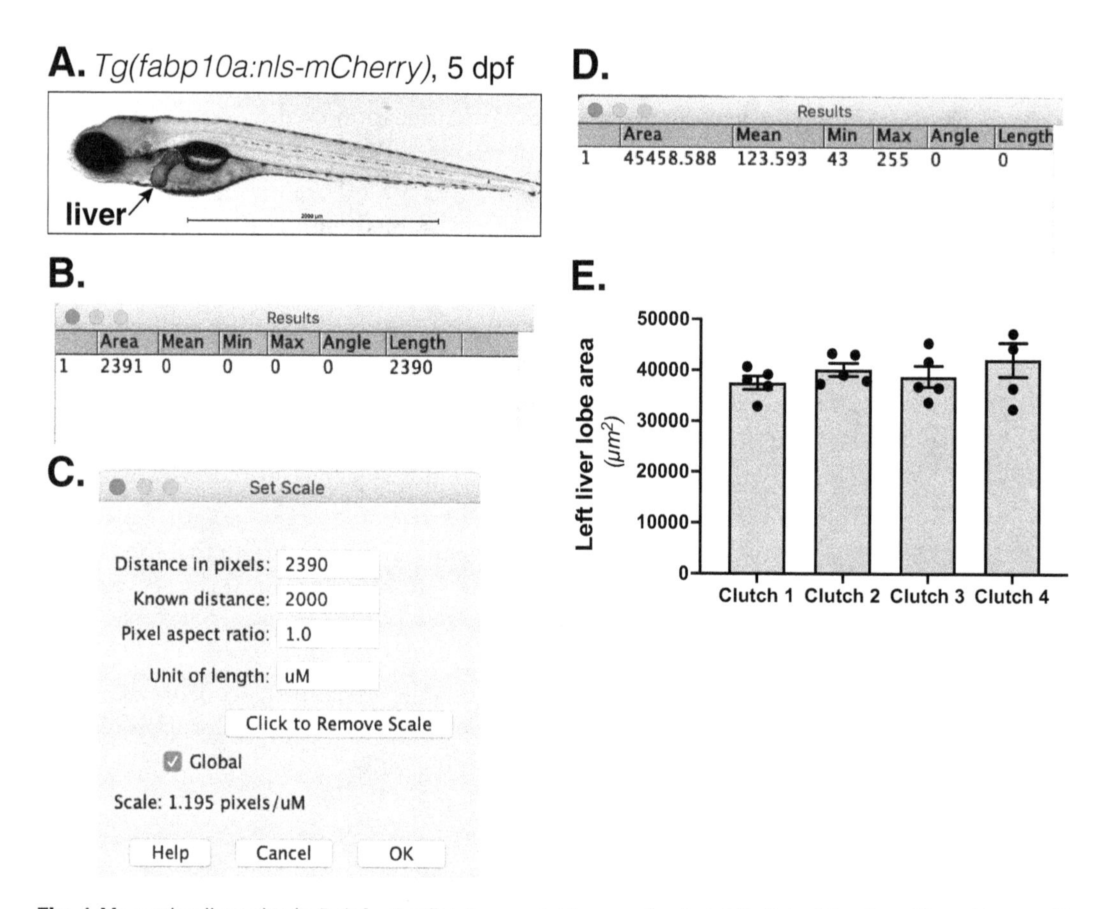

Fig. 1 Measuring liver size in 5 dpf zebrafish larvae. (**a**) Image of a 5 dpf *Tg(fabp10a:nls-mCherry)* larva with scale bar. Arrow indicates the liver and liver area is defined by the polygon tool in ImageJ. (**b**) Measure of the scale bar that is used the set the conversion of pixels in μM. (**c**) Parameters used to set the conversion of pixels in μM. (**d**) Measurements of the liver with the automatic conversion of pixels in μM. (**e**) Measurements of left liver lobe area of five individual larvae across four independent clutches

13. A separate window will pop up. Copy and paste the Area values into a software of choice (such as an MS Excel spreadsheet) for data curation and analysis.
14. Data can be displayed as individual values for each liver, or as an average for an entire clutch (*see* Fig. 1e).

3.2 Gene Expression Analysis in Microdissected Zebrafish Larval Livers

Analysis of gene expression by qPCR or RNAseq of livers is important to provide information regarding possible pathways affected by the treatment or genetic manipulation. It can also be used to assess key genes that are known to be disrupted in diseased livers or that serve as markers of disease processes, such as response to liver injury, inflammation, or activation of myofibroblasts to form fibrosis.

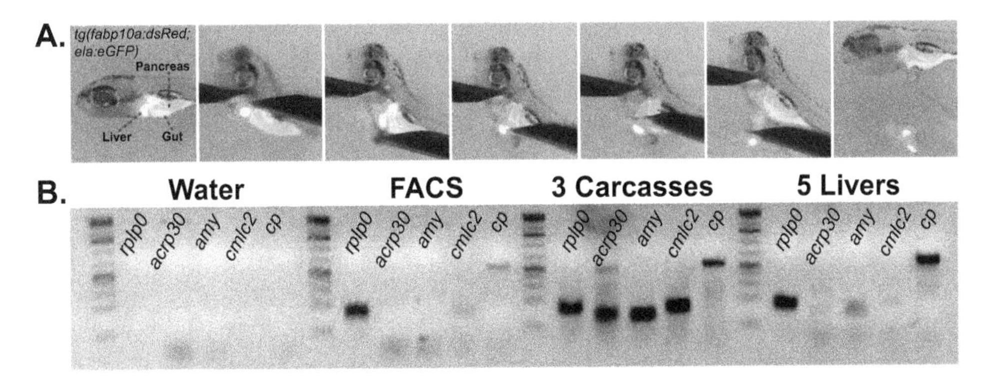

Fig. 2 Manual liver dissection. (**a**) Snapshots of manual liver dissection of a *Tg(fabp10a:dsRed;ela:EGFP)* liver from a 5 dpf larvae. (**b**) PCR of tissue-specific genes from samples collected on 5 dpf to detect contamination of gut (*acrp30*), pancreas (*amy*), and heart (*cmlc2*) in dissected livers (*cp*) and the liver-less carcass remaining after the liver was dissected. The gene expression is compared to purified hepatocytes obtained using Fluorescent Activated Cell Sorting (FACS) based on hepatocyte specific fluorescent marker expression

This protocol describes the method for liver dissection and provides tools to assess whether there is contamination of the liver sample by cells from neighboring organs (i.e., pancreas, heart, gut). We manually dissect the whole liver to obtain RNA that allows assessing transcriptome from all the cells in the liver. FACS can be used as an alternative to select a specific population of cells (e.g., hepatocytes or biliary cells), and that is described elsewhere [34, 35].

1. Set up mating as an incross of the LiPan line *Tg(fabp10a:dsRed; ela:GFP;ins:dsRed)*$^{Tg/Tg}$ [32] to distinguish the red-labeled liver from the green-labeled exocrine pancreas, or alternatively use any transgenic line that expresses a fluorescent protein in hepatocytes.
2. Collect embryos and transfer 50–60 healthy embryos (*see* **Note 6**) into 10 cm Petri dish with 30–40 mL of the culture medium of choice (*see* **Note 1**) in each plate. Maintain embryos in an incubator set to 28 °C with a 14 h light:10 h dark cycle.
3. After 5 days, anesthetize the larvae by adding 1 mL of Tricaine Solution (4 g/L) to the petri dish and wait 3–5 min until they stop swimming (*see* **Note 7**).
4. Transfer three to five larvae to a depression slide containing 1 drop of approximately 100 μL of 2% methylcellulose using a small tip plastic pipette (*see* **Note 8**).
5. Position larvae so the left lobe is facing upward (*see* Fig. 2a).
6. Use fine gauge needles to dissect the liver away from the gut and pancreas (*see* Fig. 2a). To do this place one needle on each side of the liver and pull apart. The liver usually flops out easily

by doing this method. Alternatively, pin fish still by inserting one needle anterior to the heart. Insert the second needle just posterior to first and peel back the skin around heart and liver; remove the second needle. Use the second needle to carefully pull out the liver (*see* Fig. 2a). This routinely takes less than a minute per liver.

7. Carefully pick up the liver with a pipette tip. The use of transgenic lines that express a fluorescent protein in hepatocytes enables visualization of the dissected liver using a fluorescent microscope (*see* Fig. 2a) so that it is easy to transfer (*see* **Note 10**).
8. Repeat to collect as many livers as needed based on the purpose of the experiment (*see* **Note 11**).
9. Proceed to RNA isolation protocol of choice (*see* **Note 12**) and then convert to cDNA to assess any contamination of the sample using PCR; the remaining cDNA can be used for qPCR or can then be directly used in a library preparation protocol of choice for RNAseq (*see* Fig. 2b and **Note 13**).

3.3 Assessing Tumor Suppression

A number of cellular and molecular changes occur in cells prior to malignant transformation. Tumor-suppressive mechanisms, including apoptosis and senescence, are common means to eliminate such cells from the organism, and this is observed in preneoplastic liver cancer cells. A hallmark of both processes is the withdrawal from the cell cycle and, when this process is activated in zebrafish larval livers, the outcome is reduced liver size. Over time, as successful cancer cells overcome these mechanisms, cell division resumes and, in some cases, liver size expands and ultimately tumors form.

Here, we focus on analyzing precancerous lesions in larvae aged 4–14 dpf as during this time larvae can be sustained in small volumes using a paramecia-based diet supplemented with formulated larval diet. After this time, the introduction of brine shrimp as a primary nutritional source can be more challenging to catch and eat, resulting in death in unhealthy fish. Our studies have exclusively used a transgenic line generated in our lab (*Tg(fabp10a:UHRF1-EGFP)*$^{Tg/+}$) as this routinely generates a high level of senescence in hepatocytes and HCC [17]. However, any line of interest can be used for these assays.

3.3.1 Apoptosis Using TUNEL Assay

Apoptotic cells are characterized by high levels of DNA fragmentation that can be detected using the TdT-mediated dUTP-biotin Nick End Labeling (TUNEL) assay. This is based on the incorporation of modified dUTPs by the enzyme terminal deoxynucleotidyl transferase (TdT) at the 3′-OH ends of fragmented DNA, a hallmark as well as the ultimate determinate of apoptosis. This protocol has been optimized for detecting apoptosis in livers of 4–7 dpf.

1. Collect embryos and transfer 50–60 healthy embryos (*see* **Note 6**) into 10 cm petri dish with 30–40 mL of the culture medium of choice (*see* **Note 1**) in each plate. Maintain embryos in an incubator set to 28 °C with a 14 h light:10 h dark cycle.
2. After 5 days, anesthetize the larvae by adding 1 mL of Tricaine Solution (4 g/L) to the petri dish and wait until they stop swimming (*see* **Note 7**).
3. Transfer larvae to a 2 mL tube and remove all liquid.
4. Fix embryos by adding 1 mL of 4% PFA for 2 h at room temperature or 4 °C overnight.
5. Remove all of the fixative and fill the tube with PBS to wash.
6. Gradually dehydrate embryos: remove half of the PBS from the tube, add equal volume of 100% methanol and equilibrate for 2–5 min at room temperature. Repeat this **step 3** times.
7. Remove all of the liquid from the tube and replace with 1 mL of 100% methanol.
8. Incubate in 100% methanol for at least 1 h. This is a convenient stopping point as larvae can be stored in 100% methanol at 4 °C indefinitely.
9. Rehydrate the larvae: remove half of the methanol from the tube, add equal volume of PBS, and equilibrate for 2–5 min at room temperature. Repeat this **step 3** times.
10. Remove all of the liquid from the tube and replace with 1 mL of PBS.
11. Incubate in PBS for 1 h at room temperature.
12. Dissect livers using the protocol above and transfer livers to PBS with a P20 pipette set to 2 μL (*see* **Note 14**) to 0.5 mL tube, remove all liquid.
13. Permeabilize livers by incubating in 100 μL of PBS containing 0.1% sodium citrate, 0.1% Triton-X, 20 μg/mL Proteinase K for 10 min at room temperature. This step may need optimization as too much digestion can destroy the tissue integrity.
14. Wash the larvae 3 times in PBS by carefully removing the solution from the tube and adding 100 μL of PBS (*see* **Note 15**).
15. Remove the PBS and fix the livers with 100 μL of 4% PFA for 10 min at room temperature.
16. After fixation, remove 4% PFA and add 100 μL of PBS.
17. Prepare TUNEL solution on ice by adding 1 μL of TUNEL TdT enzyme (provided in the kit) to 9 μL labeling mix (provided in the kit) per reaction mixture for up to 10 livers in a 0.5 mL tube (*see* **Note 16**).

18. Remove the PBS and add 10 μL of TUNEL solution to each tube.
19. Incubate for 1 h in the dark at 37 °C.
20. Wash the larvae 3 times in PBS by carefully removing the solution from the tube and adding 100 μL of PBS (*see* **Note 15**).
21. Remove the wash solution and stain the nuclei by adding 100 μL of Hoechst DNA staining solution and incubate for 1 h at room temperature in dark.
22. Carefully remove the Hoechst DNA staining solution from the tube and wash the livers by adding 100 μL of PBS, repeating 3 times (*see* **Note 15**).
23. To mount the livers on a glass slide, put 10 μL drop of 2% methylcellulose at the center of the slide and transfer the livers with a p20 pipette set to 2 μL (*see* **Note 14**).
24. Using the stereomicroscope to orient the livers, gently move the livers to the bottom of the slide with an insulin syringe and make them touch the slide surface. Avoid the creation of air bubbles that can interfere with the imaging. Place the coverslip on the droplet and leave undisturbed for 3–5 min. Seal the slide with clear nail polish and allow to dry for 5–10 min. The slides are ready for confocal imaging once the nail polish is dry.
25. Image the livers at the confocal microscope (*see* **Note 17**).
26. Quantification of TUNEL signal is performed with LasX or other equivalent software by counting number of spots per unit area.

3.3.2 Senescence-Associated (SA) β-Galactosidase Staining

Senescent cells are characterized by increased lysosomal activity which enables the detection of endogenous β-galactosidase activity when the substrate is provided at pH 6. Although cells have positive β-galactosidase activity in absence of senescence, it is one of the most widely used methods to detect senescence in vitro and in vivo, and it serves as a good starting point to determine whether there is senescence in your experimental condition. This protocol has been optimized to detect senescent cells in zebrafish livers up to 7 dpf, but it can be used for other organs.

1. Set up mating between your model of choice or perform treatment that you hypothesize will induce senescence in the liver. If you have a transgenic line that can mark one set—that is, the experimental set—use that so that you can mix treated and control larvae in the same tube to reduce possible technical variation.
2. Collect embryos and transfer 50–60 healthy embryos (*see* **Note 6**) into 10 cm petri dish with 30–40 mL of the culture medium of choice (*see* **Note 1**) in each plate. Maintain embryos in an incubator set to 28 °C with a 14 h light:10 h dark cycle.

3. After 5 days, anesthetize the larvae by adding 1 mL of Tricaine Solution (4 g/L) to the petri dish and wait until they stop swimming (*see* **Note 7**).
4. Collect ten larvae from the experimental group and ten control sibling larvae (*see* **Note 18**) in 2 mL round-bottom tubes (*see* **Note 19**).
5. Before starting the staining, incubate the fixative solution 10× (provided by the kit) at room temperature for 30 min and dilute 1:10 with RO water to obtain a 1× working dilution.
6. Remove the solution and wash the larvae once with 1 mL of PBS pH 7.4.
7. Remove all PBS from the tube and add 1 mL of 1× fixative solution for 15 min at room temperature.
8. Remove the fixative solution and wash the larvae twice with 1 mL of PBS pH 7.4 (*see* **Note 20**).
9. Remove the PBS and add 1 mL of SA-β-galactosidase staining solution.
10. Incubate overnight at 37 °C protected from light (*see* **Note 21**).
11. In the morning, check the wild-type larvae using the stereomicroscope for the presence of blue pigmentation in the posterior gut as indication of successful staining. If not present, leave the larvae in the SA-β-galactosidase staining solution for additional time and check every hour until it appears (*see* **Note 22**, Fig. 3).
12. When there is staining of the posterior gut of the wild-type larvae, remove the SA-β-galactosidase staining solution and add 1 mL of PBS pH 7.4. Larvae are stable for a few days (up to a week) at 4 °C protected from light.

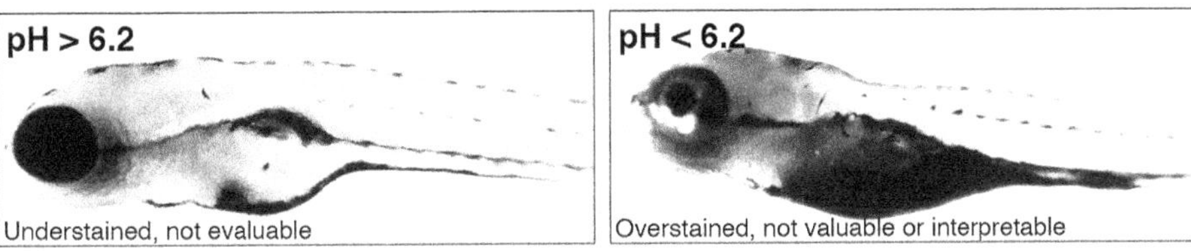

Fig. 3 SA-β-galactosidase staining of 5 dpf larvae. (**a**) Examples of wild-type sibling larva and *Tg(fabp10a: hUHRF1-EGFP)*$^{Tg/wt}$ larva properly stained with SA-β-galactosidase. Arrow indicates the posterior gut used as a positive control for successful staining. (**b**) Examples of larvae stained with staining solution at wrong pH, leading to false positive (left) or unspecific positive staining (right)

13. To determine if the experimental treatment induced senescence, the number of larvae per clutch is counted. To count how many larvae have SA-β-galactosidase-positive livers, transfer the larvae to a depression slide containing one drop of 2% methylcellulose and observe under the stereomicroscope. Positive staining is characterized by a blue haze in the liver area (*see* Fig. 3b) while negative livers appear yellowish or clear. Staining in the yolk, gut, and neighboring tissues only indicates that the staining worked and is not indicative of senescence in the liver.

3.4 Liver Cell Division Using EdU Incorporation

The EdU incorporation assay is a method to determine how many cells are undergoing DNA replication as EdU is incorporated in the newly synthesized DNA daughter strand. Different from the BrdU assay that has been used commonly in zebrafish, EdU is antibody-free since it uses click-IT chemistry. The technique is highly sensitive with minimal nonspecific signal compared to BrdU. This protocol has been optimized for zebrafish between 3.5 and 14 dpf.

1. Collect embryos and transfer 50–60 healthy embryos (*see* **Note 6**) into 10 cm petri dish with 30–40 mL of the culture medium of choice (*see* **Note 1**) in each plate. Maintain embryos in an incubator set to 28 °C with a 14 h light:10 h dark cycle.
2. After 5 days, anesthetize the larvae by adding 1 mL of Tricaine Solution (4 g/L) to the 10 cm petri dish and wait until they stop swimming (*see* **Note 7**).
3. Collect ten experimental and ten control larvae and transfer in 50 mL conical tube with a plastic Pasteur pipette in 1–5 mL of embryo medium (*see* **Note 18**).
4. Prepare 250 μL of DMSO-EdU solution for each experimental group by combining 218.75 μL of embryo medium without methylene blue, 25 μL of DMSO, 6.25 μL of EdU 20 mM.
5. Carefully remove all the embryo medium from the tube containing larvae (*see* **Note 23**), add the DMSO-EdU solution, and immediately place the tube on ice.
6. Incubate the tube on ice for 20 min.
7. After incubation, gently add embryo medium without methylene blue (*see* **Note 24**) to fill the 50 mL falcon tube.
8. Lay the tube gently on its side so the embryos are resting on a flat, wide surface and incubate at 28 °C for 30 min.
9. Remove the embryo medium and add 1 mL of freshly prepared 4% PFA directly to the embryos.
10. Fix the larvae overnight at 4 °C.
11. Transfer the larvae in 2 mL round tubes with a glass pipette and wash them twice with 1 mL of PBS using a glass pipette (*see* **Note 20**).

12. Gradually dehydrate embryos: remove half of the PBS from the tube and add equal volume of 100% methanol and equilibrate for 2–5 min at room temperature. Repeat this **step 3** times.
13. Remove all the liquid from the tube and replace with 1 mL of 100% methanol.
14. Incubate in 100% methanol for at least 1 h. This can be used as stopping point as larvae can be stored in 100% methanol at 4 °C indefinitely.
15. Rehydrate the larvae: remove half of the methanol from the tube, add equal volume of PBS, and equilibrate for 2–5 min at room temperature. Repeat this **step 3** times.
16. Remove all the liquid from the tube and replace with 1 mL of PBS.
17. Incubate in PBS for 1 h at room temperature.
18. Permeabilize larvae by incubating them for 30 min at room temperature in PBS containing 1% Triton-X and 20 μg/mL proteinase K.
19. Remove the solution and wash the larvae twice with 1 mL of PBS and transfer them in 0.5 mL tubes by using a glass pipette (*see* **Note 25**).
20. Prepare 160 μL of Click-IT Reaction for each larvae-containing tube by combining 144 μL of PBS to 16 μL of CuSO4 and 0.4 μL of Alexa Azide.
21. Remove the PBS from the larvae and add 100 μL of PBS to the tube containing larvae.
22. Add 80 μL of Click-IT Reaction to each tube containing larvae to a total of 180 μL.
23. Incubate the samples for 10 min with gentle rocking protected from light by wrapping with aluminum foil or placing in a dark container on the rocker.
24. After 10 min, add 20 μL of 0.5 M ascorbic acid to achieve a volume of 200 μL and incubate for 20 more minutes.
25. Remove the Click-IT reaction and repeat the Click-IT reaction from **step 21**.
26. Remove the solution and wash twice with 500 μL of PBST for 15 min.
27. Remove the solution and add 500 μL of PBS.
28. Manually dissect out the livers with insulin syringes (*see* Subheading 3.2) and transfer them to a 1.5 mL tube containing 100 μL of PBS with a p20 pipette set to 2 μL (*see* **Note 14**).
29. Remove the wash solution and stain the nuclei by adding 100 μL of Hoechst DNA staining solution and incubate for 1 hour at room temperature in the dark.

30. Wash the larvae 3 times in PBST by carefully removing the solution from the tube and adding 100 μL of PBST (*see* **Note 15**).
31. Wash the larvae one time in PBS by carefully removing the solution from the tube and adding 100 μL of PBS (*see* **Note 15**).
32. To mount the livers on a glass slide, put 10 μL drop of 2% methylcellulose at the center of the slide and transfer the livers with a p20 pipette set to 2 μL (*see* **Note 14**).
33. Using the stereomicroscope to orient the livers, gently move the livers to the bottom of the slide with an insulin syringe and make them touch the slide surface. Avoid the creation of air bubbles that can interfere with the imaging. Place the coverslip on the droplet and leave undisturbed for 3–5 min. Seal the slide with nail polish. The slides are ready for confocal imaging once the nail polish is dry.
34. Image the livers using confocal microscopy (*see* **Note 26**) to capture a Z-stack through the entire liver with a step size of about 0.35 μm.
35. EdU- and Hoechst-positive cells are counted with Imaris Software (Spot function) or can be manually counted. Data are presented as number of EdU-positive cells per number of total nuclei (counted using the Hoechst channel).

3.5 Assessing Hepatocyte Organelle Dynamics in Liver Disease Models by Live Imaging

Factors such as a high-fat diet, alcohol, or environmental toxicants activate stress response pathways in hepatocytes. One of the most common responses to stress is a change in the size, shape, and morphology of the affected organelles. Stress can also promote the formation of structures such as lysosomes, autophagosomes, lipid droplets, cytoplasmic globules, and stress granules. Changes in these structures can be pathognomonic for some liver diseases, such as the globules observed in patients with alpha-1-antitrypsin deficiency. Live in vivo imaging can provide qualitative and quantitative indications of how structures are altered/formed in response to stress allowing us to define cellular features during disease progression.

Here we describe a protocol for a live in vivo imaging technique that assesses cellular structures in livers of 5 dpf larvae, with particular alteration to the ER. The screening of the phenotype in live fish allows for further downstream processing for RNA or protein assays based on phenotypic segregation.

1. Set up mating between *Tg(fabp10a:ER-dTomato)*$^{Tg/WT}$ [36] and *Tg(fabp10a:CAAX-EGFP)*$^{Tg/WT}$ in order to obtain larvae with marked ER and cellular membrane (*see* **Note 27**).

2. Collect embryos and transfer 50–60 healthy embryos (*see* **Note 6**) into 10 cm petri dish with 30–40 mL of the culture medium of choice (*see* **Note 1**) in each plate. Maintain embryos in an incubator set to 28 °C with a 14 h light:10 h dark cycle.
3. Perform a treatment according to your preferred protocol (*see* **Note 28**).
4. Prepare 1% low melting agarose in embryo medium (*see* **Note 29**). Ensure the gel is approximately 37 °C before using for embedding. For this purpose, keeping the gel at a consistent temperature using a water bath or heating block set to 37 °C is the best option. Add Tricaine Solution (4 g/L) to the gel for a final concentration of 0.16 g/L and vortex gently, avoiding bubbles (*see* **Note 30**).
5. Anesthetize the zebrafish larvae with Tricaine Solution to reach 0.16 g/L final concentration (40 μL in 1 mL medium) (*see* **Note 30**) and transfer between 3 and 10 larvae to an imaging dish with glass bottom (*see* **Note 31**).
6. Using a small transfer pipette, remove all water from the imaging dish, except a small droplet that contains the fish.
7. Add cooled agarose gel to the imaging plate with a large transfer pipette, such that it covers the rim of the glass bottom, and then carefully remove surplus gel (*see* **Note 32**).
8. Using a paintbrush with a single bristle or a hair loop (*see* Fig. 4b), gently push the larvae to the bottom of the dish such that the head is oriented towards the right for imaging on an inverted confocal microscope. In this orientation, the larger lobe of the liver will be at bottom.
9. After approximately 10 min, add embryo medium with Tricaine (0.08 g/L, 20 μL in 1 mL medium) to the top of the solidified gel (*see* **Note 33**).
10. Leave the plate undisturbed for 30 min for the gel to set.
11. Image on an inverted confocal microscope (*see* **Note 34**).
12. Note the phenotype associated with each larva imaged to correlate the phenotype with the downstream assay results (*see* **Note 35**).
13. To rescue the fish, use a needle with a blunt tip (*see* Fig. 4c). Add extra embryo medium to the dish. Gently cut the gel around the larva and flip it using the needle without touching the larva. Use the hair loop or paintbrush to then free the larva gently by sliding between the larva and the gel.
14. Transfer to a labeled well in a 12-well plate containing fresh embryo medium. Repeat the same for other larvae and transfer them into individual wells (*see* **Note 36**).

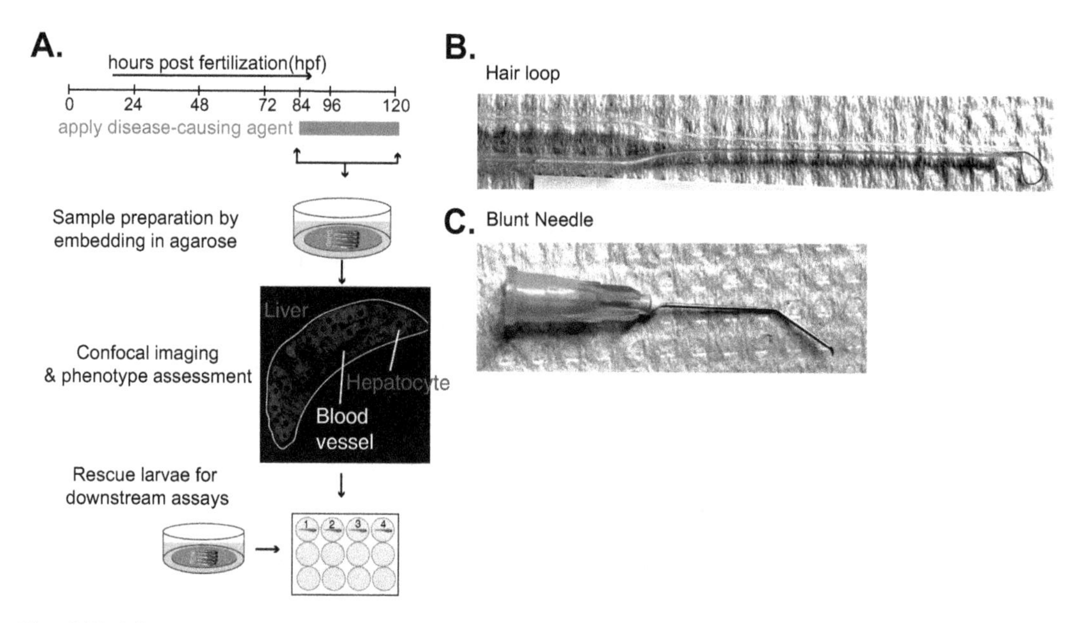

Fig. 4 Workflow and tools for studying organelle dynamics by live imaging
(**a**) Experimental setup for studying organelle dynamics. The representative image of *Tg(fabp10a:ER-tdTomato)* shows hepatocytes with a reticular ER and an absence of signal from the blood vessel. (**b**) Representative image of a hair loop used for embedding. (**c**) Representative image of blunt needle for cutting gel during rescue

15. Proceed with live larvae for other downstream analyses such as RNA or protein collection that can be correlated to imaging data.

3.6 Assessing Fibrosis

Hepatic stellate cells (HSCs) are myofibroblasts activated from a quiescent to proliferative state that secrete extracellular matrix proteins such as Collagen 1a to create the fibrotic scar that is characteristic of liver fibrosis. HSC markers useful in zebrafish include *hand2* [37], *colec11* [38], and *gfap* [39]. Immunofluorescence on cryosections can be used to visualize collagen deposition within the liver of transgenic lines where the hepatocytes are marked by a fluorescent protein (i.e. *Tg(fabp10a:dsRed)*).

1. Set up mating between *Tg(fabp10a: dsRed)*$^{Tg/WT}$ or *Tg (fabp10a:CAAX-EGFP)*$^{Tg/WT}$ in order to obtain larvae with marked hepatocytes.
2. Collect embryos and transfer 50–60 healthy embryos (*see* **Note 6**) into 10 cm petri dish with 30–40 mL of the culture medium of choice (*see* **Note 1**) in each plate. Maintain embryos in an incubator set to 28 °C with a 14 h light:10 h dark cycle.
3. Perform treatment according to preferred liver disease model.

4. Larvae are collected at the time that is optimal for assessing fibrosis based on the goal of the experiment by transferring larvae to 4% PFA overnight at 4 °C, but it should be after the liver has fully formed (i.e., after 96 hpf).
5. Transfer larvae to 30% sucrose in PBS and leave overnight at 4 °C (*see* **Note 37**).
6. Transfer fixed larvae/livers to embedding mold properly labeled and remove any extra sucrose solution.
7. Add Tissue-Tek O.C.T. compound to the top of the mold and arrange up to ten larvae/livers as needed using a fine gauge needle to position while viewing under the stereomicroscope.
8. Freeze block on dry ice until the O.C.T. turns white and is solid and store blocks at −80 °C.
9. Generate 10 μM sections onto positively charged (for tissue adhesion) microscope slides using a cryostat.
10. Wipe off excess O.C.T. around outer edges of the cryosection to leave enough room to create a barrier.
11. Use the Pap pen to create a box around the samples on each slide so as to confine the liquid to just the area on the slide that has the tissue. This will allow a small volume of liquid to cover the tissue section (200 μL).
12. Using a Pasteur pipette, add a few drops of PBS on top of the tissue for 1 min.
13. Wash the slide for 5 min with PBST. Repeat two more times.
14. Wash twice the slide with ice-cold PBS +0.4% Triton X-100 for 10 min to permeabilize.
15. Wash the slide quickly with PBST.
16. Wash slide for 5 min with PBST. Repeat two more times.
17. Move slides to the humid chamber. From this point onward all subsequent steps are done in a humid chamber (*see* **Note 38**).
18. Block slides in blocking solution (5% FBS and 2% BSA in PBST) for 1 hour and 15 min at room temperature.
19. Add 150 μL of primary antibody solution prepared by diluting Anti-Collagen 1 antibody 1:100 in blocking solution (5% FBS and 2% BSA in PBST) to the slide and incubate overnight at 4 °C.
20. After overnight incubation, wash the slide quickly with PBST.
21. Wash the slide for 10 min with 200 μL of PBST. Repeat two more times.
22. Incubate slides for 1.5 h at room temperature (*see* **Note 39**) in 150 μL of secondary antibody solution prepared diluting Goat

Anti-Rabbit AlexaFluor 1:250 in blocking solution (5% FBS and 2% BSA in PBST) (*see* **Note 40**).

23. Wash with 200 μL PBST for 15–20 s.
24. Wash the slide with 200 μL PBST for 10 min. Repeat two more times.
25. Add one drop of antifade mounting media and one drop of DAPI and mount with coverslips (*see* **Note 41**).
26. Use clear nail polish to secure the corners of the coverslip to the slide.
27. Allow to cure overnight (*see* **Note 42**).
28. After 24 h seal the coverslip to the slide with nail polish. The slides can now be stored at −20 °C indefinitely.
29. Image samples using confocal microscopy (*see* **Note 43**).
30. For image analysis, use ImageJ software to quantify Collagen 1 staining.
31. Open the image to be quantified in ImageJ.
32. Select Image > Adjust > Threshold and set the threshold to highlight the region to be measured (do not hit apply).
33. Select Edit > Selection > Create Selection.
34. Measure mean intensity and of selected region (Command + M).
35. Copy values to data storage software (such as an MS Excel spreadsheet) and average values for each liver.

4 Notes

1. There are different commonly used media that are used for rearing zebrafish embryos. The composition of these can affect the disease model under study and should be compared as an experimental parameter to assess if this is a variable that affects the liver-related outcome of the model. This is covered in detail in Ramdas Nair, Delaney et al. [40]. All the protocols described here are carried out in Embryo Medium as described [40].
2. To prepare the methylcellulose, use cold RO water and stir at 4 °C for 1–2 days until all particles are dissolved.
3. In our hands, the TMR kit is more sensitive than FITC kit.
4. This is a key step. SA-β-galactosidase staining is pH-dependent and therefore it is essential that the pH of the staining solution 1× is precise. For zebrafish, the optimal pH of the staining solution 1× is 6.2, as higher pH leads to false negative and lower pH leads to false positive (*see* Fig. 3). Since this value might slightly vary according to different pH-meters, it is

recommended to perform a pilot experiment on positive controls to determine the exact pH that gives reliable results. After the pilot study, prepare 20 mL of a 1× staining solution and adjust the pH to the right value with HCl 1 N or NaOH 1 N, and aliquot this into single-use volumes and store at −20 C until use. This is important to obtain consistency from experiment to experiment.

5. A hair loop is made by inserting a single human hair, typically an eyebrow, as a loop into the opening of a Pasteur pipette and fixed in place with super glue (*see* Fig. 4). Alternatively, it is possible to use a paint brush (size 000, Faber-Castell) that has been thinned out, leaving one single bristle.
6. If embryos are kept at higher density, embryos can be delayed or unhealthy, reducing liver size.
7. Do not use more than 3 mL Tricaine Solution or keep fish for more than 20 min in Tricaine to avoid cell necrosis.
8. Try not to dilute the methylcellulose by transferring excess embryo medium when you move the larvae over to the slide because thick methylcellulose prevents the fish from moving.
9. If you are unsure of how the magnification translates to image scale, keep in mind that the average left liver lobe area of wild-type larvae at 5 dpf is 40,000 μm^2 (*see* Fig. 1e).
10. Make sure that the liver has been transferred to the tube by checking the tip under the stereomicroscope to assure that no fluorescence remains.
11. If correctly performed, it is possible to extract 10–12 ng of total RNA from a single 5 dpf liver. This is sufficient for generating a library for RNAseq using low-input library generation protocols or to generate cDNA for qPCR analysis of 10–12 genes. If the number of genes to be analyzed exceeds this, the livers can be pooled from 5 to 10 larvae. Alternatively, 10–12 livers can be pooled to isolate RNA for library preparation using standard protocols. Note that we observe higher variability in gene expression when comparing single livers to each other than when compared across pools of livers.
12. Extraction of total RNA can be performed by using Trizol. Livers can be put directly into Trizol and dissociated by using insulin syringe by passing the solution through the needle 10 times. Chloroform is added to the Trizol as per the manufacturer's protocol and aqueous phase is collected in a new tube. Equal volume of isopropanol and 1 μL of Glycogen Blue to aid in visualizing the pellet are added and tubes are incubated overnight at −20 °C followed by centrifuging in a microcentrifuge for 1 h centrifuge at 12,000×*g* at 4 °C. Washes are performed according to manufacturer's instructions.
13. Manual liver microdissection can result in contamination by surrounding tissues. The pancreas, gut, and heart are the most

proximal tissues to the larval liver and contamination by these samples can significantly affect gene expression analysis. To control this, dissected samples are analyzed for the presence of tissue-specific genes from the pancreas, gut, and heart (*see* Table 1). Using manufacturers protocol for standard PCR reaction using primers for genes in Table 1, assess contamination based on robustness of bands for tissue-specific genes (*see* Fig. 2b). We compare this to the purity obtained by analyzing cells isolated by FACS, and typically have comparable results.

14. Fixed livers can be sticky. To prevent them from attaching to the tip, pipette PBST and expel once and then use the same tip to collect the livers.
15. As the washes are performed, inspect the livers in the 1.5 mL tube under the stereomicroscope to avoid the accidental loss of any livers.
16. If there are too many livers or if the solution is not mixed well, the labelling will be uneven and not reliable.
17. For the imaging, we used SP8 Leica Confocal microscope. Livers are located on the slide by using a 5× objective, and then a 40× oil objective was used for the acquisition of the images. Set the following parameters: size 2048 × 2048, speed 100 Hz, zoom 2.25, bidirectional and lines 3. Once it is set, acquire the whole liver with a z-stack size of 0.35 μm or three independent stacks of the livers.
18. We advise combining the experimental and the control group is the same tube to avoid possible differences in the staining due to technical reasons. In case of testing the effect of a mutation or treatment, transgenic lines that mark hepatocytes (e.g., *Tg (fabp10a:nls-mCherry)* or *Tg(fabp10a:CAAX-EGFP))* can be used to distinguish the experimental treatment group so that experimental and control samples can be mixed during processing.
19. Avoid using 1.5 mL tubes as it can lead to uneven staining among the larvae.
20. The larvae became very sticky after fixation, so it is suggested to use glass pipettes and work quickly when transferring larvae.
21. Typically, the staining should be set up between 3 and 5 pm in order to have a correct staining the following morning (i.e., 14–16 h).
22. Typically, the staining should be present after overnight incubation. Absence of staining in the posterior gut after 24 h of incubation indicates that the staining failed. This is probably due to high pH of the staining solution 1×. *See* **Note 4**.

Table 1
List of primers used for assessing fibrosis and contamination in the liver dissection

Application	Gene symbol	Ensembl ID	Forward primer (5′--> 3′)	Reverse primer (5′-->3′)
Housekeeping gene	*rplp0*	ENSDARG00000051783	CTGAACATCTCGCCCTTCTC	TAGCCGATCTGCAGACACAC
Assess fibrosis	*acta2*	ENSDARG00000045180	CGAGGCTACTCATTCGTCACCAC	AGAGGAGGAAAAGGCAGCGG
	col1a1a	ENSDARG00000012405	AATGGAGAGGATGGTGAGTCTGG	CCCTTGATGCCTGGAAGTCC
	col1a1b	ENSDARG00000035809	CCTGGTGCTGCTGGTATTGC	TCTCCATTGTTTCCTTTGGGG
	lamb4	ENSDARG00000039133	GTTGGTGACCCTGTCTCTGG	GAGTCACATTGTGGGCCTGT
	pdgfba	ENSDARG00000086778	GGACCCTCTTCCTCCATCTC	TGGGACACGTAACTGACAGC
	pdgfrb	ENSDARG00000100897	CCTGCACAAAACAAGGTCCG	TCCCACTGTCACTGTCGTTG
	tgfb1a	ENSDARG00000041502	CTGGGAACTCGCTTTGTCTCC	GCTCCAAGGTTTCTTCATCTTCTG
	timp2b	ENSDARG00000075261	TTGGTCGTGAAGAGTGTCCG	CATAAGCGTCATTGCCGGTG
Liver marker	*cp*	ENSDARG00000010312	CAGCCACACGGAGTGCA	GGACAGATGCCAAGACAC
Heart marker	*cmlc2*	ENSDARG00000019096	AGAAGGAAAAGGGCCCATAA	GGGTCATTAGCAGCCTCTTG
Gut marker	*acrp30*	ENSDARG00000100086	TCCACCTGATGACAGACAGC	CTGGTCCACATTGGTCTCCT
Pancreas marker	*amy2a*	ENSDARG00000013856	CATCAACCCTGATTCCACCT	GTCCCACCAATTGGAAAATG

23. Do not damage the larvae while removing the embryo medium since it can trigger tissue repair and EdU incorporation. This can be a great source of variation.
24. The presence of methylene blue increases the background of fluorescence so it should be avoided in this assay.
25. Efficiency of the Click-IT decreases if performed in tubes bigger than 0.5 mL.
26. For the imaging we used SP8 Leica Confocal microscope with Lightning Software. Livers are located on the slide by using a 5× objective, and then a 40× oil objective was used for the acquisition of the images. Set the following parameters: size 2048 × 2048, speed 100 Hz, zoom 2.25, bidirectional, and lines 3. Once it is set, select the Lightning software and move the cursor of speed/resolution to match the following combination: size 3264 × 3264, speed around 820 Hz. Z-stack sections are taken to acquire the whole liver with a z-stack size of 0.35 μm.
27. Although it is advisable to use transgenic reporter lines for the organelle of your interest crossed with a liver reporter line, it is also possible to use vital dyes such as Nile Red and LipidSpot to mark lipids or Lysotracker to mark lysosomes.
28. As a positive control for assessing changes to ER morphology or lipid accumulation, treatment with 0.5 μg/mL of Tunicamycin (stock: 10 mg/mL in DMSO) from 4 to 5 dpf is highly effective (*see* Fig. 4a).
29. All steps should be done close to the microscope, as the agarose gel solidifies quickly. Keeping a mini heating block close to the microscope can make the process easier. Alternatively, the gel can be prepared beforehand and transferred to 1.5 mL tube and stored at −20 °C. An hour before embedding, the tubes can be heated on a heating block at 90 °C while vortexing occasionally until the gel melts completely and then temperature can be lowered to 37 °C. Ensure the gel is cool for embedding by testing it on a petri dish. When you expel a drop on the petri dish, there should be no ring of condensation around the droplet, indicating that the gel is suitable for embedding.
30. MS-222 Tricaine batches can vary. We recommend doing different titrations (from 0.08 to 0.32 g/L Tricaine) and ensure there is no unwanted effects, such as tachycardia, that could impact health or survival after embedding and during imaging.
31. Start with embedding three larvae at a time. Try to embed them vertically with a 1 mm gap. After embedding three larvae, if the gel is still liquid enough to manipulate the larvae, you may increase the number of fish for next round of

embedding. It is advisable to embed a few well than too many improperly as that may result in none being able to be imaged effectively.

32. Ensure the gel touches the rim of the glass bottom. In the presence of gaps, the gel along with samples, can float during imaging, ruining the acquisition.
33. Gel will have a bluish-white hue when it has solidified. Adding embryo medium before the gel has solidified will cause the gel to dissolve or float.
34. For the imaging, we use the Leica SP8 STED 3× Microscope. Livers are located on the slide by using the 10× dry objective and then either a 40× or 63× objective that uses water or glycerol as the refractive medium was used for the acquisition of the images. Water objective will give the best quality images. Set the following parameters: pixel size 1024 × 1024, scanning speed 100 Hz bidirectionally, zoom 1× and line average 2×. Z-stack sections are set from the top of the liver to the middle of the liver with a z-stack size of 1.0 μm (approximately 40–50 slices).
35. Having the larvae lined up vertically will help at this stage. Alternatively draw a map of the dish, with each larvae numbered such that they can easily be identified during rescue.
36. This is a critical step, as it is very easy injure the larvae. The success of this stage depends almost exclusively on the quality of the gel and the embedding of the larvae. Ensure all the gel has been removed from the body and fins. The larvae should start moving within 15–30 min. Mutant/treated larvae tend to be more sensitive to the rescue, so plan to image more mutant/treated larvae compared to controls.
37. Can be washed in 30% sucrose longer if larvae are still floating in sucrose solution.
38. Humid chamber can be created by placing damp Kim wipes in a container that includes a surface for the slides to rest on and a top that seals tightly.
39. High background in the staining can be eliminated by overnight incubation of secondary antibody at 4 °C instead of room temperature.
40. From this step forward wrap humid chambers in tin foil to keep slides in the dark.
41. ProLong Gold is used for standard procedures default, but the use of ProLong Diamond is recommended for samples with fluorescent proteins.
42. Samples can be imaged as early as 3 h after adding Prolong Gold.

43. We use Leica SP5 DM with setting the parameters for this analysis, use control to set brightness for each type of fluorescence. Check negative control condition to look for nonspecific staining and lower brightness if necessary. Confirm that sufficient signal is still obtained from control samples. Image all samples using these conditions.

Acknowledgments

The authors are grateful to Shashi Ranjan for expert fish care and Joshua Morrison for the optimization of the IF protocol. This work is supported by the Al Jalila Foundation (AJF2018098 to KSE), NYUAD Research Enhancement Fund (RE188 to KSE), and R01 DK121154, R01 DK121154-01A1S1 to JC.

References

1. Asrani SK, Devarbhavi H, Eaton J et al (2019) Burden of liver diseases in the world. J Hepatol 70:151–171
2. Collaborators GBDC (2020) The global, regional, and national burden of cirrhosis by cause in 195 countries and territories, 1990–2017: a systematic analysis for the Global Burden of Disease Study 2017. Lancet Gastroenterol Hepatol 5:245–266
3. Di Sessa A, Cirillo G, Guarino S et al (2019) Pediatric non-alcoholic fatty liver disease: current perspectives on diagnosis and management. Pediatric Health Med Ther 10:89–97
4. Roehlen N, Crouchet E, Baumert TF (2020) Liver fibrosis: mechanistic concepts and therapeutic perspectives. Cell 9
5. Goessling W, Sadler KC (2015) Zebrafish: an important tool for liver disease research. Gastroenterology 149:1361–1377
6. Cox AG, Goessling W (2015) The lure of zebrafish in liver research: regulation of hepatic growth in development and regeneration. Curr Opin Genet Dev 32:153–161
7. Wilkins BJ, Pack M (2013) Zebrafish models of human liver development and disease. Compr Physiol 3:1213–1230
8. Wang S, Miller SR, Ober EA et al (2017) Making it new again: insight into liver development, regeneration, and disease from zebrafish research. Curr Top Dev Biol 124:161–195
9. Morrison J, DeRossi C, Alter I et al (2021) Single-cell transcriptomics reveals conserved cell identities and fibrogenic phenotypes in zebrafish and human liver. bioRxiv. https://doi.org/10.1101/2021.08.06.455422
10. Cheng D, Morsch M, Shami GJ et al (2020) Observation and characterisation of macrophages in zebrafish liver. Micron 132:102851
11. Wrighton PJ, Oderberg IM, Goessling W (2019) There is something fishy about liver cancer: zebrafish models of hepatocellular carcinoma. Cell Mol Gastroenterol Hepatol 8: 347–363
12. Evason KJ, Francisco MT, Juric V et al (2015) Identification of chemical inhibitors of beta-catenin-driven liver tumorigenesis in zebrafish. PLoS Genet 11:e1005305
13. Lu JW, Yang WY, Tsai SM et al (2013) Liver-specific expressions of HBx and src in the p53 mutant trigger hepatocarcinogenesis in zebrafish. PLoS One 8:e76951
14. Nguyen AT, Emelyanov A, Koh CH et al (2012) An inducible krasV12 transgenic zebrafish model for liver tumorigenesis and chemical drug screening. Dis Model Mech 5:63–72
15. Li Z, Huang X, Zhan H et al (2012) Inducible and repressable oncogene-addicted hepatocellular carcinoma in Tet-on xmrk transgenic zebrafish. J Hepatol 56:419–425
16. Spitsbergen JM, Tsai HW, Reddy A et al (2000) Neoplasia in zebrafish (Danio rerio) treated with N-methyl-N′-nitro-N-nitrosoguanidine by three exposure routes at different developmental stages. Toxicol Pathol 28:716–725

17. Mudbhary R, Hoshida Y, Chernyavskaya Y et al (2014) UHRF1 overexpression drives DNA hypomethylation and hepatocellular carcinoma. Cancer Cell 25:196–209
18. Passeri MJ, Cinaroglu A, Gao C et al (2009) Hepatic steatosis in response to acute alcohol exposure in zebrafish requires sterol regulatory element binding protein activation. Hepatology 49:443–452
19. Howarth DL, Passeri M, Sadler KC (2011) Drinks like a fish: using zebrafish to understand alcoholic liver disease. Alcohol Clin Exp Res 35:826–829
20. Delaney P, Nair AR, Palmer C et al (2020) Arsenic induced redox imbalance triggers the unfolded protein response in the liver of zebrafish. Toxicol Appl Pharmacol 409:115307
21. Bambino K, Zhang C, Austin C et al (2018) Inorganic arsenic causes fatty liver and interacts with ethanol to cause alcoholic liver disease in zebrafish. Dis Model Mech 11:dmm031575
22. Carlson P, Van Beneden RJ (2014) Arsenic exposure alters expression of cell cycle and lipid metabolism genes in the liver of adult zebrafish (Danio rerio). Aquat Toxicol 153: 66–72
23. Lam SH, Winata CL, Tong Y et al (2006) Transcriptome kinetics of arsenic-induced adaptive response in zebrafish liver. Physiol Genomics 27:351–361
24. North TE, Babu IR, Vedder LM et al (2010) PGE2-regulated wnt signaling and N-acetylcysteine are synergistically hepatoprotective in zebrafish acetaminophen injury. Proc Natl Acad Sci U S A 107:17315–17320
25. DeRossi C, Bambino K, Morrison J et al (2019) Mannose phosphate isomerase and mannose regulate hepatic stellate cell activation and fibrosis in zebrafish and humans. Hepatology 70:2107–2122
26. Michael C, Martinez-Navarro FJ, de Oliveira S (2021) Analysis of liver microenvironment during early progression of non-alcoholic fatty liver disease-associated hepatocellular carcinoma in zebrafish. J Vis Exp. https://doi.org/10.3791/62457
27. Yang Q, Salim L, Yan C et al (2019) Rapid analysis of effects of environmental toxicants on tumorigenesis and inflammation using a transgenic zebrafish model for liver cancer. Mar Biotechnol (NY) 21:396–405
28. Salmi TM, Tan VWT, Cox AG (2019) Dissecting metabolism using zebrafish models of disease. Biochem Soc Trans 47:305–315
29. Fei F, Wang L, Sun S et al (2019) Transgenic strategies to generate heterogeneous hepatic cancer models in zebrafish. J Mol Cell Biol 11:1021–1023
30. Kendall GC, Amatruda JF (2016) Zebrafish as a model for the study of solid malignancies. Methods Mol Biol 1451:121–142
31. Ellis JL, Yin C (2017) Histological analyses of acute alcoholic liver injury in zebrafish. J Vis Exp. https://doi.org/10.3791/55630
32. Korzh S, Pan X, Garcia-Lecea M et al (2008) Requirement of vasculogenesis and blood circulation in late stages of liver growth in zebrafish. BMC Dev Biol 8:84
33. Zheng W, Li Z, Nguyen AT et al (2014) Xmrk, kras and myc transgenic zebrafish liver cancer models share molecular signatures with subsets of human hepatocellular carcinoma. PLoS One 9:e91179
34. Huo X, Li H, Li Z et al (2019) Transcriptomic profiles of tumor-associated neutrophils reveal prominent roles in enhancing angiogenesis in liver tumorigenesis in zebrafish. Sci Rep 9: 1509
35. Stuckenholz C, Lu L, Thakur P et al (2009) FACS-assisted microarray profiling implicates novel genes and pathways in zebrafish gastrointestinal tract development. Gastroenterology 137:1321–1332
36. DeRossi C, Vacaru A, Rafiq R et al (2016) trappc11 is required for protein glycosylation in zebrafish and humans. Mol Biol Cell 27: 1220–1234
37. Yin C, Evason KJ, Maher JJ et al (2012) The basic helix-loop-helix transcription factor, heart and neural crest derivatives expressed transcript 2, marks hepatic stellate cells in zebrafish: analysis of stellate cell entry into the developing liver. Hepatology 56:1958–1970
38. Yang W, He H, Wang T et al (2021) Single-cell transcriptomic analysis reveals a hepatic stellate cell-activation roadmap and myofibroblast origin during liver fibrosis in mice. Hepatology 74:2774–2790
39. Yang Q, Yan C, Gong Z (2018) Interaction of hepatic stellate cells with neutrophils and macrophages in the liver following oncogenic kras activation in transgenic zebrafish. Sci Rep 8:8495
40. Ramdas Nair A, Delaney P, Koomson AA et al (2021) Systematic evaluation of the effects of toxicant exposure on survival in zebrafish embryos and larvae. Curr Protoc 1:e231

Chapter 4

Developmental Toxicity Assessment Using Zebrafish-Based High-Throughput Screening

Subham Dasgupta, Michael T. Simonich, and Robyn L. Tanguay

Abstract

Zebrafish-based high-throughput screening has been extensively used to study toxicological profiles of individual chemicals and mixtures, identify novel toxicants, and study modes of action to prioritize chemicals for further testing and policy decisions. Within this chapter, we describe a protocol for automated zebrafish developmental high-throughput screening in our laboratory, with emphasis on exposure setups, morphological and behavioral readouts, and quality control.

Key words High-throughput screening, Zebrafish, Morphology, Behavior, Chemicals, Automation, Embryos, Development

1 Introduction

Zebrafish has been widely used as a model for toxicity evaluations of environmental chemicals, particularly for understanding human health effects. While rodent models have closer relevance to human health, their large size, in utero development, comparatively low reproductive capacity, and high husbandry costs preclude their use in high-throughput studies. In vitro models circumvent these limitations but lack sufficient complexity and the metabolic capacity for translation into human health risk [1, 2]. The developmental zebrafish model would seem to address all the above limitations with whole animal complexity and metabolism, *ex utero* development, comparatively high reproductive capacity, and low husbandry costs. The developmental zebrafish is thus amenable to high-throughput assessments for discovery, structure-bioactivity relationships, and understanding mechanisms of toxicity through molecular studies. Its translatability to human health effects is facilitated by a remarkable similarity in anatomy, molecular landscape, signaling mechanisms, cellular structure, and physiology

James F. Amatruda et al. (eds.), *Zebrafish: Methods and Protocols*, Methods in Molecular Biology, vol. 2707,
https://doi.org/10.1007/978-1-0716-3401-1_4,

between zebrafish and higher vertebrates; indeed, ~80% of the zebrafish genome is similar to the human genome [3].

Our lab routinely uses zebrafish as a model to study the developmental bioactivity of hundreds of thousands of chemicals, profile their hazard potential, calculate their effective concentrations, and study the mechanisms of toxic effects. Our protocols are modified from the Fish Embryo Toxicity (FET) test, recommended by the OECD Guidelines for the Testing of Chemicals [4]. The short developmental trajectory (~24 h from fertilization to onset of organogenesis) effectively means that nearly every molecular target the animal has is transcriptionally and translationally active during an easily dosed experimental window. The capacity for a female fish to produce hundreds of embryos daily enables design of studies with large concentration ranges and replicate numbers with robust statistical power. Our typical screening paradigm includes exposure to seven concentrations on a whole or fractional log scale, with at least 32 embryos per concentration. Singulated, age-matched embryos are arrayed in 96-well plates, typically in a 100 μL volume of embryo medium. Thus, chemical use can be minimized, the experimental footprint is small, and as long as evaporation is prevented, zebrafish development is robust under these conditions through 6 days postfertilization (dpf) at 28 °C. We typically initiate chemical exposures at 6 h postfertilization (hpf)—the gastrulation stage. This timepoint enables dechorionation *en masse* at 4 hpf, the earliest point at which chorion removal is well tolerated, and sufficient time after dechorionation for automated singulation of embryos into plate wells [5]. We continue the static exposures until 120 hpf (5 dpf) to encompass organogenesis. The feeding stage, beginning at 6 dpf, is typically avoided to eliminate any compounding effects of food intake on biology and metabolism. At 24 and 120 hpf, we record the incidences of embryo mortality and malformation visualized under the microscope. We also perform an automated embryo photomotor behavior assay at 24 hpf and a larval photomotor and startle response assay at 120 hpf. All these assays are performed noninvasively over the course of development.

2 Materials

1. System water: 0.3 g/L Instant Ocean salts in reverse osmosis water; pH of ~7.
2. Recirculating System for zebrafish.
3. Fish flakes.
4. Strainers.
5. 96-well plates.

6. Pronase from *S. griseus*.
7. Automated dechorionator.
8. 4-axis SCARA robotic platform (Denso SCARA arm, Cognex vision, custom software).
9. HP D300 Digital Dispenser (Palo Alto, CA).
10. Zebrabox behavior chamber.
11. Stereomicroscopes.
12. Tricaine methanesulfonate (MS-222).

3 Methods

3.1 Fish Husbandry and Feeding

Our adult zebrafish are reared under standard laboratory conditions of 28 °C in system water on a 14 h light/10 h dark cycle. The fish are housed in 1.8, 3, 6, or 9 L tanks on an Aquaneering recirculating system (San Diego, CA) at approximately 6 fish/L in groups of both males and females [6]. Alternatively, groups of 250 or 500 males and similar numbers of female zebrafish are housed in our 190 and 380 L Mass Embryo Production System (MEPS) tanks, respectively (*see* Fig. 1). Fish are fed with ground fish flakes twice a day.

Fig. 1 Zebrafish housed within the Mass Embryo Production System (MEPS)

Fig. 2 Zebrafish embryos collected from spawning chambers (left) and MEPS (right)

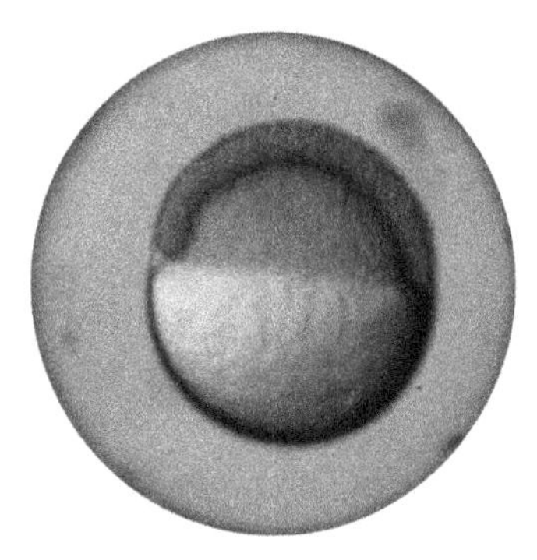

Fig. 3 Zebrafish embryos at 6 hpf

3.2 Embryo Spawning

On the day before screening, fish reared in small individual tanks are placed in spawning baskets, with males and females (3:2 ratio) separated by a barrier (*see* **Note 1**). Zebrafish will typically spawn in the morning when the barrier is removed, and the lights come on after the 10 h dark period (*see* Fig. 2). Newly fertilized eggs are collected using a strainer, rinsed, and stored in system water at 28 °C for sorting. For MEPS, embryos are collected at the outlet of the MEPS egg lift (air/water lift) plumbing (*see* Fig. 2). Embryos from multiple baskets or MEPSs are mixed together for experiments to avoid clutch-specific effects.

3.3 Embryo Sorting

Prior to initiation of exposure at 6 hpf, embryos that are unfertilized, dead, or at a too advanced (off-spawned) developmental stage, are discarded. All embryos are staged at ~50% epiboly (*see* Fig. 3), based on [7].

3.4 Dechorionation

We developed and published the design of a partially automated dechorionator [5]. Though the design has been improved since then, the operating principle has remained the same: to gently shake a dish of approximately 1000 embryos in a stuttering circular motion and to deliver a gentle overflowing quantity of rinse water with precise timing to avoid overdigestion of the chorion/embryo. The automated dechorinator has a holder stage or four petri dishes enabling dechorionation of up to 4000 embryos at once with 80–90% efficiency (successfully dislodged from the chorion without further manipulation). Enzymatic digestion is accomplished with pronase (from *S. griseus*) dissolved to make a stock of 224 U/mL and stored as single-use aliquots at −80 °C. 83 μL of the pronase are added to a dish of 1000 embryos in approximately 30 mL of EM. The automated dechorionator is started and gently shakes the embryo/pronase mixture for 6.5 min, gently rinses with EM by overflowing the dish for 10 min, mixes gently with a stuttering motion for 10 min to help dislodge the partially digested chorions, pauses for a 30-min rest and recovery of the embryos, then overflow rises more vigorously for 5 min to wash away the dislodged chorions.

3.5 Loading into 96-Well Plates

Dechorionated embryos are singulated into 96-well plates, each well containing 100 μL EM, using a custom four-axis SCARA robotic platform (Denso SCARA arm, Cognex vision, custom software) [5] (*see* Fig. 4).

3.6 Waterborne Exposure of Chemicals

Chemicals are loaded into barcoded 96-well plates using an HP D300 Digital Dispenser (Palo Alto, CA). Stock concentrations of chemicals are dissolved in dimethyl sulfoxide (DMSO) and loaded into cassettes of the D300e which only dispenses in sequential jets of 13 pL drops. The stock concentration on the cassette, desired test concentrations in the wells, and location on the plate are all user-specified in the controlling software. The chemical concentration map is assigned to the plate barcode and input to a custom LIMS platform that interfaces with a MySQL database. All the concentration math is performed by the software and there is no need for serial dilution (*see* Fig. 5). After dispensing is complete, plates are covered with sheet of parafilm sandwiched between the plate surface and the lid and placed on a shaker at 225 rpm in a 28 °C dark room until morphology and embryo photomotor response assessment at 24 hpf.

3.7 24 h Mortality and Developmental Progression Assessment

Chemical exposures encompass 6–120 hpf (*see* Fig. 6). At 24 hpf, embryos are assessed for viability and developmental progression. Developmental effects are recorded as mortality or >12 h delay in developmental progression. Each observation of an abnormality is manually input into the LIMS via touchscreen tablet PC as binary data (live/dead or presence/absence) for each well location [8].

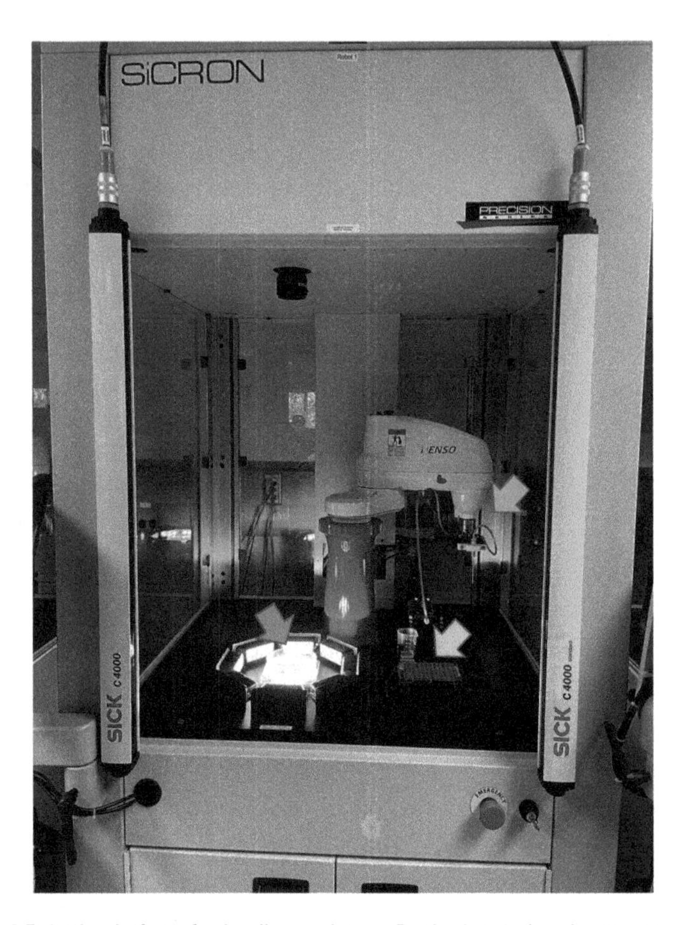

Fig. 4 Robotic platform for loading embryos. Dechorionated embryos are gently aspirated from a high-contrast environment dish (orange arrow) and transferred by capillary action to the liquid surfaces of plate wells (yellow arrow) by a robotic arm (blue arrow)

3.8 Embryonic Photomotor Response (EPR)

At 22–24 hpf, embryos are assessed in plate for photomotor response using a custom photomotor response analysis tool. The EPR is based on detection of significant departures from the canonical burst of embryo tail flexions in response to a bright pulse of visible light, but no response to a second pulse (*see* Fig. 7). This is an early but sensitive indication of chemical impact and a robust predictor of the likelihood of later-manifesting chemical impacts [9]. For every exposure plate, 850 frames of digital video are recorded at 17 frames s^{-1} from beneath a custom 96-well plate mount, and lighted from above with white LED and infrared (IR) LED lights. The light cycle consists of 30 s of dark (IR) background (prior to the first light pulse), a short pulse of white light, and 9 s later, a second pulse of white light, and then 10 more seconds of dark (IR). (*See* **Note 2**).

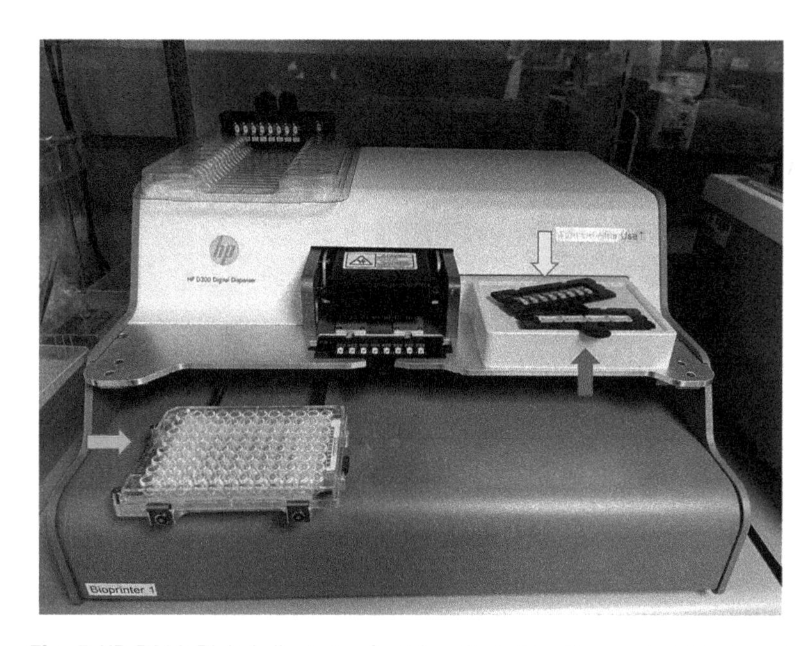

Fig. 5 HP D300 Digital dispenser for 96-well dosing. Four and eight-channel cartridges (orange and yellow arrow, respectively) are preloaded with concentrated (mM) stock solutions of chemicals in DMSO and rapidly dispensed as jets of 13 pL droplets into standard plate wells (green arrow)

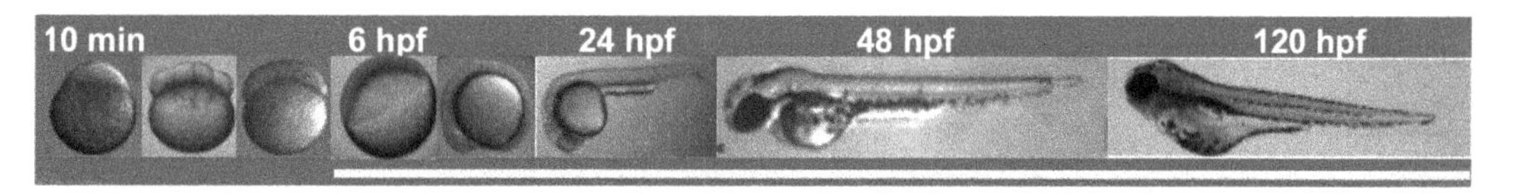

Fig. 6 Developmental timeline to 120 hpf. Yellow bar represents our typical exposure window enabling routine dechorionation at 4 hpf with high efficiency and survival

3.9 Larval Photomotor Response (LPR) and Larval Startle Response (LSR)

At 120 hpf (5 dpf) zebrafish are free-swimming larvae and the photomotor response assays are executed as total movement (swim distance) in response to multiple light-dark transitions. The LPR is based on detection of significant departures from the canonical pattern of two- to tenfold more swimming activity in the dark (IR) phases than in the visible light phases. A ZebraBox behavior chamber (ViewPoint Life Sciences, Montreal, CAN) with an infrared backlit stage tracks total movement in 96 wells during a 24-min assay. HD video is captured at 15 frames s^{-1} and processed in real time by the manufacturer's software. A representative wild-type versus treated LPR data set over four light-dark cycles is shown in Fig. 8 (*see* **Note 3**).

The larval startle response (LSR) is assessed on the same ZebraBox platform as the LPR and immediately afterward. The assay consists of a 1 s, 600 Hz audio tone at 100 dB. The audio speaker

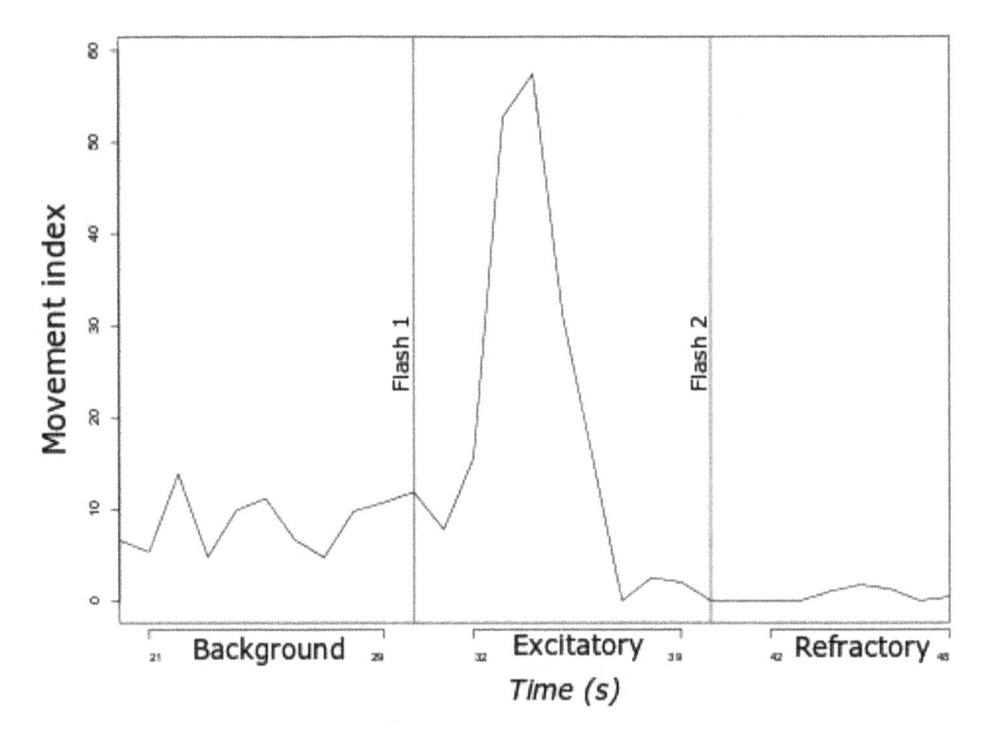

Fig. 7 Wild-type embryo photomotor response. Red lines indicate a bright 1 s flash of visible light. The assay duration is 60 s trimmed in the analysis to second 21–29 (baseline phase), 32–39 (excitatory phase) and 42–48 (refractory phase). The Y-axis values are a unitless index of motion (tail flexons) recorded as pixel changes between successive video frames. Embryos are tested 1 per well of a 96-well plate

is directly coupled to the plate stage and results in a 4 mm/s vibration rate at the center of the plate. The tone is repeated every 10 s of the 2-min assay. Again, the goal of the assay is to detect significant departures from the canonical startle spike and decay in swimming activity in response to the sudden vibration shown in Fig. 9.

3.10 Mortality and Morphology Assessments at 120 hpf

At 120 hpf, plates are evaluated for larval mortality and a battery of morphological endpoints under a stereoscope. Incidence of chemical effect in each well are again captured by the LIMS as binary (presence/absence) data. The morphological endpoints typically recorded include (*see* Fig. 10):

1. Cranial defects: abnormality within the craniofacial area.
2. Axis defects: deformed body axis.
3. Edema: presence of an edema. Typically pericardial or yolk sac, but also possible within other areas.
4. Somites: number of somites present.
5. Trunk: defects or deformities within the trunk.
6. Brain: brain deformities or enlarged brain.
7. Skin: defects within skin, skin pigmentation.

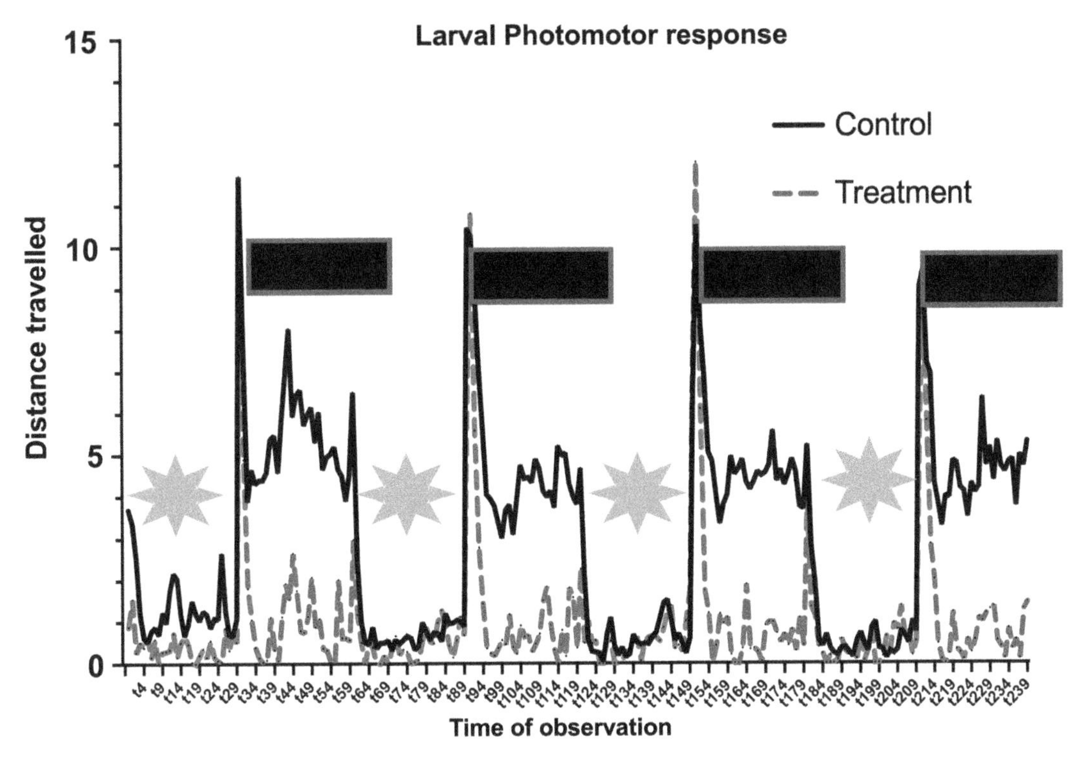

Fig. 8 LPR trace for control and treated (a representative flame retardant) embryos. Yellow stars represent visible light period, black bars represent dark IR light period. The canonical wild-type larval zebrafish behavior pattern is two- to tenfold more swimming activity in the dark than in the light

8. Notochord: defects within notochord or bent notochord.
9. Touch response: escape response when larvae is touched with a probe.

3.11 Statistics

Statistical analyses are performed by custom R-scripts running in the LIMS background. For mortality and morphology data, lowest effect concentrations (LECs) and statistical significance are estimated based on a one-sided Fisher's exact test. For behavioral assays, statistical estimations and LECs are estimated from a Kolmogorov-Smirnov test comparison of area under curves and peak heights. More details are provided in [10].

3.12 Quality Control

QC for each experiment is estimated by custom R-scripts in the LIMS background whenever a plate barcode is retrieved and exported from the MySQL database for analysis. The QC for negative control behavior data requires meeting a minimum response threshold of swimming activity following the stimuli. The QC for teratogenicity data is two parts: (1) A stand-alone, daily positive teratogenicity control plate consisting of a seven-concentration curve of ethyl parathion plus untreated control animals ($n = 12$ per concentration) set up each day as an internal check

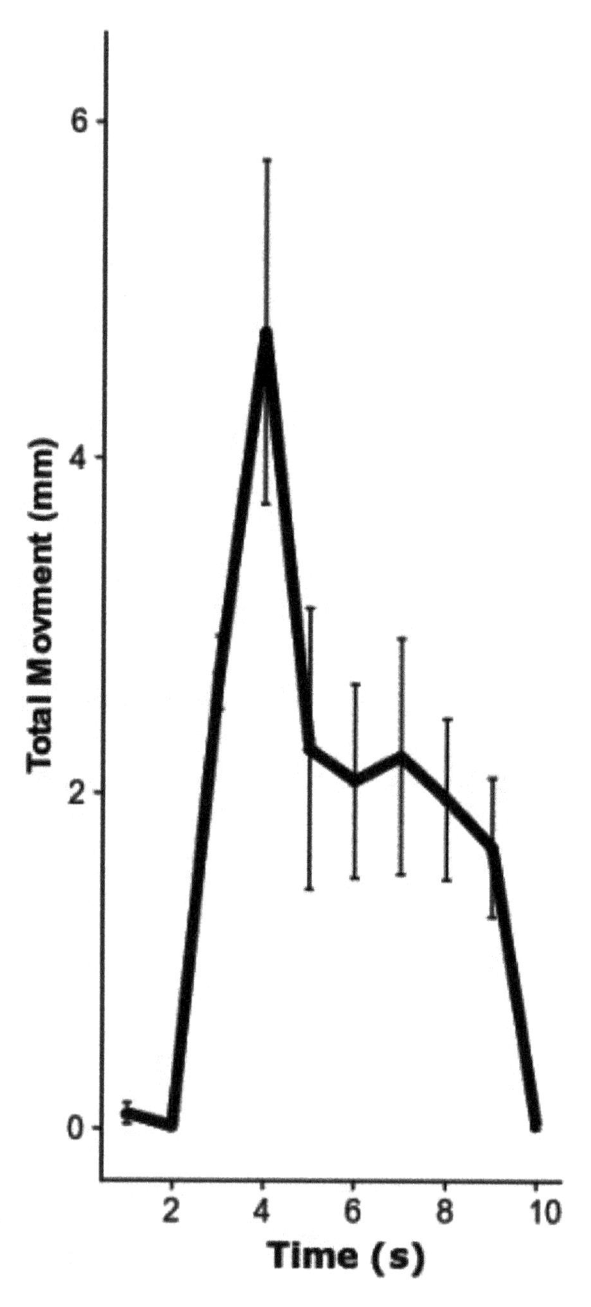

Fig. 9 Typical larval startle response of singulated larvae in a 96-well plate to an audible tone where the speaker is coupled to the plate stage

for response consistency in animals from different hatches, parathion mortality/morbidity concentration response must exhibit an EC_{50} of 29.1 ± 4 μM. (2) On plate experiment negative controls (n = 12) must exhibit less than 20% cumulative mortality and morbidity across the replicate plates of an experiment, typically three plates (n = 36), so seven or fewer affected negative control animals (*see* **Note 4**).

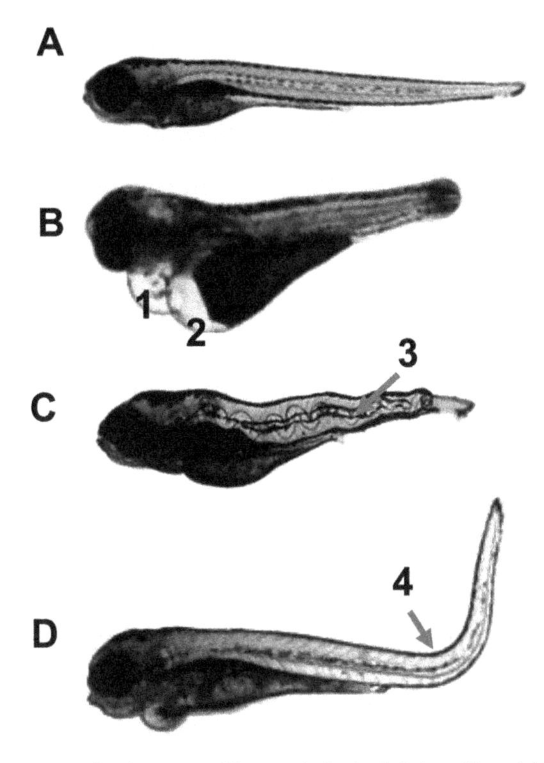

Fig. 10 Representative images with morphological deformities. (**a**) Normal larva at 120 hpf. (**b**) Larvae with pericardial (*1*) and yolk sac (*2*) edema. (**c**) Larvae with undulated notochord (*3*). (**d**) Larvae with bent axis (*4*)

4 Notes

1. The adult fish are not fed in the afternoon prior to spawning. It is always advisable to spawn multiple pairs of fish together to maintain genetic diversity.
2. Animals recorded as dead, delayed, or malformed at the 24 hpf visual evaluation are automatically filtered from subsequent analysis of the behavior data.
3. Additional animals dead or malformed (above what was recorded at 24 hpf) at the 120 hpf timepoint are automatically filtered from the subsequent larval behavior analysis.
4. Disposal of embryos: After analysis of morphology and behavior, plates are overdosed with tricaine methane sulfonate (MS-222). Euthanized larvae in the plates are suctioned out to liquid chemical waste using a Tecan Hydrospeed 96 channel plate washer and disposed of as low-level liquid chemical waste.

References

1. Truong L, Tanguay RL (2017) Evaluation of embryotoxicity using the zebrafish model. In: Gautier J-C (ed) Drug safety evaluation: methods and protocols. Springer, New York, pp 325–333. https://doi.org/10.1007/978-1-4939-7172-5_18
2. Truong L, Tanguay RL (2017) Evaluation of embryotoxicity using the zebrafish model. Methods Mol Biol 1641:325–333. https://doi.org/10.1007/978-1-4939-7172-5_18
3. Garcia GR, Noyes PD, Tanguay RL (2016) Advancements in zebrafish applications for 21st century toxicology. Pharmacol Ther 161: 11–21
4. OECD (2019) Test no. 203: fish, acute toxicity test. https://doi.org/10.1787/9789264069961-en
5. Mandrell D, Truong L, Jephson C et al (2012) Automated zebrafish chorion removal and single embryo placement: optimizing throughput of zebrafish developmental toxicity screens. J Lab Autom 17(1):66–74. https://doi.org/10.1177/2211068211432197
6. Barton CL, Johnson EW, Tanguay RL (2016) Facility design and health management program at the Sinnhuber Aquatic Research Laboratory. Zebrafish 13(Suppl 1):S39–S43. https://doi.org/10.1089/zeb.2015.1232
7. Kimmel CB, Ballard WW, Kimmel SR et al (1995) Stages of embryonic development of the zebrafish. Dev Dyn 203(3):253–310. https://doi.org/10.1002/aja.1002030302
8. Truong L, Reif DM, St Mary L et al (2014) Multidimensional in vivo hazard assessment using zebrafish. Toxicol Sci 137(1):212–233. https://doi.org/10.1093/toxsci/kft235
9. Reif DM, Truong L, Mandrell D et al (2016) High-throughput characterization of chemical-associated embryonic behavioral changes predicts teratogenic outcomes. Arch Toxicol 90(6):1459–1470. https://doi.org/10.1007/s00204-015-1554-1
10. Truong L, Marvel S, Reif DM et al (2020) The multi-dimensional embryonic zebrafish platform predicts flame retardant bioactivity. Reprod Toxicol 96:359–369. https://doi.org/10.1016/j.reprotox.2020.08.007

Chapter 5

Cancer Modeling by Transgene Electroporation in Adult Zebrafish (TEAZ)

Emily Montal, Shruthy Suresh, Yilun Ma, Mohita M. Tagore, and Richard M. White

Abstract

Transgenic expression of genes is a mainstay of cancer modeling in zebrafish. Traditional transgenic techniques rely upon injection into one-cell embryos, but ideally these transgenes would be expressed only in adult somatic tissues. We provide a method to model cancer in adult zebrafish in which transgenes can be expressed via electroporation. Using melanoma as an example, we demonstrate the feasibility of expressing oncogenes such as BRAFV600E as well as CRISPR/Cas9 inactivation of tumor suppressors such as PTEN. These approaches can be performed in any genetic background such as existing fluorophore reporter lines or the *casper* line. These methods can readily be extended to other cell types allowing for rapid adult modeling of cancer in zebrafish.

Key words Electroporation, Melanoma, Cancer modeling, Somatic transgenesis, CRISPR/Cas9, Oncogenes, Tumor suppressors

1 Introduction

The ability to manipulate the genome has proven to be one of the most powerful techniques across model organisms. In zebrafish, transgenic methods have proven to be especially powerful since they allow for expression of marker fluorophores or genes of interest in a cell-type specific manner [1]. This has allowed for the development of many stable lines of fish that mark cells such as endothelial cells [2], red blood cells [3], melanocytes [4], and many others. Over the past decade, these types of approaches have greatly expanded into the realm of cancer modeling. Building off of pioneering work done in cell culture and mouse models, expression of oncogenes under cell-type-specific promoters has enabled the generation of numerous cancer models in zebrafish. For example, expression of the MYC oncogene under the rag2 promoter leads to leukemia [5], and expression of the BRAFV600E oncogene

James F. Amatruda et al. (eds.), *Zebrafish: Methods and Protocols*, Methods in Molecular Biology, vol. 2707, https://doi.org/10.1007/978-1-0716-3401-1_5,

under the mitfa promoter leads to melanoma in a p53−/− background [6]. Inactivation of tumor suppressors in a cell-type specific manner can also now be achieved using CRISPR/Cas9, such as the loss of SPRED1 in melanoma using an mitfa:Cas9 approach [7].

Despite these advances, most of the prior approaches to modeling cancer in zebrafish have relied upon injection of transgenes into one-cell stage embryos. In some cases, the animals are then raised to adulthood, crossed, and the offspring screened for transgenes that have passed through the germline. This allows for stable founder lines which then spontaneously give rise to cancer in adults, but is extremely time-consuming and most animals wind up uninformative. A different approach has been to study the tumors that arise in the injected (F0) animals [8]. This is more efficient, but can be highly variable due to mosaic expression of the transgene. In addition, a major issue with these one-cell embryo approaches is that the transgenes are typically expressed throughout life, including embryonic and larval periods. This is not what happens in human cancer, where somatic mutations in genes like BRAFV600E do not occur during infancy, but rather occur somatically later in life. Attempts to circumvent this issue using inducible Cre/Lox approaches are available [9, 10], but these can be challenging to work with due to variability in induction by tamoxifen or mifepristone, although this is still an important method that could be improved upon. A final difficulty with one-cell embryo injection approaches is that many thousands of cells are transformed at the same time, which again likely does not happen in humans. This makes it difficult to discern a multifocal primary tumor versus true metastatic lesions.

For these reasons, an alternative approach in which transgenes could be expressed in a specific cell type at the appropriate adult stage of development would be useful. Several approaches to such "somatic transgenesis" have been applied, including direct viral injection or electroporation of vectors into the tissue of interest. While lentiviruses, retroviruses, and AAV vectors all work very efficiently in mouse models of cancer, they are far less well developed in zebrafish and would require significant optimization for this species. Based on this, we developed a highly efficient method for delivery of transgenes in adult zebrafish using electroporation (TEAZ). In the original method, we demonstrated its feasibility using a BRAFV600E-driven melanoma model that includes inactivation of RB1 via CRISPR/Cas9 [11]. However, here we provide additional data that the tumors are far more robust and easy to generate with inactivation of the PTEN tumor suppressor rather than RB1. Because TEAZ uses cell-type specific drivers and different transgenes can be combined, this opens up the possibility of modeling many different cancer types with oncogenes and tumor suppressors appropriate to each type of tumor. In addition, because these tumors can be generated in any existing genetic background, this will allow investigators to take advantage of the wide variety of fluorescent reporter lines already available (i.e., fli-GFP, etc.) [2] or more optically transparent lines such as casper [12], which will enhance imaging capabilities.

2 Materials

2.1 Zebrafish Lines

1. AB strain.
2. Casper strain [12].
3. Casper triples strain [13].

2.2 Instrumentation

1. Electroporator (e.g., BTX ECM 830),
2. Electrodes (e.g., BTX 3 mm platinum Tweezertrodes 45-0487),
3. Microinjection apparatus.
4. Needle puller.
5. Stereo dissecting microscope.

2.3 Plasmid Preparation Materials

1. Entry and destination vectors required are part of the Tol2Kit [14] and include 5′ entry vector p5E, middle entry vector pME, 3′ entry vector p3E, and pDestTol2pA2 destination vector.
2. Gateway BP Clonase II enzyme mix (Thermo Fisher).
3. Gateway LR Clonase II Plus enzyme mix (Thermo Fisher).
4. High-fidelity Taq polymerase.
5. zU6a, zU6b, and zU6c promoters: Addgene plasmid numbers 64245, 64247, 64248.
6. Oligonucleotides for the desired gRNA sequence (*see* Subheading 3.1.2).
7. 1X NEB Buffer (New England Biolabs).
8. Gel purification kit.
9. Competent cells (e.g., top10 chemically competent cells, Thermo Fisher).
10. Kanamycin (for entry clones).
11. Ampicillin (for expression clones).
12. LB broth.
13. LB Agar plates with ampicillin or kanamycin.
14. Plasmid purification kit capable of large-scale (e.g., Maxi) plasmid preparation.
15. Nucleospin cleanup kit.
16. MiniCOOPR-GFP (Addgene plasmid #118850).

2.4 Electroporation

1. Agarose plate made in 100 mm × 20 mm petri dish.
2. E3 medium. For 2 L of 60× E3 stock, dissolve 34.4 g NaCl, 1.52 g KCl, 5.8 g CaCl2.2H2O, and 9.8 g MgSO4.7H2O, add double distilled water up to 2000 mL.
3. Quartz needles.
4. Razor.
5. Tricaine MS-222.
6. Plasmid mix for injection.
7. Food coloring (e.g., McCormick green FC, containing propylene glycol).

2.5 Microscopy

1. Fluorescence stereo microscope (e.g., Zeiss Axio V16).

2.6 CRISPR-SEQ

1. 15 mL conical screw-cap tube .
2. Liberase (e.g., Millipore Sigma # 05401020001).
3. Fetal bovine serum (FBS).
4. 40 μm tip filter.
5. Kit for preparation of genomic DNA from tissues.

3 Methods

3.1 Plasmid Cloning to Overexpress or Knockout Genes

3.1.1 Overexpression Plasmids: Construct Plasmids Using the Tol2Kit [14] by Gateway Cloning

1. For cell-type specific expression, clone selected promoter sequence into the p5E vector, the overexpressed gene ORF into the pME vector, and SV40 into the p3E vector, using BP Clonase II.
2. Perform LR cloning with three entry vectors into the pDest-Tol2pA2 destination vector, using LR Clonase II.
3. Transform the cloning reaction into competent cells under Ampicillin selection and use the Qiagen HiSpeed or Plasmid Plus MaxiPrep kit or equivalent to extract plasmids.

3.1.2 CRISPR Plasmids: Generate Cell-Type Specific Cas9 Plasmid and gRNA Plasmids

1. Cell-type specific Cas9 plasmids: For Cas9 overexpression plasmids, use the above Tol2Kit protocol with the desired promoter (such as *mitf*) as the p5E vector, Cas9 in the pME vector, and SV40 into the p3E vector.
2. Plasmid with single gRNA cassette: We generated a single gRNA cassette within Tol2 arms by infusion cloning the zU6a-sgRNA [15] sgRNA guide cassette with the pDest-Tol2pA2 destination vector (*see* Fig. 1). Use this plasmid to generate single gRNA constructs (soon to be available on Addgene). Cloning strategy for this plasmid is adapted from Yin et al. [15].

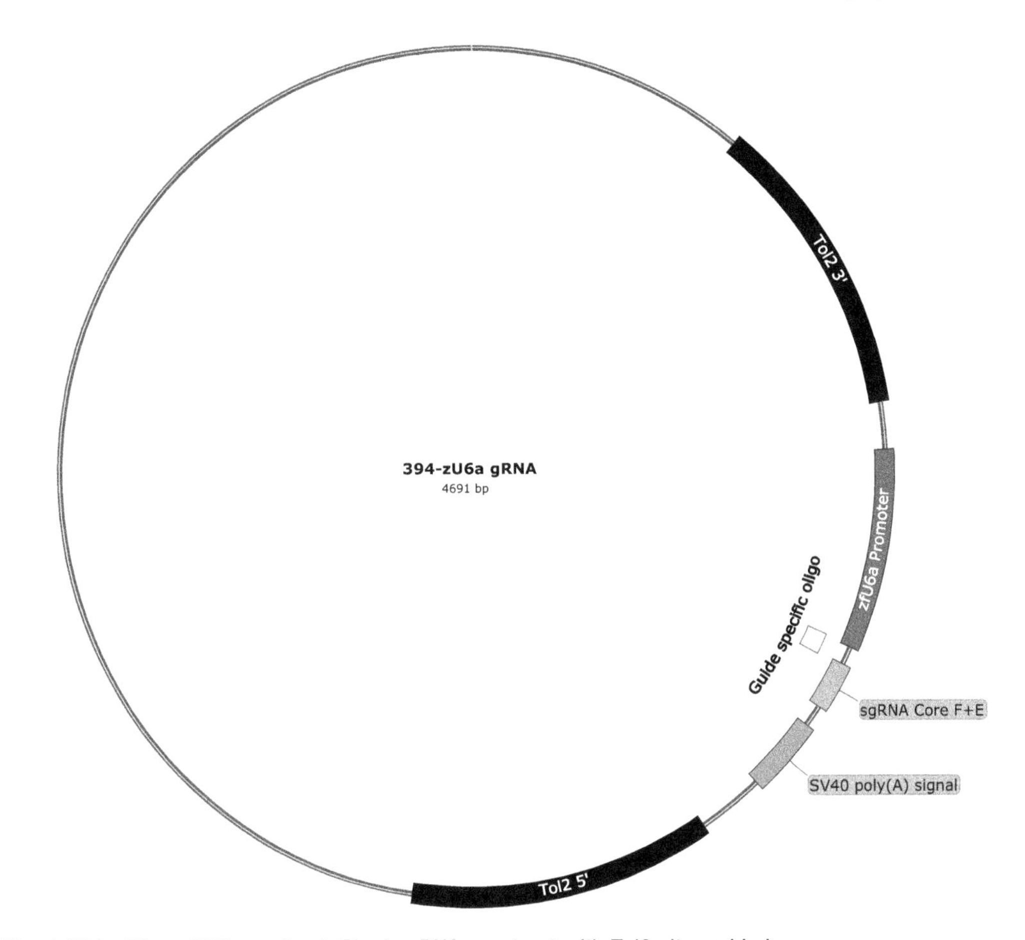

Fig. 1 394-zU6a-sgRNA construct. Single gRNA construct with Tol2 sites added

3. Design a pair of oligos for the desired gRNA sequence by adding TTCG on the forward primer and AAAC on the reverse primer. For example, with the target sequence AAGAGAGCAA CAAAATCGCG, the forward oligo would be **TTCG**AAGA GAGCAACAAAATCGCG and the reverse oligo would be **AAAC**CGCGATTTTGTTGCTCTCTT.
4. Combine the oligos (1 μL of 100 μM stock) with 18 μL of 1x NEB buffer 2.1. Anneal by incubating on a thermocycler with the following protocol:
   (a) 95 °C for 5 min
   (b) Ramp down to 50 °C at 0.1 °C/s.
   (c) Incubate at 50 °C for 10 min.
   (d) Chill to 4 °C at normal ramp speed.

5. Ligate the annealed oligos with the 394-U6 vector. Prepare a PCR reaction:

10× CutSmart buffer	1 μL
T4 DNA ligase buffer	1 μL
394-U6a plasmid (about 100 ng)	0.3 μL
Annealed Oligos	1 μL
T4 DNA ligase	0.3 μL
BsmB1 (Eps1)	0.3 μL
DNAse-free water	6.1 μ

6. Incubate on thermocycler as follows:
   (a) 3 cycles of:
      1. 37 °C for 20 min
      2. 16 °C for 15 min
   (b) Then:
      1. 37 °C for 10 min
      2. 55 °C for 15 min
      3. 80 °C for 15 min
      4. 4 °C hold.
7. Transform into competent cells and spread onto ampicillin plates.
8. Plasmid with three gRNAs cassette: We generated a three-gRNA cassette within tol2 arms by infusion cloning the zU6a, zU6b, and zU6c promoters into gateway compatible entry vectors, using zU6 promoter sequences described earlier (Addgene plasmid numbers are 64245, 64247, 64248) (Fig. 2). Individual gRNA plasmids can be cloned into each gateway compatible entry vector using blunt-end cloning (described below) followed by LR cloning into the pDest-Tol2pA2 destination vector.
   (a) Blunt-end ligation cloning for individual gRNAs:
      (i) Linearize the individual sgRNA promoter empty plasmid vectors (p5E-zU6a-empty, pME-zU6b-empty and p5E-zU6c-empty) using 15 bp forward and 15 bp reverse primers, annealing to the vector and design individual 20 bp sgRNAs. Split each gRNA sequence down the middle such that 10 bp is included in the forward primer and 10 bp is included in the reverse primer, such that each forward and reverse

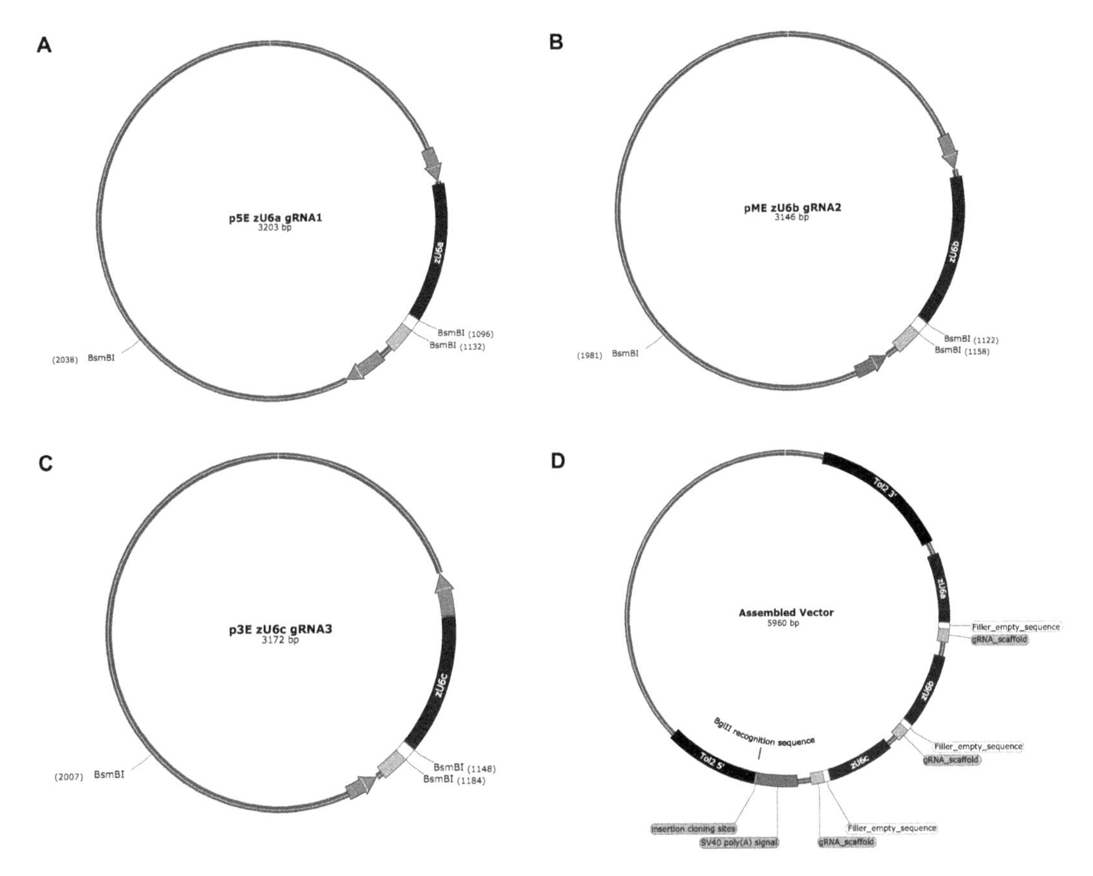

Fig. 2 394-zU6-3X-sgRNA construct. Triple gRNA construct with Tol2 sites. (**a–c**)The individual p5E-zU6a, pME-zU6b, and p5E-zU6c vectors. (**d**) Assembled triple gRNA plasmid

primer length is 25 bp. Set up PCR with Q5 HiFi DNA polymerase in a 50 μL reaction using the individual p5E-zU6a-empty, pME-zU6b-empty, and p5E-zU6c-empty vectors to linearize the plasmids and insert specific guide RNA sequences..

(ii) After PCR is done, add 1 μL of DpnI directly to the Q5 PCR mix and incubate at 37 °C for 1 h.

(iii) Do a PCR purification using Nucleospin cleanup kit or equivalent, eluting in 33 μL of DNAse-free water.

(iv) Prepare the components of the T4 PNK reaction.

T4 DNA ligase buffer	2 μL
T4 PNK	0.5 μL
DNA	500 ng
DNAse-free water	Fill to 20 μL total volume

(v) Incubate at 37 °C for 30 min and then 10 min at 65 °C.

(vi) Add 2 μL of HiFi T4 DNA ligase and incubate at room temperature for ~1 h.

(vii) Take 3 μL of the ligation mix and transform into TOP10 or similar competent cells.

(b) Upon confirmation of successful sgRNA ligated entry clones, perform LR cloning with the three entry vectors into the pDestTol2pA2 destination vector.

See **Note 1**.

3.2 Electroporation

3.2.1 Preparation of Injection Mix

1. Dilute and combine plasmids of interest with water. The concentration of the desired plasmids should be equimolar in concentration to the fluorescent plasmid of interest (e.g., 250 ng/uL MiniCoopr-GFP). For more efficient tumor generation, add 10% of the tol2 plasmid (not mRNA) into the injection mix.
2. To generate BRAFV600E *p53−/− pten−/−* melanomas in the **casper triple strain,** use the following plasmids:
   (a) MiniCoopR-GFP [16] (could substitute for MiniCoopR driving any fluorophore).
   (b) *mitfa*-Cas9 (equimolar to MiniCoopR plasmid),
   (c) 394-sgptena (equimolar to cas9 plasmid)
   (d) 394-sgptenb (equimolar to cas9 plasmid)
   (e) Tol2 (10% of total injection concentration).
3. To generate BRAFV600E *p53−/− pten −/−* melanomas in **casper or AB strains,** use the following plasmids:
   (a) MiniCoopR-GFP (could substitute for MC driving any fluorophore).
   (b) mitf-BRAFV600E,
   (c) 394-sgp53
   (d) mitf-Cas9,
   (e) 394-sgptena (equimolar to cas9 plasmid)
   (f) 394-sgptenb (equimolar to cas9 plasmid)
   (g) Tol2 (10% of total injection concentration).
4. Add food coloring to the injection mix at 1:50 to visualize the injection site for subsequent electroporation.

3.2.2 Prepare Injection Plate

1. Make a 2% 10 cm agarose dish by combining approximately 1.2 gm agarose with 60 mL E3 water. Pour into a 100 mm x 20 mm plastic petri dish and let it solidify.
2. Cut a rectangle with two slots, to accommodate placement of electrodes, as seen in Fig. 3 into the 2% agarose plate.
3. Store agarose plates at 4 °C for reuse between experiments.

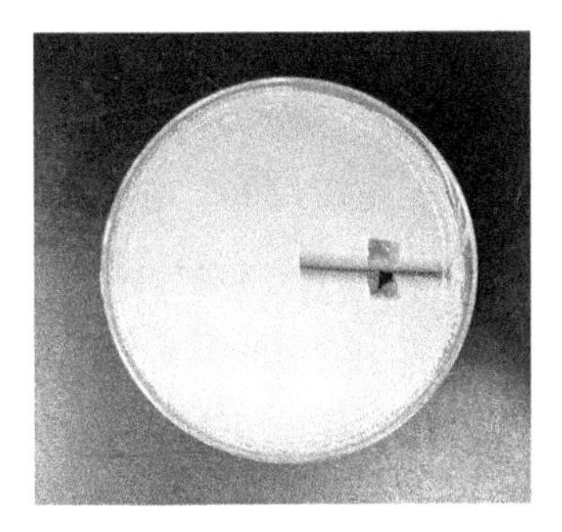

Fig. 3 Example TEAZ electroporation plate

3.2.3 Set Up Workspace

1. Turn on instruments (electroporator, nitrogen gas, microinjector, microscope).
2. Check settings on the electroporator:
   (a) Mode: LV.
   (b) Voltage: 40 V.
   (c) # of pulses: 5
   (d) Length of pulse: 60 ms.
   (e) Pulse interval: 1 s.
3. Dilute tricaine to 0.16 mg/mL in a petri dish with system water.
4. Fill the recovery tank with fresh system water.
5. Place thick paper towels to set up the workspace area.

3.2.4 Prepare Injection Needle

1. Using a needle puller, pull several quartz needles.
   (a) Settings:
      (i) Heat: 900.
      (ii) Fil: 4.
      (iii) Vel: 55.
      (iv) Del: 130.
      (v) Pul: 55.

      See **Note 2**.
2. Use a razor blade or a fine tip jewelers tweezer to cut an opening into the quartz glass pulled needle. The edge should be sharp and beveled.
3. Attach needle to microinjector holder.
4. Measure out 1.5 μL of injection mix with a p10 micropipette and eject droplet onto a petri dish lid or parafilm.
5. Fill the needle from the droplet, pulling up as much as possible.

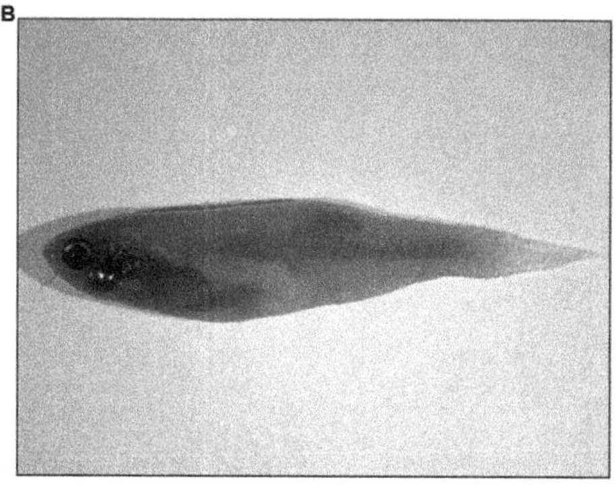

Fig. 4 Injection procedure. Proper injection technique (**a**). Image of fish with food coloring indicating injection site (**b**)

6. Mark the needle with a marker to indicate the top of the injection mix to ensure injection of the same amounts of DNA across all animals.
7. Set the pressure on the microinjector between 5 and 15 PSI. It should take a total of five to ten pedal steps to eject total volume.

3.2.5 Microinjection of DNA into Fish

1. Anesthetize fish in 0.16 mg/mL tricaine.
2. Dry fish slightly on a paper towel.
3. Place the fish flat on its side on the injection plate with the head facing the opposite hand that is holding the injector (*see* Fig. 4a).
4. Hold the fish steady with two fingers of the nondominant hand.
5. Locate the desired spot for injection under the microscope at 2× or 2.5× (right below the dorsal fin Fig. 4b).
6. Holding the needle parallel to the fish, displace scales and pierce fish just below the skin.

   See **Note 3**.
7. Inject the volume up until the needle starts to taper so as not to inject air into the fish and for total injection to be approximately 1 μL.

   See **Note 4**.

3.2.6 Electroporation of Fish

1. Place fish with the injected spot towards the top into the designated chamber on the plate filled with tricaine (*see* Fig. 5a).
2. Place electrodes on either side of the fish covering the injection site (should be visible by food dye).The positive side of the electrode should be on the side of the injection (*see* Fig. 5b).

   See **Note 5**.

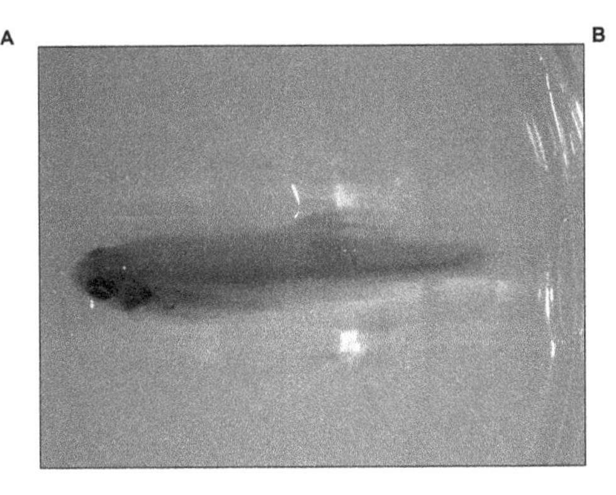

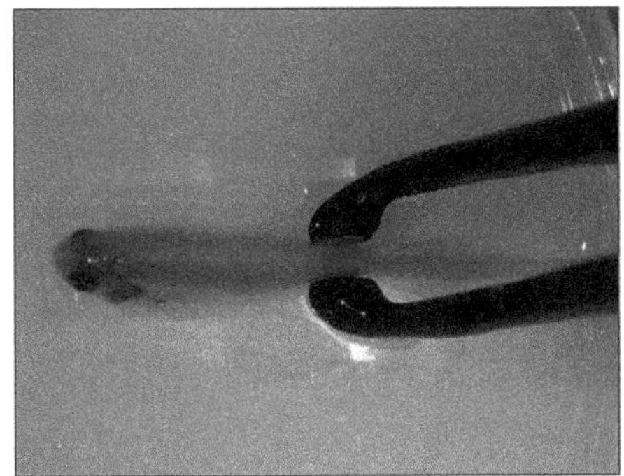

Fig. 5 Fish placed in TEAZ electroporation plate. Correct orientation indicated (**a**). Proper tweezertrode placement (**b**)

3. Use one hand to hold the electrodes by the wires or the tweezer (DO NOT TOUCH ELECTRODES DIRECTLY) and the other to press pulse on the electroporator.
4. Place the fish into a recovery tank and confirm the fish recovers breathing shortly after.

   See **Note 6**.
5. Repeat above steps, filling the injection needle to the previously labeled position with injection mix, for subsequent fish.

3.3 CRISPR-Sequencing

3.3.1 Primer Design

1. Use a primer design tool such as NCBI Primer Design to design primers to amplify an amplicon of 200–280 bp in length.
2. The mutation site that is targeted by the sgRNA should be within 100 bp of the beginning or end of the amplicon.

3.3.2 Sample Preparation

1. Euthanize fish of interest in tricaine and carefully dissect the tumor under the microscope to isolate as much of the tumor as possible.
2. Tumors can be isolated 7–10 weeks post TEAZ for successful CRISPR Seq without purification of tumor cells via flow cytometry.

Note: It is best to have three independent tumors from each experimental group. Optional: Tail tissue of fish can be used as a negative control and CRISPR Seq for *ptena* and *ptenb* could serve as positive controls.

3. To CRISPR Seq cells at earlier time points or in a nontumor context, dissect regions of interest and perform flow cytometry to isolate tumor cells (using fluorophores of interest).
   (a) Dissection: mark region of interest under fluorescent microscope using a scalpel or razor. Move fish to dissection microscope and remove the tissue. Dice the tissue

into small pieces that can fit through a p1000 wide-bore pipette. Deposit in 15 mL conical screw-cap tube.

(b) Dissociation: Vary based on cell type. For skin cells, treat samples with 187.5uL of 2.5 mg/mL liberase (e.g., Millipore Sigma # 05401020001) in PBS and incubate at room temperature for 30 min with agitation to keep cells suspended in the solution. At 15 min and 30 min of liberase treatment, pipette the solution using a p1000 wide-bore pipette up and down for 90 s to obtain single cells. After final pipetting, add 250 μL of FBS to stop the dissociation. Remove the liberase solution by pelleting cells at 500 g for 5 min. Replace the supernatant with suitable medium, PBS or DMEM, and resuspend cells. Filter cells through a 40 μm tip filter to obtain a single cell suspension for FACS.

4. Isolate genomic DNA from tumor samples using Qiagen DNeasy Blood and Tissue kit or equivalent as per kit manual.

3.3.3 PCR to Generate CRISPR Seq Amplicons

1. Use a Taq polymerase (e.g., Promega GoTaq Green Master Mix) to amplify the amplicon of interest.
2. For a 50 μL reaction volume using the GoTaq Green Master Mix, set up PCR reaction as follows:

GoTaq green master mix.	25 μL
Forward primer 10 μM.	5 μL
Reverse primer 10 μM.	5 μL
Genomic DNA (50–100 ng).	1–5 μL
DNAse-free water.	Up to 50 μL.

3. Perform a standard PCR and run 5 μL of the PCR product on a 2% agarose DNA gel to visualize amplicons.
4. Purify PCR product using a PCR purification kit and submit samples for CRISPR Sequencing. These samples can also be used as starting material for Surveyor nuclease assays.

   Optional: As positive control, the following primers can be used for CRISPR Seq at the *pten* locus:

   ptena_F: GAAGTGTTTTGAACTGCTGT,

   ptena_R: GGAAGTCGTATTGTTACAGCT,

   ptenb_F: CCTTCTGAGGAATAAGCTGGAG,

   ptenb_R: GCAAGCTCATACCAGGTGTAAA,

   See comment 7 in the NOTES section.

5. Analyze CRISPR Seq results using CRISPRESSO analysis or CRISPR Variants R package to assess gene editing efficiency.

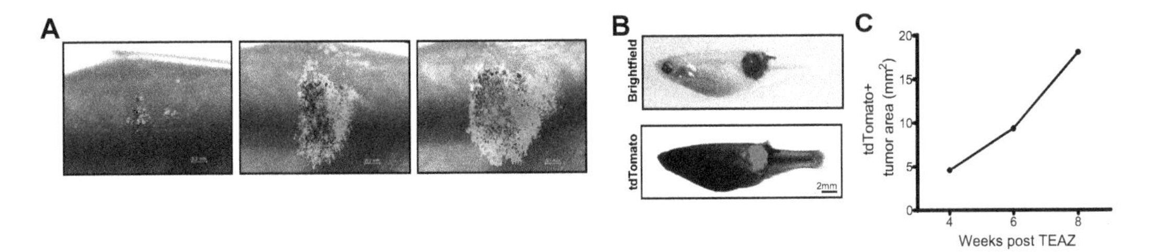

Fig. 6 Example of tumor development over time. The development of melanoma generated through TEAZ from weeks 2 to 6 (**a**). Representative images (top: brightfield, bottom: fluorescent image) of BRAFV600E *pten−/−* melanomas at week 12 (**b**). Primary tumor area progression over time (**c**)

3.4 Microscopy

1. Validate TEAZ by fluorescence imaging (e.g., Zeiss Axio V16). Promoters such as *ubb* and *mitf* may be seen as early as 24 h post-electroporation. For melanoma modeling, confirm transformation of precursor cells by fluorescence of individual cells or groups of cells ~7 days post-electroporation.
   (a) Image at 3.5× for view of the whole animal.
   (b) Image at 15–25× to track tumor growth.
   (c) Image at 40–60× to assess single cells or patches.
2. For TEAZ experiments in *caspers* or *casper triples*, melanocyte-specific transformation can be confirmed by brightfield imaging of melanin starting at 1–2 weeks post-electroporation.
3. Typical tracking of melanoma growth should be performed weekly or biweekly. An example can be seen in Fig. 6.
4. Metastasis in BRAFV600E *p53−/− pten−/−* tumors is observed as early as 7–8 weeks post-electroporation and can be monitored via fluorescent imaging (3.5× and 10×) in the kidney marrow region of the animal, as in Fig. 6b.
5. Quantify tdTomato+ area by either FIJI or Matlab (*see* Fig. 6c).

4 Notes

1. Plasmids containing both the Cas9 and gRNA in a single construct driven by a cell-type specific promoter such as MAZERATI (Zon Lab) [7] can be used as well. However, we have found that the efficiency of generating tumors is much higher when using multiple smaller plasmids. It is recommended to prepare plasmid stocks every 4–6 months from fresh bacteria.
2. Needles can be pulled prior to the procedure and stored in an appropriate container for future use.

3. When piercing the skin with the needle, try not to go too deep into the muscle. It is helpful if the needle is parallel to the fish and the bevel is facing upwards.
4. When injecting, inject a little of the mix (one to two pedal steps) and ensure the plasmids (+ food coloring) are on the intended side of injection. If you see more of the mix on the other side, withdraw the needle and reinsert gently to prevent going into the muscle.
5. When electroporating the fish, the electrodes should be as close as possible to the fish without touching it so as not to burn the fish.
6. When the fish is put into fresh system water, it is helpful to use a pipette to gently wake up the fish.
7. A typical CRISPR Sequencing run will aim to sequence 75,000–100,000 reads.

References

1. Stuart GW, McMurray JV, Westerfield M (1988) Replication, integration and stable germ-line transmission of foreign sequences injected into early zebrafish embryos. Development 103:403–412
2. Lawson ND, Weinstein BM (2002) In vivo imaging of embryonic vascular development using transgenic zebrafish. Dev Biol 248:307–318. https://doi.org/10.1006/dbio.2002.0711
3. Long Q, Meng A, Wang H et al (1997) GATA-1 expression pattern can be recapitulated in living transgenic zebrafish using GFP reporter gene. Development 124:4105–4111
4. Lister JA, Robertson CP, Lepage T et al (1999) Nacre encodes a zebrafish microphthalmia-related protein that regulates neural-crest-derived pigment cell fate. Development 126: 3757–3767
5. Langenau DM, Traver D, Ferrando AA et al (2003) Myc-induced T cell leukemia in transgenic zebrafish. Science 299:887–890. https://doi.org/10.1126/science.1080280
6. Patton EE, Widlund HR, Kutok JL et al (2005) BRAF mutations are sufficient to promote nevi formation and cooperate with p53 in the genesis of melanoma. Curr Biol 15:249–254. https://doi.org/10.1016/j.cub.2005.01.031
7. Ablain J, Xu M, Rothschild H et al (2018) Human tumor genomics and zebrafish modeling identify SPRED1 loss as a driver of mucosal melanoma. Science 362:1055–1060. https://doi.org/10.1126/science.aau6509
8. Langenau DM, Keefe MD, Storer NY et al (2008) Co-injection strategies to modify radiation sensitivity and tumor initiation in transgenic zebrafish. Oncogene 27:4242–4248. https://doi.org/10.1038/onc.2008.56
9. Nguyen AT, Koh V, Spitsbergen JM, Gong Z (2016) Development of a conditional liver tumor model by mifepristone-inducible Cre recombination to control oncogenic kras V12 expression in transgenic zebrafish. Sci Rep 6: 19559. https://doi.org/10.1038/srep19559
10. Langenau DM, Feng H, Berghmans S et al (2005) Cre/lox-regulated transgenic zebrafish model with conditional myc-induced T cell acute lymphoblastic leukemia. Proc Natl Acad Sci U S A 102:6068–6073. https://doi.org/10.1073/pnas.0408708102
11. Callahan SJ, Tepan S, Zhang YM et al (2018) Cancer modeling by transgene electroporation in adult zebrafish (TEAZ). Dis Model Mech 11. https://doi.org/10.1242/dmm.034561
12. White RM, Sessa A, Burke C et al (2008) Transparent adult zebrafish as a tool for in vivo transplantation analysis. Cell Stem Cell 2:183–189. https://doi.org/10.1016/j.stem.2007.11.002
13. Heilmann S, Ratnakumar K, Langdon E et al (2015) A quantitative system for studying metastasis using transparent zebrafish. Cancer Res 75:4272–4282. https://doi.org/10.1158/0008-5472.CAN-14-3319
14. Kwan KM, Fujimoto E, Grabher C et al (2007) The Tol2kit: a multisite gateway-based construction kit for Tol2 transposon transgenesis

constructs. Dev Dyn 236:3088–3099. https://doi.org/10.1002/dvdy.21343

15. Yin L, Maddison LA, Li M et al (2015) Multiplex conditional mutagenesis using transgenic expression of Cas9 and sgRNAs. Genetics 200: 431–441. https://doi.org/10.1534/genetics.115.176917
16. Ceol CJ, Houvras Y, Jane-Valbuena J et al (2011) The histone methyltransferase SETDB1 is recurrently amplified in melanoma and accelerates its onset. Nature 471:513–517. https://doi.org/10.1038/nature09806

Chapter 6

Lineage Tracing of Bone Cells in the Regenerating Fin and During Repair of Bone Lesions

Wen Hui Tan and Christoph Winkler

Abstract

Small teleost fishes such as zebrafish and medaka show remarkable regeneration capabilities upon tissue injury or amputation. To elucidate cellular mechanisms of teleost tissue repair and regeneration processes, the Cre/LoxP recombination system for cell lineage tracing is a widely used technique. In this chapter, we describe protocols used for inducible Cre/LoxP recombination-mediated lineage tracing of osteoblast progenitors during medaka fin regeneration as well as during the repair of osteoporosis-like bone lesions in the medaka vertebral column. Our approach can be adapted for lineage tracing of other cell populations in the regenerating teleost fin or in other tissues undergoing repair.

Key words Lineage analysis, Fin regeneration, Osteoporosis, Bone repair, Osteoblast progenitors, Medaka

1 Introduction

Bone is a dynamic organ that possesses an intrinsic ability to heal upon injury. Bone-resorbing osteoclasts facilitate the removal of aged and damaged bone while bone-forming osteoblasts secrete bone matrix, mediating the formation of new bone [1, 2]. Across vertebrates, bone repair and regeneration capabilities vary [3, 4]. Mammalian bone repair is limited to fractures below a critical size [5], and healing efficiency decreases with age [6]. Moreover, upon amputation of complex tissues such as limbs, mammals are incapable of regeneration. Instead, wound healing occurs, leading to the formation of fibrous scars [7]. On the other hand, small teleost species such as zebrafish (*Danio rerio*) and medaka (*Oryzias latipes*) possess remarkable bone healing abilities [8–12]. After amputation of the adult fin, which is mainly composed of bony fin rays, almost perfect regeneration occurs within two weeks [9, 10]. In addition, efficient bone repair is triggered after mechanically induced fractures in zebrafish [13] and after ectopic induction

James F. Amatruda et al. (eds.), *Zebrafish: Methods and Protocols*, Methods in Molecular Biology, vol. 2707,
https://doi.org/10.1007/978-1-0716-3401-1_6,

of osteoporosis-like lesions in the medaka vertebral column [14]. For the latter, upon transgenic overexpression of Receptor Activator of NF-κB Ligand (Rankl) and degradation of bone matrix in the medaka vertebral column, osteoblast progenitors get recruited to bone lesion sites and differentiate into mature osteoblasts, facilitating de novo mineralization [14, 15].

A technique commonly used to uncover cellular mechanisms governing tissue repair and regeneration is the Cre/LoxP recombination system for genetic cell lineage tracing [13, 16–19]. Using a Cre recombinase fused to a mutant estrogen receptor ligand-binding domain (ERT2), recombination between LoxP sites can be induced upon tamoxifen treatment [20]. By permanently labeling a specific cell population and their progeny with a fluorescence marker via Cre/LoxP recombination, contribution of various cell lineages to the restoring tissue can be tracked in vivo [20]. Using this technique, it has been shown that teleost fin regeneration is highly lineage-restricted, with no evidence of transdifferentiation between cell lineages [17, 18, 21]. In addition, several cellular sources contributing to the regenerating bony fin rays have been identified with this method. These include *osterix* (*osx*)-expressing osteoblasts [16, 21], *matrix metalloproteinase 9* (*mmp9*)-expressing osteoblast progenitor cells [19], as well as *collagen10a1* (*col10a1*)-expressing osteoblast progenitors and joint cells [17]. More recently, also using the Cre/LoxP recombination system for cell lineage tracing, we showed that resident *col10a1*-expressing osteoblast progenitors in the medaka vertebral column represent a cell pool that can be mobilized for bone repair after the formation of Rankl-induced osteoporotic lesions [24]. Identification of cellular sources contributing to teleost bone repair and regeneration will provide valuable insights into the high bone healing abilities in these species and aid the development of methods to stimulate bone repair and regeneration in mammals.

Here, we outline procedures for amputation of the adult medaka fin and induction of osteoporosis-like bone lesions in the medaka vertebral column, to trigger bone regeneration and repair, respectively. In addition, we describe protocols used for in vivo Cre/LoxP recombination-mediated lineage tracing of *col10a1*-expressing cells as bone regeneration and repair take place. These methods can be modified and applied to the lineage analyses of other cell populations in the study of various repair and regeneration processes in medaka and zebrafish.

2 Materials

2.1 Medaka Lines

1. *col10a1*:CreERT2-p2a-mCherry transgenic Cre driver line [17] (available from authors upon request).

2. *col10a1*$^{p2a\text{-}CreERT2;cmlc2:EGFP}$ knock-in Cre driver line ([24]; available from authors upon request; *see* **Note 1**).
3. GaudiBBW2.1 transgenic LoxP reporter line [22] (available at NBRP Medaka, strain ID TG1065, http://shigen.nig.ac.jp/medaka/).
4. *rankl*:HSE:CFP transgenic line [15] (available at NBRP Medaka, strain ID TG1135. http://shigen.nig.ac.jp/medaka/).

2.2 Reagents

1. Embryo medium: 17.4 mM NaCl, 0.12 mM KCl, 0.12 mM $MgSO_4$, 0.19 mM $Ca(NO_3)_2$, 1.5 mM HEPES. Adjust pH to 7 using NaOH.
2. Tricaine (MS-222): Prepare 0.4% stock solution by dissolving 0.2 g of tricaine powder in 50 mL of deionized water. Adjust pH to 7 using Tris buffer (1 M, pH 9). Store at 4 °C up to 6 months. To prepare working solution for anesthetizing larval and adult medaka, dilute tricaine stock solution to 0.016% in embryo medium and 0.03% in fish system water, respectively.
3. 4-hydroxytamoxifen (4-HT): Prepare 1 mM stock solution by dissolving 5 mg of 4-HT powder in 12.5 mL of 100% ethanol. Aliquot the stock solution in 1.5 mL Eppendorf tubes and store the aliquots at −20 °C in the dark. To prepare 4-HT working solution for treatment of larval and adult medaka, dilute stock solution to 10 μM in embryo medium and 0.5 μM in fish system water, respectively.
4. Low melting agarose: Prepare a 1.2% solution by dissolving 0.24 g of low melting agarose powder in 20 mL of embryo medium in a glass flask. Melt the powder by microwaving. Aliquot 1 mL of solution in 1.5 mL Eppendorf tubes and store the aliquots at room temperature.
5. Calcein (C0875, Sigma): Prepare a 1% stock solution by dissolving 0.2 g of calcein powder in 20 mL of embryo medium. Store at room temperature in the dark. For use, dilute the stock solution to 0.01% in embryo medium.

2.3 Equipment and Consumables

1. Plastic fish tanks (500 mL).
2. 30 °C incubator.
3. Water bath set at 39 °C.
4. Thermomixer set at 65 °C.
5. Polystyrene petri dish (60 × 15 mm).
6. Glass bottom imaging dish (170 μm, 35 × 10 mm).
7. P100 and P1000 pipette and tips.
8. P10 pipette tips.
9. 6-well plates.

10. 1.5 mL Eppendorf tubes.
11. Disposable polyethylene transfer pipettes (3 mL).
12. Stainless steel fine forceps.
13. Sterile scalpel blade.
14. Plastic spoon.
15. Glass coverslips (22 mm × 50 mm).
16. Aluminum foil or opaque box.
17. Fluorescence stereo microscope with CFP, GFP, and mCherry filter cubes.
18. Fluorescence confocal microscope with 488 nm, 514 nm, and 561 nm lasers.

3 Methods

3.1 Lineage Tracing of col10a1 Cells During Bone Repair in the Medaka Vertebral Column

3.1.1 Preparation of Transgenic Embryos for Lineage Tracing

For this experiment, use medaka heterozygous for the GaudiBBW2.1 transgenic LoxP reporter [22], the *col10a1*$^{\text{p2a-CreERT2;}\textit{cmlc2}\text{:EGFP}}$ knock-in Cre driver [24] and the *rankl*:HSE:CFP transgene [15]. In GaudiBBW2.1 (kindly provided by Lazaro Centanin, University of Heidelberg) [22], a *ubiquitin* promoter drives the expression of a floxed ECFP, which can be switched to dTomato, YFP, or nlsEGFP upon Cre/LoxP recombination (*see* Fig. 1a). In *col10a1*$^{\text{p2a-CreERT2;}\textit{cmlc2}\text{:EGFP}}$, a 4-HT-inducible Cre recombinase is knocked in and expressed under control of the endogenous *col10a1* promoter (*see* Fig. 1b). In *rankl*:HSE:CFP transgenic medaka, heat-shock treatment drives Rankl expression, which induces ectopic formation of osteoclasts and osteoporosis-like bone lesions in the medaka vertebral column [15].

1. Cross the *col10a1*$^{\text{p2a-CreERT2;}\textit{cmlc2}\text{:EGFP}}$ Cre driver medaka line with the GaudiBBW2.1 LoxP reporter medaka line.
2. Collect embryos. Remove debris and transfer clean embryos to a 60 × 15 mm petri dish filled with embryo medium.
3. Raise embryos in a 30 °C incubator. Check and discard dead or abnormally developed embryos every other day.
4. At 5 dpf, screen embryos for the GaudiBBW2.1 LoxP reporter transgene via ubiquitous ECFP fluorescence using a fluorescence stereo microscope (*see* Fig. 1c, c').
5. At 10 dpf, screen larvae for *col10a1*$^{\text{p2a-CreERT2;}\textit{cmlc2}\text{:EGFP}}$ via EGFP fluorescence in the heart using a fluorescence stereo microscope (*see* Fig. 1d, d').
6. Raise GaudiBBW2.1; *col10a1*$^{\text{p2a-CreERT2;}\textit{cmlc2}\text{:EGFP}}$ double heterozygous medaka in a recirculating system at 27 °C with a controlled 14 h light/ 10 h dark cycle.

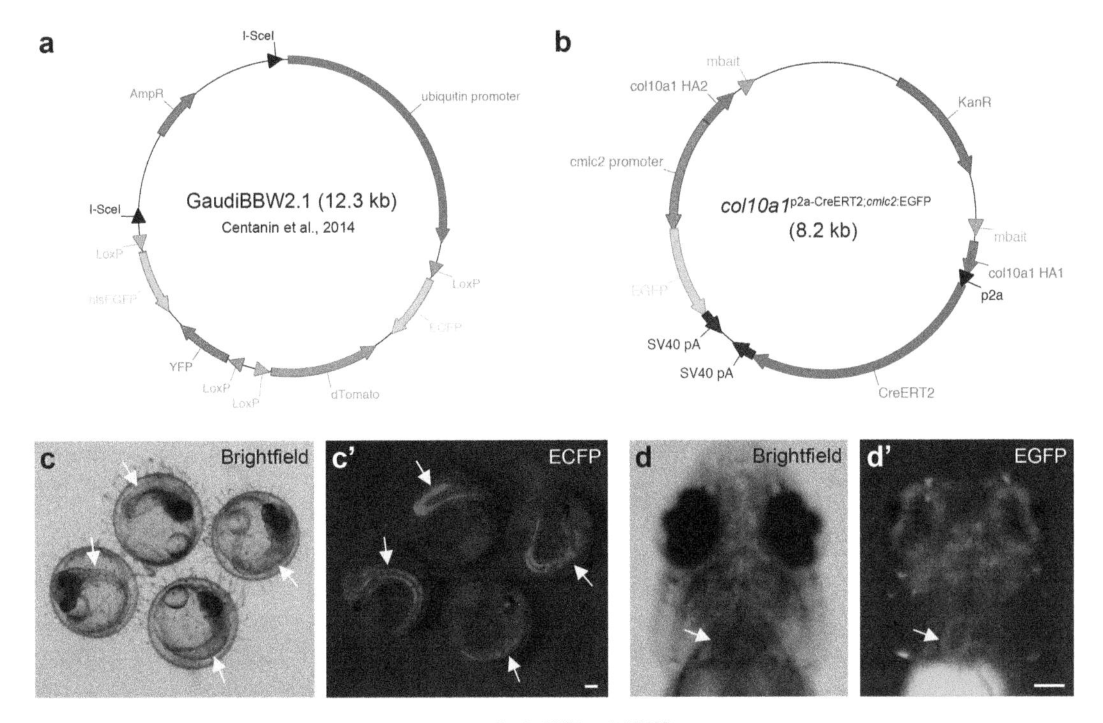

Fig. 1 Preparation of GaudiBBW2.1 and *col10a1*$^{\text{p2a-CreERT2;}\textit{cmlc2}\text{:EGFP}}$ medaka. (**a**) Simplified map of the plasmid used for generation of GaudiBBW2.1 transgenic medaka (adapted from [22]). ECFP is ubiquitously expressed by default and can be switched to dTomato, YFP, or nlsEGFP upon Cre/LoxP recombination. (**b**) Simplified map of the donor plasmid used for generation of *col10a1*$^{\text{p2a-CreERT2;}\textit{cmlc2}\text{:EGFP}}$ knock-in medaka. Mbait guide RNA sequences flank the knock-in cassette, which consists of *col10a1* homology arms HA1 (292 bp) and HA2 (215 bp), a p2a self-cleaving peptide linking endogenous *col10a1* to CreERT2 and a *cmlc2*:EGFP secondary fluorescence reporter. pA, polyA. AmpR, ampicillin resistance gene. KanR, kanamycin resistance gene. (**c, c'**) GaudiBBW2.1-expressing embryos were screened via ubiquitous ECFP expression (yellow arrows) at 5 dpf using the CFP filter cube of a fluorescence stereo microscope. (**d, d'**) *col10a1*$^{\text{p2a-CreERT2;}\textit{cmlc2}\text{:EGFP}}$-expressing larvae were screened via *cmlc2*:EGFP expression in the heart (yellow arrows) at 10 dpf using the GFP filter cube of a fluorescence stereo microscope. Scale bars: 100 μm

7. At 3 months old, cross GaudiBBW2.1; *col10a1*$^{\text{p2a-CreERT2;}\textit{cmlc2}\text{:EGFP}}$ double heterozygous medaka with homozygous *rankl*:HSE:CFP medaka to obtain embryos heterozygous for GaudiBBW2.1, *col10a1*$^{\text{p2a-CreERT2;}\textit{cmlc2}\text{:EGFP}}$ and *rankl*:HSE:CFP. Screen embryos for GaudiBBW2.1 and *col10a1*$^{\text{p2a-CreERT2;}\textit{cmlc2}\text{:EGFP}}$ as described above.

3.1.2 Induction of Cre/LoxP Recombination by 4-Hydroxytamoxifen: Labeling of col10a1-Positive Osteoblast Progenitors

1. Raise embryos collected from the crossing of GaudiBBW2.1; *col10a1*$^{\text{p2a-CreERT2;}\textit{cmlc2}\text{:EGFP}}$ and *rankl*:HSE:CFP in embryo medium in a 30 °C incubator (*see* **Note 2**).
2. To induce Cre/LoxP recombination and labeling of *col10a1* cells, incubate 7 dpf embryos in 10 μM of 4-HT in embryo medium in a 60 × 15 mm petri dish (*see* **Note 3**). Conduct the drug treatment for 6 h in the dark in the 30 °C incubator (*see* **Notes 4** and **5**).

3. Wash 4-HT-treated embryos at least three times with embryo medium. Use a new 60 × 15 mm petri dish and remove as much solution as possible for each wash (*see* **Notes 6** and 7).
4. Place the 4-HT-treated embryos back in the 30 °C incubator.

3.1.3 Induction of Osteoporosis-Like Bone Lesions in the Medaka Larval Vertebral Column

1. At 9 dpf, heat-shock 4-HT-treated embryos for Rankl induction by placing the 60 × 15 mm petri dish in a 39 °C water bath for 2 h (*see* **Note 8**).
2. Place heat-shock-treated embryos back in the 30 °C incubator. Embryos will start to hatch at this stage. Feed hatchlings twice daily with larval feed.

3.1.4 In Vivo Tracking of Labeled Cells During Repair of Bone Lesions

To track the contribution of labeled *col10a1* positive osteoblast progenitors to the repair of Rankl-induced bone lesions in the vertebral column, live fluorescence imaging should ideally be performed at two time points. At 11 dpf when bone lesions form, perform imaging to determine the location and number of labeled cells in the vertebral column. At 13 dpf when bone repair is initiated, perform imaging for analysis of migration and proliferation of labeled cells.

1. Melt 1 mL 1.2% low melting agarose aliquots using a thermomixer set at 65 °C.
2. Anesthetize 11 dpf larvae using 0.016% tricaine in embryo medium for 5 min.
3. Cool the 1.2% low melting agarose to approximately 40 °C. Add 40 μL of 0.4% tricaine to 1 mL of low melting agarose to obtain a final concentration of 0.016% tricaine.
4. Use a transfer pipette to transfer anesthetized larvae to the center of a 35 × 10 mm glass bottom imaging dish. Remove as much solution as possible using the transfer pipette.
5. Add approximately 700 μL of 1.2% low melting agarose with 0.016% tricaine to the larvae on the glass bottom imaging dish.
6. Use a P10 pipette tip to push all larvae to the bottom of the imaging dish and orientate the larvae in a lateral view (*see* Fig. 2a, b). We do not recommend mounting more than 8 larvae per imaging dish as the low melting agarose will begin solidifying in approximately 40 s. Be careful to avoid damaging any larval tissues.
7. Wait 5 min to let the agarose solidify completely. Add 2.5 mL of embryo medium on top of the solidified agar.
8. Image the larval vertebral column using a fluorescence confocal microscope. Cre/LoxP recombination-labeled cells expressing nlsEGFP, YFP or dTomato can be identified using the 488 nm, 514 nm or 561 nm laser, respectively.

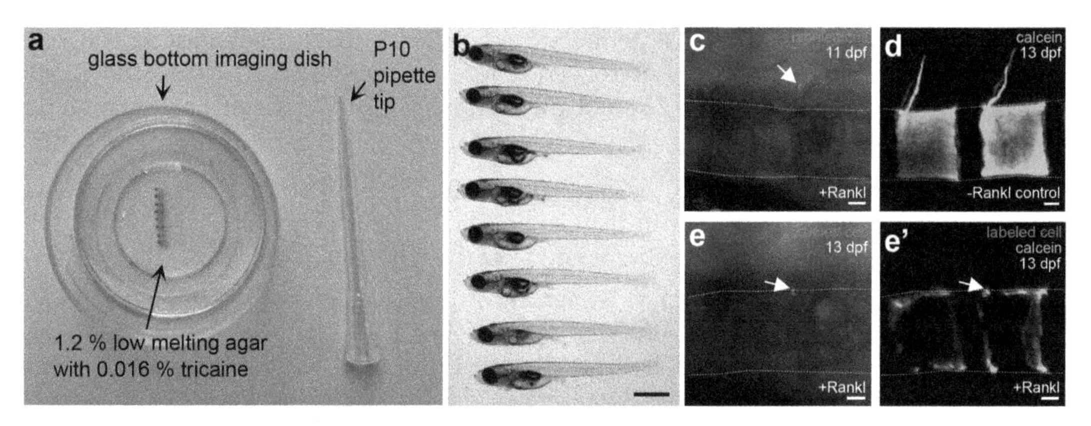

Fig. 2 Live imaging for in vivo tracking of labeled cells after induction of osteoporosis-like bone lesions in the larval vertebral column. (**a, b**) Anesthetized larvae were mounted in 1.2% low melting agarose containing 0.016% tricaine in a glass bottom imaging dish as shown in (**a**). Using a P10 pipette tip, larvae were arranged in a lateral view as shown in (**b**). (**c–e'**) Imaging of Cre/LoxP recombination-labeled cells was performed using a fluorescence confocal microscope. Yellow dotted lines demarcate the vertebral bodies. At 11 dpf, a labeled cell was observed at a neural arch above the vertebral body (white arrow in **c**). At 13 dpf, in the same fish, the labeled cell migrated from the neural arch to the vertebral body where bone matrix is degraded (white arrows in **e, e'**). Compared to -Rankl control (**d**), bone matrix in the vertebral column was degraded in the +Rankl larva (**e'**). For this experiment, GaudiBBW2.1; *col10a1*$^{\text{p2a-CreERT2;}\textit{cmlc2}\text{:EGFP}}$; *rankl*:HSE:CFP larvae were used. Calcein was used to stain the bone matrix. Scale bars: (**b**) 1000 μm, (**c–e'**) 20 μm

9. Note the location and number of labeled cells in the vertebral column of each 11 dpf larva.
10. Using forceps, free the larvae from the low melting agarose. Place each larva in individually labeled wells of a 6-well plate. Raise larvae in a 30 °C incubator and feed them twice daily with larval feed.
11. At 13 dpf, repeat mounting and imaging of larvae as described in the steps above. By comparing the position and number of labeled cells at 11 and 13 dpf in each individual larva, the migration pattern and proliferation rate of labeled cells can be determined (*see* Fig. 2c, e'). Optional: calcein staining can be performed for visualization of bone lesions in the vertebral column in vivo after Rankl induction. For this, incubate larvae in 0.01% calcein in embryo medium for an hour in the dark at 28 °C. Before imaging, wash stained larvae with embryo medium three times, for 15 min per wash at room temperature. Calcein staining can be detected using the GFP filter on the fluorescence stereo microscope or the 488 nm laser on the fluorescence confocal microscope.

3.2 Lineage Tracing of col10a1 Cells During Medaka Adult Fin Regeneration

For this experiment, use adult medaka (3–5 months old; *see* **Note 9**) heterozygous for the GaudiBBW2.1 transgenic LoxP reporter [22] and the *col10a1*:CreERT2-p2a-mCherry transgenic Cre driver [17] (*see* **Note 1**).

3.2.1 Preparation of Transgenic Embryos for Lineage Tracing

1. Cross the *col10a1*:CreERT2-p2a-mCherry Cre driver medaka line with the GaudiBBW2.1 LoxP reporter medaka line. Collect embryos and raise them in a 30 °C incubator.
2. Using a fluorescence stereo microscope, screen for GaudiBBW2.1 and *col10a1*:CreERT2-p2a-mCherry via ubiquitous ECFP fluorescence at 5 dpf and mCherry fluorescence in osteoblast progenitors at 7 dpf, respectively.
3. Raise GaudiBBW2.1; *col10a1*:CreERT2-p2a-mCherry fish in a recirculating system at 27 °C with a controlled 14 h light/ 10 h dark cycle to 3–5 months.

3.2.2 Induction of Cre/LoxP Recombination-Labeling of col10a1 Cells in Adult Medaka

1. For induction of Cre/LoxP recombination and labeling of *col10a1* cells in the adult medaka fin, treat 3–5 months old GaudiBBW2.1; *col10a1*:CreERT2-p2a-mCherry fish with 0.5 μM of 4-HT in fish system water in a plastic fish tank for 72 h in the dark (*see* **Notes 5** and **10**).
2. Rinse 4-HT-treated medaka at least three times with fish system water. Use a new fish tank and remove as much solution as possible for each rinse (*see* **Notes 6** and 7).

3.2.3 Fin Amputation and In Vivo Tracking of Labeled Cells in the Regenerating Fin

1. Anesthetize adult medaka using 0.03% tricaine in fish system water in a fish tank. Wait approximately 1 min for the fish to be anesthetized.
2. Using a plastic spoon, transfer the anesthetized fish to the center of the lid of a 60 × 15 mm petri dish. Use a P10 pipette tip to position the fish in a lateral view with its caudal fin fanned out taut.
3. With a sterile scalpel blade, amputate the caudal fin at approximately one segment before the first fin bifurcation of the seventh dorsal-most fin ray (*see* **Note 11**; Fig. 3a, b). Use sufficient force to ensure a clean amputation with one cut. Be careful not to damage other regions of the fin.
4. Transfer the fish to a 22 × 50 mm glass coverslip using a plastic spoon. With a P10 pipette tip, position the fish in a lateral view (*see* Fig. 3c). Be gentle as the glass coverslip is fragile.
5. Using a transfer pipette, add five drops of 0.03% tricaine in fish system water on the gills of the fish to prevent them from drying out.

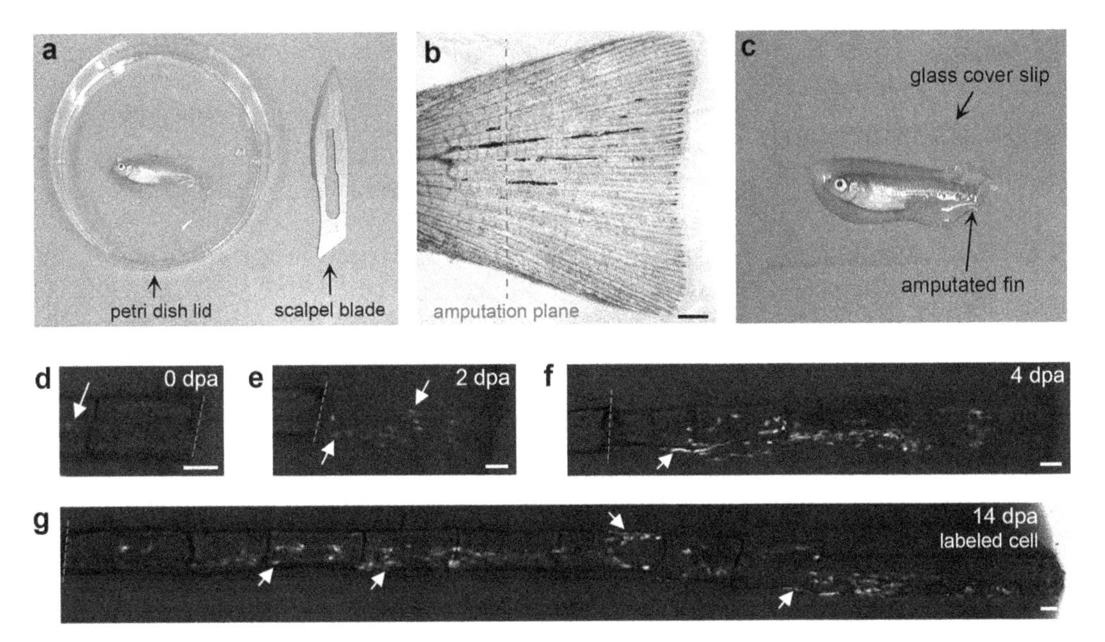

Fig. 3 Lineage tracing of labeled cells in the regenerating adult medaka fin. (**a**) Fin amputation was performed using a sterile scalpel blade and on the lid of a 60 × 15 mm petri dish. (**b**) Amputation was performed approximately one segment before the first fin bifurcation of the seventh dorsal-most fin ray. Red dotted line demarcates the amputation plane. (**c**) For live imaging of the regenerating fin under the confocal microscope, the anesthetized fish was placed on a glass coverslip. (**d**) Imaging was performed immediately after fin amputation and at every 2 days up to 14 days as the fin regenerated. Labeled cells that were observed proximal to the amputation plane at 0 dpa (white arrow in **d**) contributed to the blastema at 2 dpa (white arrows in **e**) and gave rise to regenerating bone-lining cells at 4 and 14 dpf (white arrows in **f** and **g**). dpa, days post amputation. For this experiment, GaudiBBW2.1; *col10a1*:CreERT2-p2a-mCherry adult medaka were used. Scale bars: (**b**) 500 μm, (**d**, **g'**) 50 μms

6. Image the fish at the fin amputation site immediately under a fluorescence confocal microscope. Using the 488, 514, or 561 nm laser, identify the respective nlsEGFP, YFP, or dTomato-labeled cells. Image as quickly as possible. Complete imaging in less than 15 min to avoid harm to the anesthetized fish.
7. Return the fish back to a fish tank with fish system water. Use an air bubbler to increase aeration. The fish should recover within 2 min.
8. Repeat anesthesia and imaging as described above every second day up to 14 days post fin amputation (*see* Fig. 3d, g'). Based on the position and/or morphology of labeled cells in the fin regenerate, lineage contributions of *col10a1* cells to the regenerating fin can be inferred. To further confirm the identity of labeled cells in the fin regenerate, immunostaining with established cell lineage markers can be performed.

4 Notes

1. In *col10a1*:CreERT2-p2a-mCherry or *col10a1*$^{p2a\text{-}CreERT2;cmlc2:EGFP}$, CreERT2 is expressed under control of a cloned 5.8 kb *col10a1* promoter or the endogenous *col10a1* promoter, respectively. The *col10a1*:CreERT2-p2a-mCherry transgenic line recapitulates *col10a1* expression in the adult medaka fin but not in the trunk region. The *col10a1*$^{p2a\text{-}CreERT2;cmlc2:EGFP}$ knock-in line recapitulates *col10a1* expression in all medaka tissues. For precise cell lineage analysis, it is important to ensure that the Cre driver line accurately recapitulates endogenous gene expression in tissue or organ of interest.
2. To ensure development of embryos at similar rates, we recommend keeping the ratio of the number of embryos to the volume of embryo medium consistent. We raise 25 embryos in 12 mL of embryo medium in 60 × 15 mm petri dishes.
3. To ensure equal drug distribution to all embryos for 4-HT treatments, we recommend keeping the ratio of the number of embryos treated to the volume of 4-HT consistent. We use 12 mL of 4-HT solution for 25 embryos.
4. At a 4-HT concentration of 10 μM and a treatment duration of 6 h, we obtained one to five labelled cells per medaka vertebral column. If greater proportion of cell labeling is desired, 4-HT concentration can be increased to 15 μM and treatment duration can be increased to 48 h without affecting embryonic development.
5. 4-HT is light-sensitive. Ensure all 4-HT treatments are performed in the dark. Use an opaque box or aluminum foil to cover the petri dishes or fish tanks.
6. After 4-HT treatment, it is important to conduct washes thoroughly. Incomplete 4-HT washout may result in delayed and undesired Cre/LoxP recombination. This may lead to wrong conclusions in cell lineage analyses.
7. 4-HT is hazardous. Handle with care and discard all 4-HT waste according to institutional guidelines.
8. Ensure the petri dish is not filled to the brim with embryo medium to avoid mixing of embryo medium in the petri dish with water in the water bath.
9. Fin regeneration rate is dependent on the age of the fish. Older fish regenerate at slower rates. Use fish of similar age if fin regeneration rate is a factor for comparison.
10. After treatment with 0.5 μM of 4-HT in fish system water for 72 h, cell labeling was observed in 61.5% of 4-HT-treated fish.

To increase cell labeling efficiency, 4-HT concentration and treatment duration can be increased.

11. Fin regeneration rate is influenced by the proximal-distal level of amputation [23]. For studies in which regeneration rate is a factor for comparison, ensure fin amputation at the same proximal-distal level across all fish.

Acknowledgments

We thank Lazaro Centanin (Centre for Organismal Studies, University of Heidelberg) for sharing the GaudiBBW2.1 transgenic line. We also thank the Centre for Bioimaging Sciences confocal unit and the fish facility at Department of Biological Sciences, National University of Singapore, for continued support. This work was supported by the Singapore Ministry of Education (MOE2016-T2-2-086) and the National Research Foundation Singapore (NRF2017-NRF-ISF002-2671). The authors declare no competing interests.

References

1. Hadjidakis DJ, Androulakis II (2006) Bone remodeling. Ann N Y Acad Sci 1092:385–396. https://doi.org/10.1196/annals.1365.035
2. Raggatt LJ, Partridge NC (2010) Cellular and molecular mechanisms of bone remodeling. J Biol Chem 285(33):25103–25108. https://doi.org/10.1074/jbc.R109.041087
3. Daponte V, Tylzanowski P, Forlino A (2021) Appendage regeneration in vertebrates: what makes this possible? Cell 10(2):242. https://doi.org/10.3390/cells10020242
4. Stoick-Cooper CL, Moon RT, Weidinger G (2007) Advances in signaling in vertebrate regeneration as a prelude to regenerative medicine. Genes Dev 21(11):1292–1315. https://doi.org/10.1101/gad.1540507
5. Schemitsch EH (2017) Size matters: defining critical in bone defect size! J Orthop Trauma 31(Suppl 5):S20–s22. https://doi.org/10.1097/bot.0000000000000978
6. Clark D, Nakamura M, Miclau T, Marcucio R (2017) Effects of aging on fracture healing. Curr Osteoporos Rep 15(6):601–608. https://doi.org/10.1007/s11914-017-0413-9
7. Jaźwińska A, Sallin P (2016) Regeneration versus scarring in vertebrate appendages and heart. J Pathol 238(2):233–246. https://doi.org/10.1002/path.4644
8. Takeyama K, Chatani M, Takano Y et al (2014) In-vivo imaging of the fracture healing in medaka revealed two types of osteoclasts before and after the callus formation by osteoblasts. Dev Biol 394(2):292–304. https://doi.org/10.1016/j.ydbio.2014.08.007
9. Azevedo AS, Grotek B, Jacinto A et al (2011) The regenerative capacity of the zebrafish caudal fin is not affected by repeated amputations. PLoS One 6(7):e22820. https://doi.org/10.1371/journal.pone.0022820
10. Nishidate M, Nakatani Y, Kudo A et al (2007) Identification of novel markers expressed during fin regeneration by microarray analysis in medaka fish. Dev Dyn 236(9):2685–2693. https://doi.org/10.1002/dvdy.21274
11. Kaliya-Perumal A-K, Ingham PW (2021) Musculoskeletal regeneration: A zebrafish perspective. Biochimie 196:171. https://doi.org/10.1016/j.biochi.2021.10.014
12. Sousa S, Valerio F, Jacinto A (2012) A new zebrafish bone crush injury model. Biol Open 1(9):915–921. https://doi.org/10.1242/bio.2012877
13. Geurtzen K, Knopf F, Wehner D et al (2014) Mature osteoblasts dedifferentiate in response to traumatic bone injury in the zebrafish fin and skull. Development 141(11):2225–2234. https://doi.org/10.1242/dev.105817

14. Renn J, Buttner A, To TT et al (2013) A col10a1:nlGFP transgenic line displays putative osteoblast precursors at the medaka notochordal sheath prior to mineralization. Dev Biol 381(1):134–143. https://doi.org/10.1016/j.ydbio.2013.05.030
15. To TT, Witten PE, Renn J et al (2012) Rankl-induced osteoclastogenesis leads to loss of mineralization in a medaka osteoporosis model. Development 139(1):141–150. https://doi.org/10.1242/dev.071035
16. Singh SP, Holdway JE, Poss KD (2012) Regeneration of amputated zebrafish fin rays from de novo osteoblasts. Dev Cell 22(4):879–886. https://doi.org/10.1016/j.devcel.2012.03.006
17. Dasyani M, Tan WH, Sundaram S et al (2019) Lineage tracing of col10a1 cells identifies distinct progenitor populations for osteoblasts and joint cells in the regenerating fin of medaka (Oryzias latipes). Dev Biol 455(1):85–99. https://doi.org/10.1016/j.ydbio.2019.07.012
18. Tu S, Johnson SL (2011) Fate restriction in the growing and regenerating zebrafish fin. Dev Cell 20(5):725–732. https://doi.org/10.1016/j.devcel.2011.04.013
19. Ando K, Shibata E, Hans S et al (2017) Osteoblast production by reserved progenitor cells in zebrafish bone regeneration and maintenance. Dev Cell 43(5):643–650.e643. https://doi.org/10.1016/j.devcel.2017.10.015
20. Kretzschmar K, Watt FM (2012) Lineage tracing. Cell 148(1–2):33–45. https://doi.org/10.1016/j.cell.2012.01.002
21. Knopf F, Hammond C, Chekuru A et al (2011) Bone regenerates via dedifferentiation of osteoblasts in the zebrafish fin. Dev Cell 20(5):713–724. https://doi.org/10.1016/j.devcel.2011.04.014
22. Centanin L, Ander J-J, Hoeckendorf B et al (2014) Exclusive multipotency and preferential asymmetric divisions in post-embryonic neural stem cells of the fish retina. Development 141(18):3472–3482. https://doi.org/10.1242/dev.109892
23. Uemoto T, Abe G, Tamura K (2020) Regrowth of zebrafish caudal fin regeneration is determined by the amputated length. Sci Rep 10(1):649. https://doi.org/10.1038/s41598-020-57533-6
24. Tan WH, Winkler C (2022) A non-disruptive and efficient knock-in approach allows fate tracing of resident osteoblast progenitors during repair of vertebral lesions in medaka. Development 149(12). https://doi.org/10.1242/dev.200238

Part II

Neuroscience

Chapter 7

Optimized Primary Culture of Neuronal Populations for Subcellular Omics Applications

Richard Taylor and Corinne Houart

Abstract

Primary cell culture is an invaluable method frequently used to overcome challenges associated with in vivo experiments. In zebrafish research, in vivo live imaging experiments are routine owing to the high optical transparency of embryos, and, as a result, primary cell culture has been less utilized. However, the approach still boasts powerful advantages, emphasizing the importance of sophisticated zebrafish cell culture protocols. Here, we present an enhanced protocol for the generation of primary cell cultures by dissociation of 24 hpf zebrafish embryos. We include a novel cell culture medium recipe specifically favoring neuronal growth and survival, enabling relatively long-term culture. We outline primary zebrafish neuronal culture on glass coverslips, as well as in transwell inserts which allow isolation of neurite tissue for experiments such as investigating subcellular transcriptomes.

Key words Zebrafish, Cell culture, Embryo dissociation, Transwell Insert, RNA extraction, RNAseq, Compartmentalized culture

1 Introduction

Cell culture is a laboratory method frequently employed by a wide range of biomedical researchers to overcome challenges associated with in vivo experiments, such as live imaging of cells in complex 3D environments with very low optical clarity. This problem is faced extensively in research carried out using mammalian model organisms, where plating cells in culture allows high-resolution imaging of cell behavior and the dynamics of subcellular structures. Over many years of development, culture techniques and media recipes have undergone extensive modification to optimize conditions to promote the viability and growth of many different cell types [1, 2]. Contrastingly, in zebrafish research, although primary culture of cells derived from embryos has been described by multiple labs [3–9], its use has been relatively limited. This is in large part because of the excellent degree of optical clarity zebrafish embryos

James F. Amatruda et al. (eds.), *Zebrafish: Methods and Protocols*, Methods in Molecular Biology, vol. 2707, https://doi.org/10.1007/978-1-0716-3401-1_7,

possess. At 24 hpf, embryos are almost completely transparent providing the opportunity to track in real time, in vivo cell fate and subcellular structures. As a result, efforts have not focused on a similar level of protocol optimization.

However, there remains a strong need for sophisticated zebrafish cell culture techniques. Rapid advances made in improving imaging technologies now allows the visualization of ever smaller structures inside cells, such as individual RNAs localizing to specific subcellular compartments. However, fluorescent signals associated with such tiny structures are typically very dim. Signal-to-noise ratio is also dampened by cell autofluorescence, exacerbated in thicker tissues. Imaging in 24 hpf embryos is therefore tricky. In addition to this, cell culture approaches are also important for other types of experiment, such as those requiring the separation of subcellular compartments (e.g., neuronal axons/dendrites from somas), achieved by use of compartmentalized chambers [10]. These can facilitate the treatment of specific cell compartments with chemicals/drugs, or allow for the collection of material such as RNA for compartment-specific RNAseq experiments. Data generated from zebrafish in vitro experiments will be closely aligned with existing in vivo data, and can feed cohesively back into further downstream in vivo experiments.

Here, we present an updated protocol for zebrafish embryo dissociation and primary cell culture, generated by collating aspects of existing protocols [3–9]. Despite the cells initially plated reflecting the many cell types that make up 24 hpf whole embryos, over time neuronal fates are favored [11]. This owes to our cell culture media recipe optimization, incorporating factors used in mammalian neuronal culture that promote neuronal viability and neurite outgrowth [12, 13]. This process of deriving neuronal cultures is preferable over methods such as fluorescent-activated cell sorting of transgenic lines, which causes substantial cell stress and death. Our protocol is highly scalable as it can be performed quickly using many embryos. It is also compatible with relatively long-term zebrafish cell culture, owing to measures made in tackling contamination issues and in media optimization [12, 13]. We followed cultures up to 6 days after plating (*see* Fig. 1); however, there is great potential to extend beyond this period. We outline primary neuronal culture on coverslips, as well as performing compartmentalized culture using transwell inserts (*see* Fig. 2). We go on to describe the subsequent extraction of RNAs from isolated neurite tissue, allowing neurite-specific RNAseq.

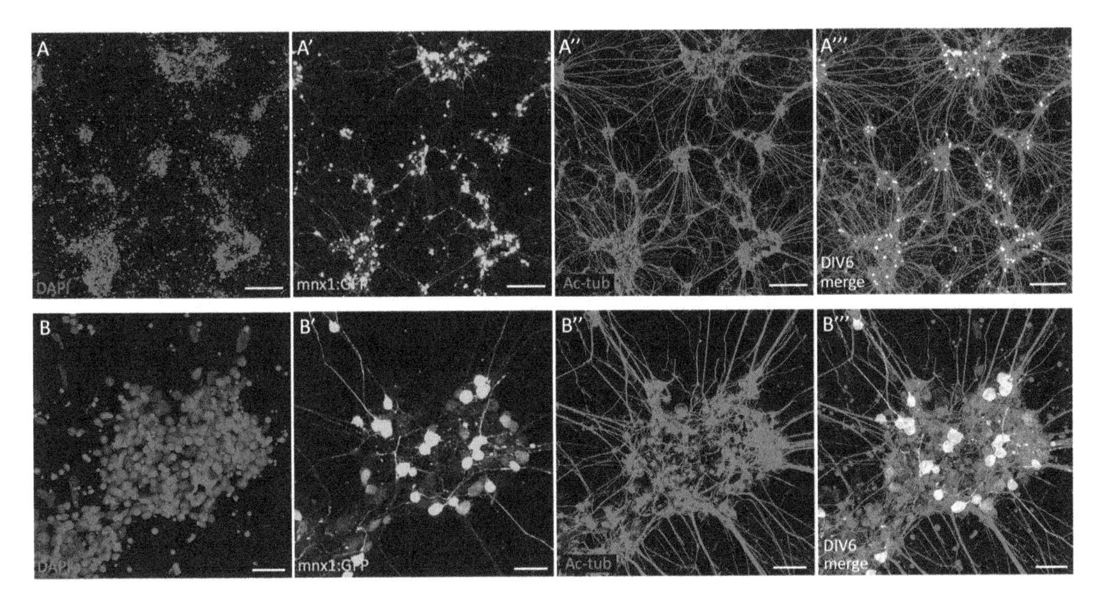

Fig. 1 Relatively long-term primary cultured neurons from dissociated zebrafish embryos. (A-A′′′) Confocal images of day 6 in vitro (DIV6) primary cell cultures from dissociated 24 hpf embryos. Neurites are immunolabelled using acetylated-tubulin antibody (red). DAPI labels nuclei (blue). Dissociated embryos were of the tg(mnx1:GFP), labelling motor neurons and some interneurons with GFP. Scale bar: 100 μm. (B-B′′′) High magnification images of cultures described in (A-A′′′). Scale bar: 20 μm

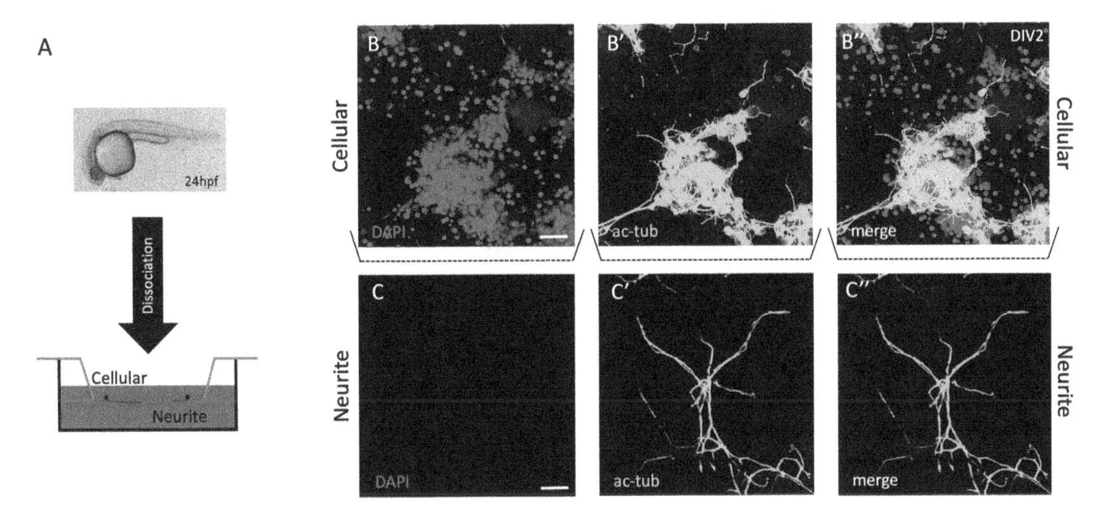

Fig. 2 Cellular and neurite compartments of transwell inserts allow for the isolation of neurite tissue from primary neuron cultures. (A) Following 24 hpf zebrafish embryo dissociation, primary cells are plated on to the upper/cellular membrane of transwell inserts. Extending neurites can pass through 1-μm pores in the membrane and adhere to the lower membrane on which they can continue to extend. (B-C′′) Cells culturing in a transwell insert day 2 in vitro (DIV2). Scale bars 20 μm. (B-B′′) Cellular/upper surface: Nuclei labelled by DAPI (A) indicate cell somas are restricted to this membrane surface. Neurites labelled by acetylated-tubulin (A′). (C-C′′) Neurite/lower surface: No nuclei indicate that somas do not pass through 1-μm pores. After passing through 1-μm pores neurites continue their extension (B′)

2 Materials

2.1 Embryo Dissociation

1. AFW: Autoclaved fish facility system water, 0.01% methylene blue, 1× Penicillin-Streptomycin, 50 μg/mL Gentamicin.
2. Bleaching solution: 0.003% Fisher Scientific sodium hypochlorite in AFW.
3. Egg strainer.
4. Petri dishes.
5. Pronase in AFW (0.6 mg/mL).
6. 3 mL plastic dropping pipettes.
7. 1.5 mL Eppendorf tubes.
8. Dissociation solution: 25% Cell Dissociation Buffer (enzyme-free), PBS, 3.25×10^{-3}M EDTA (pH 8.0).
9. 40 μm cell strainers.
10. Embryo wash solution: 70% ethanol in autoclaved ddH_2O.
11. 30 mm diameter petri dishes.
12. Bunsen burner.
13. 150 mm and 230 mm cotton-plugged glass Pasteur pipettes (*see* **Note 1**).
14. PVC teat for glass pipettes.
15. PBS.
16. 50 mL Falcon tubes.
17. Primary tissue culture hood.
18. IMS solution.
19. Centrifuge.

2.2 Primary Cell Culture

1. Zebrafish Neural Medium (ZNM): Leibovitz L-15 medium, 1× N-2 Supplement, 1× NeuroBrew-21 Supplement, 10 ng/mL BDNF, 2% FBS, 1× Penicillin-Streptomycin, 50 μg/mL Gentamicin.
2. Pipette controller and 10 mL stripettes.
3. Poly-D-Lysine/Laminin-coated glass coverslips.
4. 1000 μL and 200 μL filter tips.

2.3 Primary Culture Using Transwell Inserts

1. Transwell inserts with a 1 μm pore diameter.
2. 6-well tissue culture plates.
3. Poly-D-Lysine (15 μg/mL) in autoclaved ddH_2O.
4. Laminin (3 μg/mL) in autoclaved ddH_2O.

2.4 Tissue Isolation from Transwell Inserts and RNA Extraction

1. RNase Zap.
2. Cell scrapers.
3. Swabs.
4. 35 mm diameter culture dishes.
5. 1.5 mL RNase free Eppendorf tubes.
6. Sterile scalpel.
7. Qiagen RNeasy Micro Kit.

3 Methods

3.1 Embryo Dissociation

1. Collect fertilized embryos and put them into fresh AFW before incubating them overnight at 28.5 °C.
2. When embryos are 24 hpf, place an egg strainer into a new petri dish containing AFW. Pipette healthy embryos into the egg strainer (*see* **Note 2**).
3. Prepare 100 mL bleaching solution and divide it between two petri dishes (labelled Bleach-1 and Bleach-2). Prepare two petri dishes with AFW (labelled Wash-1 and Wash-2).
4. Immerse the egg strainer containing the embryos in dish, Bleach-1, and leave the embryos here for 5 min, before transferring the strainer to dish, Wash-1, for 5 min.
5. Repeat the bleaching and washing steps above by immersing the egg strainer in Bleach-2 and Wash-2 for 5 min each (*see* **Note 3**).
6. Turn out the egg strainer to deposit the embryos in a new petri dish containing pronase in AFW. Leave the embryos here for 15 min.
7. Pass the embryos rapidly up and down a 3 mL plastic dropping pipette to agitate them. After ten or so times of doing this, most embryos will be removed from their chorions.
8. Quickly transfer the dechorionated embryos to a clean petri dish with fresh AFW to remove pronase from the embryos. Repeat this step.
9. Transfer embryos into a 1.5 mL Eppendorf tube (*see* **Note 4**) and remove as much AFW from as possible using an autoclaved 150 mm cotton-plugged glass Pasteur pipette.
10. Add 1 mL of PBS to the embryos. When the embryos settle to the bottom of the tube, remove as much PBS from the tube as possible and replace with 1 mL fresh PBS.
11. Incubate the tube on ice. Take the sample to a primary tissue culture hood to complete the protocol whilst minimizing the risk of contamination (*see* **Note 5**).

12. Before starting, disinfect the hood and all instruments with IMS solution.
13. Prepare two 35 mm culture dishes; one with 4 mL embryo wash solution and the other with 4 mL ZNM.
14. Pipette the 1 mL of embryos in PBS into a 40 μm cell strainer. Agitate the strainer to drain PBS.
15. Immerse the strainer of embryos into the dish containing embryo wash solution, agitating for a maximum of 5 s. Immediately remove the strainer and embryos from this solution and shake the strainer to drain Embryo wash solution. Quickly immerse the strainer in the dish containing ZNM (*see* **Note 6**).
16. Allow the embryos to sit in ZNM for at least 10 s, then, using a 1000 μL filter tip, transfer the embryos in ZNM to a new sterile 1.5 mL Eppendorf tube. Allow embryos to settle to the bottom of the tube and pipette away as much ZNM and debris as possible.
17. Resuspend the embryos in 1 mL dissociation solution and place the tube on ice for 2 min.
18. Immerse a PVC pipette teat in IMS for 10 s to thoroughly disinfest. Dry with tissue paper and then insert it onto the end of an autoclaved 150 mm cotton-plugged glass Pasteur pipette (*see* **Note 1**, Fig. 3).
19. Begin the dissociation by passing the embryos in dissociation solution rapidly up and down the pipette making sure not to contact/wet the cotton plug. After a couple of minutes or so of pipetting, place the tube on ice.
20. Swap the previous autoclaved cotton-plugged glass Pasteur pipette for the next size down in tip bore size diameter (middle size; *see* **Note 1**; Fig. 3). Pipette up and down until the embryos fragment into smaller pieces. Place the tube back on ice for 2 min.
21. Swap the previous autoclaved glass Pasteur pipette for the one with the smallest tip diameter (smallest size; *see* **Note 1**; Fig. 3). Pipette up and down until no large embryo fragments remain visible. Place the tube back on ice for 2 min.
22. Place a fresh 40 μm cell strainer into an unskirted 50 mL Falcon tube and pipette the dissociated cell solution into the strainer.
23. Use a filter tip to add 1 mL fresh dissociation solution to the 1.5 mL Eppendorf to remove remaining cells stuck to the surfaces and pass this through the cell strainer.
24. Use a new filter tip to add a further 1 mL fresh dissociation solution directly to the cell strainer to remove remaining cells stuck to it. Shake the cell strainer to ensure dissociated cell solution in the insert drops into the Falcon.

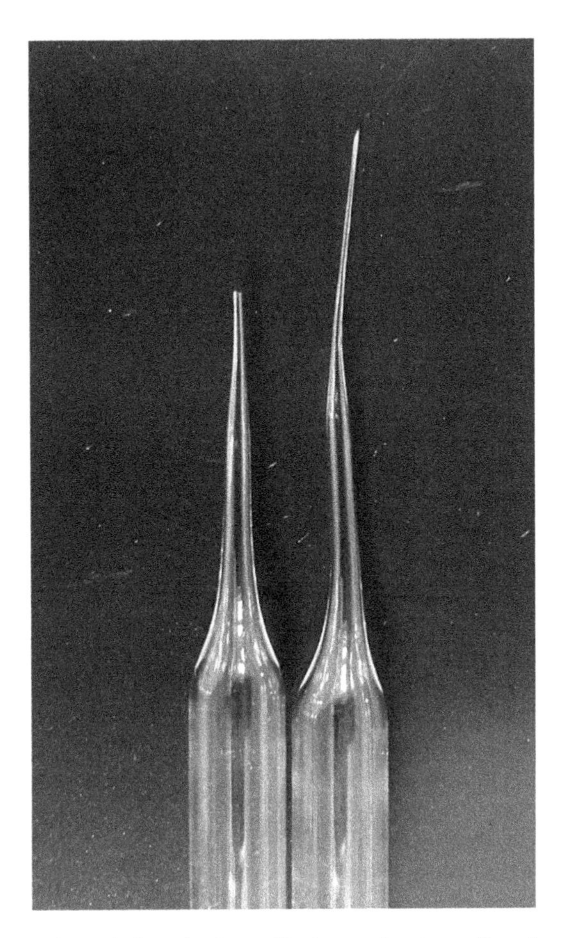

Fig. 3 Embryo dissociation pipettes with decreasing bore diameter size made using Bunsen burner. Embryos are passed through glass pipettes with three different bore sizes, with the smallest two shown (*see* **Note 1**)

25. Discard the cell strainer and centrifuge the Falcon containing the dissociated cell solution at 300 g for 7 min at 4 °C. Upon removal, the cells will be pelleted at the bottom of the tube.
26. Use a filter tip to remove as much as possible of the supernatant from the tube without disrupting the pellet.
27. Resuspend the cell pellet by adding 1 mL ZNM before pipetting vigorously up and down (*see* **Note 7**).

3.2 Primary Cell Culture

1. In a primary tissue culture hood, to minimize risk of contamination, prepare a sterile 24 well plate by placing a 12 mm Poly-D-Lysine/Laminin-coated coverslip in each well using decontaminated forceps (*see* **Note 2**).
2. Following embryo dissociation, cell filtering, centrifugation, and resuspension of cell the pellet in ZNM (see **Embryo dissociation** protocol above), pipette a 200 μL droplet of cell solution onto the coverslip in each well (*see* **Note 8**).

3. Cover the plate and keep it still for 90 min to allow the cells to settle and adhere.
4. Top up each well with 2 mL additional ZNM. Recover the plate and wrap with parafilm to seal (*see* **Note 9**). Store in a decontaminated container at room temperature (*see* **Note 10**).
5. Perform a 50% media change after 72 h of culture or when ZNM starts to turn yellow.
6. Cells may be fixed with 4% paraformaldehyde and processed for downstream applications such as in situ hybridization or immunohistochemistry. Prior to fixation, remove ZNM and perform one PBS wash.

3.3 Primary Culture Using Transwell Inserts

1. In a primary tissue culture hood to minimize the risk of sample contamination, place transwell inserts into a sterile 6-well plate using decontaminated forceps (*see* **Note 2**).
2. Use a filter tip to add 5 mL of 15 μg/mL Poly-D-Lysine to each well with an insert, pipetting 2 mL in the insert on to the upper membrane, and 3 mL into the well below the lower membrane of the insert. Incubate at room temperature for 2 h.
3. Remove the Poly-D-Lysine from below the insert of each well and replace with 3 mL of 3 μg/mL laminin. Incubate at room temperature for a further 90 min.
4. Remove the solutions from below and above the insert and replace with PBS for 2 min.
5. Remove PBS from both compartments. Replace with 2.5 mL ZNM both above and below the insert membrane.
6. Following embryo dissociation, cell filtering, centrifugation, and resuspension of the pellet in ZNM (see **Embryo dissociation** protocol above), pipette an appropriate volume of cell solution on to the upper membrane of the insert already covered with ZNM (*see* **Note 11**).
7. Place the lid on the plate and wrap with parafilm to seal (*see* **Note 9**). Store in a decontaminated container at room temperature (*see* **Note 10**).
8. Perform a 50% media change both above and below the insert membranes after 72 h of culture, or when ZNM starts to turn yellow (*see* **Note 12**).

3.4 Neurite Isolation and RNA Extraction

1. Decontaminate the workbench and all pipettes and other equipment you will use, first with IMS solution and then with RNase zap.
2. Obtain two nuclease free tubes from Qiagen's RNeasy Micro Kit and use filter tips to add 350 μL Buffer RLT to one of the

tubes. Label this tube as the upper compartment sample and the empty tube as the lower compartment sample.

3. Prepare a sterile 35 mm culture dish with 350 μL Buffer RLT. This dish will be used for collection of the lower compartment sample. Once collected, the solution will be transferred to the empty tube labelled as for the lower compartment sample.
4. When you are ready to collect tissue culturing on the insert, unwrap the 6-well plate and remove ZNM from the upper and lower membrane surfaces.
5. Wash the sample by incubating 2 mL of PBS on both the upper and lower surfaces of the membrane for 2 min.
6. Remove PBS from both sides of the membrane. Avoiding the membrane drying out, quickly but carefully remove the transwell insert from the well using your fingers (*see* **Note 13**).
7. Swiftly but gently swab the upper surface of the membrane to collect cellular material (*see* **Note 14**). Wipe repeatedly in the same direction.
8. Immerse the swab in the tube of 350 μL Buffer RLT labelled as for the upper compartment sample. Leave to soak.
9. Whist the swab is soaking, take a cell scraper and thoroughly scrape tissue from the lower surface of the membrane. Transfer the material to Buffer RLT by immersing the scraper head in the 35 mm culture dish prepared.
10. After thoroughly dabbing the scraper head in the 350 μL Buffer RLT, transfer it to the empty nuclease free tube labelled as for the lower compartment sample.
11. Use a scalpel to cut the handle from the swab head soaking in Buffer RLT.
12. Using decontaminated forceps, grab the head of the swab not immersed in Buffer RLT and press it against the side of the tube to rinse. Discard the swab head.
13. Perform the remaining steps of the Qiagen RNeasy Micro Kit protocol, making a few minor modifications (*see* **Note 15**).
14. Following elution, RNA sample concentration can be measured using the Qubit RNA HS Kit. Using 2 μL of each Upper and Lower compartment sample.
15. Samples should be stored at −70 °C prior to use in any downstream applications.

4 Notes

1. To dissociate, embryos should be passed up and down cotton-plugged glass pipettes with decreasing bore diameters. Three diameters are used. First use 150 mm glass pipettes without

modification. Next, use 230 mm glass pipettes that you have narrowed the tip of using a Bunsen burner. Two gradations of narrowed pipettes should be prepared: one with a medium-sized aperture of roughly 0.5 mm and the other with a smaller-sized aperture of roughly 0.25 mm (*see* Fig. 3). Prepare these by holding a 230 mm pipette with one hand at each end. Carefully hold it so that the middle, narrow part, of the pipette sits in a blue Bunsen burner flame, until the glass begins to deform. When this happens, pull your hands away from each other so that the middle of the pipette lengthens and narrows. Snap the pipette and discard the distal part. Carefully shorten the new end of the pipette to desired length and bore diameter, by snapping the narrowed part of the pipette in folded sheets of tissue paper. When complete, autoclave the three different-sized pipettes before use.

2. One should be as sterile as possible for the remainder of this section of the protocol to reduce the risk of bacterial contamination of cultures following embryo dissociation. Regularly spray gloves with IMS and use sterile/decontaminated tools.
3. Whilst embryos are in bleach and wash steps, prepare the pronase solution so you can immediately move to the dechorionation step after the final wash.
4. A maximum of 200 embryos should be put into each tube for dissociation. More than this will cause frequent blocking of the glass pipettes, decreasing the efficiency of dissociation.
5. When you go to tissue culture, do not forget to take with you: autoclaved pulled glass pipettes with different bore diameters (*see* **Note 1**), PVC pipette teat, dissociation solution, 2 × 40-μ m cell strainers, unskirted falcon tubes, autoclaved 1.5 mL Eppendorf tubes, filter tips, embryo wash solution, ZNM & 35 mm culture dishes.
6. This step needs to be quick! Inevitably there will always be some amount of cell death, but too long spent in this embryo wash solution will cause excessive damage to embryos, leaving very little viable material left for culture.
7. Minimize bubbles in the solution by preventing air from entering the tip whilst pipetting.
8. The volume of ZNM used for resuspension will vary based on how many embryos were used for dissociation and the density of culture you wish to achieve. The resuspended solution can then be evenly divided between the number of wells prepared. Plating cells from 25 to 50 dissociated embryos on to a 12 mm coverslip provides good density. Testing this could be a good place to start. One can then scale up or down the starting number of embryos, accordingly, based on whether cultures are at too low or too great density. Cells plated at too low a density will not grow well.

9. Decontaminate strips of parafilm prior to use by spraying with IMS solution.
10. Check the plate under a scope 24 h after plating cells. Be sure to thoroughly decontaminate gloves and the scope with IMS prior to doing this. You should see cell somas adhered to the surface of the coverslip or upper membrane of the transwell insert.
11. A rough guide to achieve good density is to plate cells from 200 to 300 dissociated embryos onto the upper membrane of each transwell insert.
12. To confirm that neurites pass through 1 μm pores of the transwell insert membrane and adhere and extend along the lower membrane surface, fix the sample, and perform immunohistochemistry targeting neurites using an acetylated-tubulin antibody, and nuclei with DAPI. Following immunohistochemistry, use a scalpel to cut the membrane from the insert and mount it on a slide with mounting medium, and cover with a large coverslip.
13. Make sure to avoid touching the membrane when removing the insert from the well. Hold the insert by its walls (*see* Fig. 4).

Fig. 4 Isolation of tissues from cellular and neurite membranes of transwell inserts. Steps summarizing the process of tissue extraction from the upper and lower transwell membrane surfaces

14. Make sure the swab head does not touch anything following removal from its wrapping to avoid contamination. Whilst still in the wrapper, bend the head of the swab as this will make it easier to collect tissue. Wipe the transwell membrane applying minimal pressure to avoid piercing the membrane. Also, ensure the swabs used have absorbent heads. We use swabs with polyester heads that prove to be fine alternatives to buccal swabs.
15. To ensure removal of all traces of Buffer RW1 and good purity of RNA samples, tubes were rotated after adding 500 μL Buffer RPE to each column prior to centrifugation. A second 500 μL Buffer RPE step was also performed, repeating the rotation prior to centrifugation.

References

1. Price PJ (2017) Best practices for media selection for mammalian cells. In Vitro Cell Dev Biol Anim 53(8):673–681. https://doi.org/10.1007/s11626-017-0186-6
2. Gordon J, Amini S, White MK (2013) General overview of neuronal cell culture. Neuronal Cell Cult Methods Proto:1–8. https://doi.org/10.1007/978-1-62703-640-5_1
3. Chen Z, Lee H, Henle SJ, Cheever TR, Ekker SC, Henley JR (2013) Primary neuron culture for nerve growth and axon guidance studies in zebrafish (Danio rerio). PLoS One 8(3): e57539. https://doi.org/10.1371/journal.pone.0057539
4. Sassen WA, Lehne F, Russo G, Wargenau S, Dübel S, Köster RW (2017) Embryonic zebrafish primary cell culture for transfection and live cellular and subcellular imaging. Dev Biol 430(1):18–31. https://doi.org/10.1016/j.ydbio.2017.07.014
5. Myhre JL, Pilgrim DB (2010) Cellular differentiation in primary cell cultures from single zebrafish embryos as a model for the study of Myogenesis. Zebrafish 7(3):255–266. https://doi.org/10.1089/zeb.2010.0665
6. Acosta JR et al (2018) Neuronal cell culture from transgenic zebrafish models of neurodegenerative disease. Biol Open. https://doi.org/10.1242/bio.036475
7. Fassier C et al (2010) Zebrafish atlastin controls motility and spinal motor axon architecture via inhibition of the BMP pathway. Nat Neurosci 13(11):1380–1387. https://doi.org/10.1038/nn.2662
8. Sakowski SA, Lunn JS, Busta AS, Palmer M, Dowling JJ, Feldman EL (2012) A novel approach to study motor neurons from zebrafish embryos and larvae in culture. J Neurosci Methods 205(2):277–282. https://doi.org/10.1016/j.jneumeth.2012.01.007
9. Ciarlo CA, Zon LI (2016) Embryonic cell culture in zebrafish. Methods Cell Biol:1–10. https://doi.org/10.1016/bs.mcb.2016.02.010
10. Vogelaar CF et al (2009) Axonal mRNAs: characterisation and role in the growth and regeneration of dorsal root ganglion axons and growth cones. Mol Cell Neurosci 42(2): 102–115. https://doi.org/10.1016/j.mcn.2009.06.002
11. Taylor R et al (2022) Prematurely terminated intron-retaining mRNAs invade axons in SFPQ null-driven neurodegeneration and are a hallmark of ALS. Nat Commun 13(1):6994. https://doi.org/10.1038/s41467-022-34331-4
12. Shetty AK, Turner DA (1998) In vitro survival and differentiation of neurons derived from epidermal growth factor-responsive postnatal hippocampal stem cells: inducing effects of brain-derived neurotrophic factor. J Neurobiol 35(4):395–425
13. Peljto M, Dasen JS, Mazzoni EO, Jessell TM, Wichterle H (2010) Functional diversity of ESC-derived motor neuron subtypes revealed through Intraspinal transplantation. Cell Stem Cell 7(3):355–366. https://doi.org/10.1016/j.stem.2010.07.013

Chapter 8

Holographic Optogenetic Activation of Neurons Eliciting Locomotion in Head-Embedded Larval Zebrafish

Xinyu Jia and Claire Wyart

Abstract

Understanding how motor circuits are organized and recruited in order to perform complex behavior is an essential question of neuroscience. Here we present an optogenetic protocol on larval zebrafish that allows spatial selective control of neuronal activity within a genetically defined population. We combine holographic illumination with the use of effective opsin transgenic lines, alongside high-speed behavioral monitoring to dissect the motor circuits of the larval zebrafish.

Key words Optogenetics, Computer-generated holography, Opsins, Spatial light modulator, Behavioral monitoring, Light patterning, Stable zebrafish transgenic lines, Locomotion

1 Introduction

The neuronal circuits controlling locomotion are highly conserved in vertebrate species [1, 2]. Due to its small size, genetic amenability, and transparency, larval zebrafish is an ideal model organism to investigate the function of motor circuits using optogenetic techniques [3]. The use of genetically encoded calcium indicators combined with two-photon fluorescence imaging has succeeded in establishing critical links between the single-cell level neuronal activity and the generation of behaviors. With the discovery of light-gated channels and ion pumps [4, 5], optical manipulations of neuronal activity now allow us to causally examine an intact neuronal circuit in an awake and behaving animal with high temporal precision.

Previous optogenetic studies achieved spatial selectivity through the use of fiber optics [6, 7], pinhole aperture [8], digital micromirror devices (DMDs) [9–11], or laser spiral-scanning approaches [12]. However, it remains challenging to manipulate single-cell activity independently and simultaneously at different spatial locations within a dense population of neurons. Computer-

James F. Amatruda et al. (eds.), *Zebrafish: Methods and Protocols*, Methods in Molecular Biology, vol. 2707,
https://doi.org/10.1007/978-1-0716-3401-1_8,

generated holography (CGH), based on the use of a spatial light modulator (SLM), offers a unique opportunity to create any arbitrary intensity profile in two or three dimensions with unparalleled spatiotemporal resolution [13–16]. A Fourier transform-based iterative algorithm is used to compute the phase hologram to be displayed at the SLM such that the desired patterns of illumination can be reproduced at the target positions of the sample [17, 18].

Here, we took advantage of our calibrated optogenetic toolbox of transgenic zebrafish lines with stable opsin expression to overcome the variability of opsin expression and efficacy in vivo [19]. We provide an overview of the steps required to prepare, perform, and control for a CGH-based optogenetic experiment coupled with high-speed behavioral monitoring. Our protocol achieves 2D light patterning with high lateral precision by projecting visible laser light onto a SLM. Two-photon illumination with CGH, although more costly, can further improve the axial confinement of the optogenetic stimulation to achieve single-cell activation.

This protocol emphasizes the importance of the selection and calibration of the stimulation patterns in vivo, as well as the identification of the proper controls for the experiment.

2 Materials

2.1 Preparations of Larval Zebrafish

1. 5–6 days postfertilization (dpf) zebrafish larvae expressing an optogenetic actuator of choice (*see* **Note 1**) in neurons of specific molecular identity with the GAL4/UAS system (*see* **Note 2**) or in a direct line such as *Tg(elavl3:CoChR-EGFP)* [20], combined with a mutation that removes pigments from the fish skin.
2. 50 mm wide diameter MatTek Petri dish with a 14 mm wide diameter microwell equipped with a glass bottom coverslip (https://www.mattek.com/).
3. E3 embryo medium for raising larvae and experiment: 5 mM NaCl, 0.17 mM KCl, 0.33 mM $CaCl_2$, 0.33 mM $MgSO_4$. To make a buffered E3 solution, add Tris solution, adjust to pH 7.0 with HCl, to a final concentration of 1 mM.
4. Blue water: 0.6 g Instant Ocean synthetic salts, 2 mL of stock methylene blue solution (1 g/L), 10 L of osmosis water.
5. 0.2% Tricaine (MS-222, Sigma-Aldrich, Saint Louis, Memphis, USA)
6. aCSF solution for enucleation: 134 mM NaCl, 2.9 mM KCl, 1.2 mM $MgCl_2$, 10 mM HEPES, 10 mM glucose and 2.1 mM $CaCl_2$; 290 mOsm.kg-1, adjusted to pH 7.7–7.8 with NaOH.

7. Low melting point agarose: 1.5% and 3% dissolved in E3 medium, 2.5% dissolved in aCSF. Keep melted in a heat block at 42 °C.
8. Stereo microscope (e.g., LEICA M165 FC).
9. Loader tip (e.g., 20 μL GELoader, Eppendorf).
10. Tungsten wire for enucleation (*see* Fig. 2a).
11. 3D-printed mounting claw.
12. Forceps (Fine Science Tools).
13. Fine scalpel (e.g., Sharpoint, 5 mm Depth 15° Stab).
14. Disposable transfer pipettes.

2.2 Light Patterning Protocol

We use CGH combined with visible light to achieve stimulation of a subset of neurons in vivo. This approach requires two-photon fluorescent imaging to locate the position of the target neuron, and a calibration procedure to register the coordinates of the stimulation to that of the imaging system. We list here all the necessary equipment required to perform light patterning calibration and experiments for a one-photon holography setup (*see* Fig. 1):

1. Two-photon laser scanning upright microscope equipped with a water dipping objective (typically 20×, NA = 1) in order to image cells expressing the opsin in vivo.
2. A fluorescent slide for calibration (*see* **Note 3**).
3. Distilled water.
4. Software to define the required illumination pattern based on the fluorescent imaging and control accordingly the phase displayed on the SLM.
5. Universal serial bus (USB) power meter (e.g., Thorlabs, PM16-130).
6. A visible light laser line that allows an efficient activation of the opsins expressed in the transgenic larva (*see* **Note 4**).

2.3 Behavioral Monitoring

1. Infrared light emitting devices (LEDs) (e.g., Intelligent LED solution, ILS, ILH-IW01-85SL-SC211-WIR200).
2. Low magnification objective (e.g., ZEISS FLUAR 5×/0.25).
3. High-speed camera (e.g., Basler acA2000-340 km).
4. Silver front-faced mirror (e.g., Thorlabs CM1-P01).
5. C-mount lens (e.g., NAVITAR HR F2.8/50 mm).
6. Bandpass filter centered on LED wavelength (e.g., FB850–40 – Ø1" bandpass filter, CWL = 850 ± 8 nm, FWHM = 40 ± 8 nm).

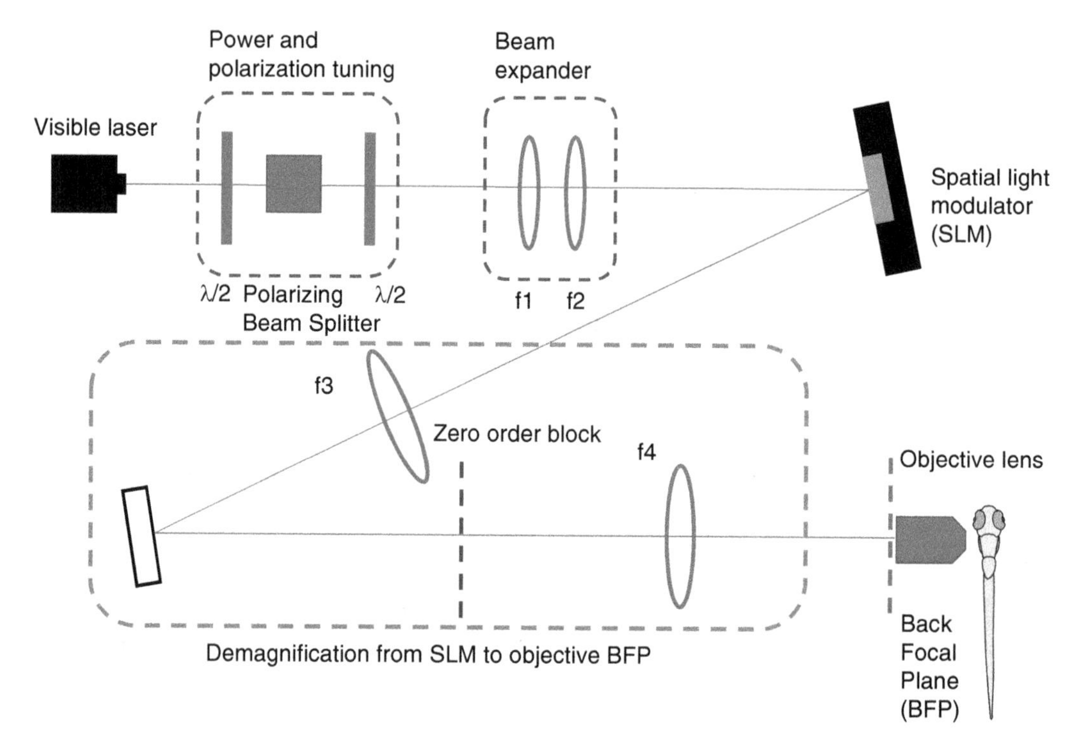

Fig. 1 Light path for one-photon holography. A set of half-wave plates and a polarizing beam splitter are used to control the power of the visible laser and the orientation of the polarization arriving on the spatial light modulator (SLM). The latter is critical since a SLM is polarization-sensitive and only a specific linear polarization undergoes phase modulation. A set of lenses (f1 and f2) is used as a beam expander in order to fully illuminate the SLM display so that all available pixels on the SLM can be used for light modulation. Another set of lenses (f3 and f4) is used to demagnify or magnify the beam and to project the image of the SLM onto the back focal plane (BFP) of the objective. The choice of the magnification factor matches the size of the beam with the size of the BFP. Typically, at the focal plane of lens f3 along the light path, a zero-order block is positioned in order to prevent the fraction of non-modulated light to reach the sample

3 Methods

3.1 Expression of Opsins and Selection of Transgenic Larvae

The optimal expression level of opsins in stable transgenic fish lines is challenging but key for success. We recently established an optogenetic toolbox including nine stable transgenic fish UAS lines selected for high expression and wide coverage [19]. The benefit of the system is its flexible deployment of appropriate opsins expressed in the neuronal population of specific genetic identity [21–23].

1. Set up crosses of the chosen GAL4 transgenic fish line with the UAS opsin line typically 6–7 days before the experiment (*see* **Note 5**). Collect eggs the next morning and raise them in E3 or blue water at 28.5 °C. Keep the embryos out of light

exposure to avoid any unwanted stimulation of the opsin during development.

2. Identify the optimal stage of sorting based on the GAL4/UAS combination of choice for the experiment (*see* **Note 6**). To sort larvae with opsin expression in neurons within motor circuits, identify the larvae who exhibit locomotor response to opsin stimulation and express the fluorescent protein fused to the opsin.
3. Turn on the white background light to track the position of the larva. Select the optimal excitation filter set to activate the opsin and turn on the fluorescent lamp. Adjust the field of view of the sorting scope to follow the larva after its locomotor response to light. Opsin-expressing larvae in GAL4 lines targeting excitatory neurons in the motor circuits in the hindbrain or spinal cord typically exhibit coiling or swimming patterns temporally matched to light illumination of correct wavelength. However, the transgenic larvae should not respond to a wavelength that is nonspecific to the opsin used. Similarly, the sibling negative larvae should not exhibit any locomotor response to light.

3.2 Preparation of the Larvae Before Experiment

To eliminate any spurious visual response to the light used, we perform enucleation at 3–4 dpf (*see* Fig. 2a), and allow enough time for recovery from the procedure.

1. Place a larva in E3 water in the center well of the 50 mm-wide petri dish. Add a droplet of 0.2% Tricaine to anesthetize the larva. Carefully suck off the E3 water in the dish. Promptly add 2.5% agarose dissolved in artificial cerebrospinal fluid (aCSF) solution to cover the larva. Keep the larvae in the center of the dish and quickly manipulate its orientation using a plastic loader tip to keep the larva dorsal side up.
2. Wait for 5–10 min until the agarose is set and add aCSF solution to cover the bottom of the petri dish.
3. Once the aCSF agarose solidifies, perform enucleation using a fine tungsten wire bent at the tip to form a hook about the size of the eye. Gently scoop down with the wire to cut apart the eyecup from the forebrain (*see* Fig. 2b). Verify that the lens is fully detached from the brain before freeing the 4 dpf old larva using a forceps and a scalpel.
4. Allow the enucleated larvae to recover in the aCSF solution for 20–30 min before transferring it back to E3 medium using a disposable transfer pipette.
5. On the day of the experiment, mount the 5–6 dpf enucleated larva in a 35 mm petri dish with 3% agarose. Mounting procedure is similar to that described in the enucleation section. In order to hold the agarose in place during the experiment, we

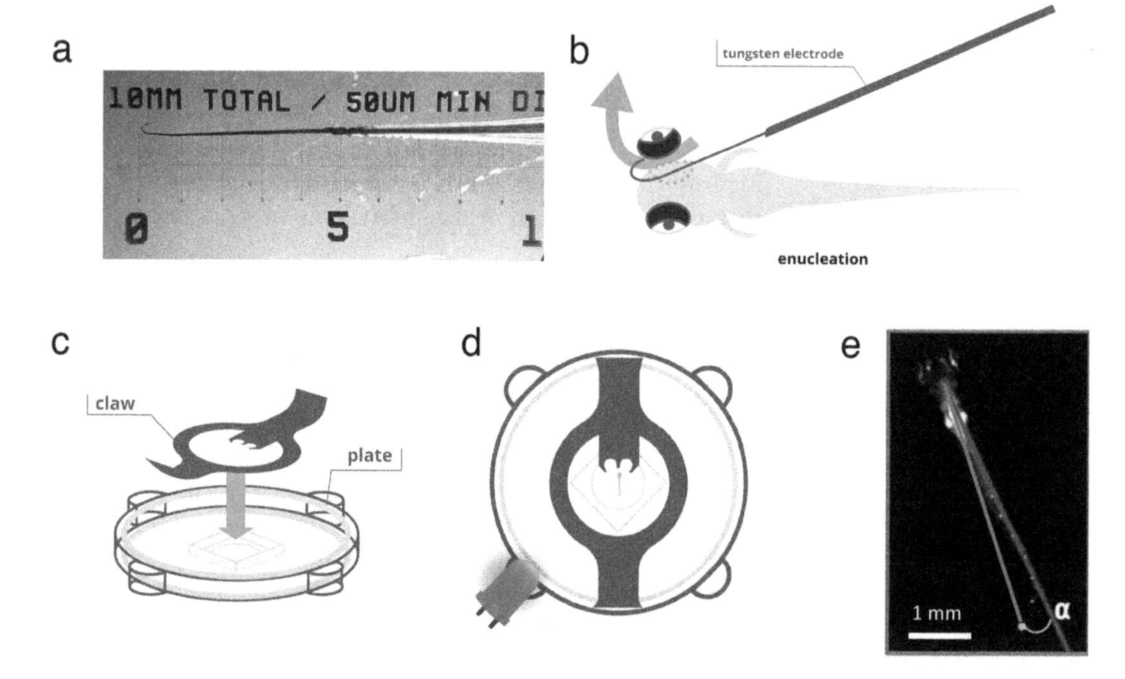

Fig. 2 Preparation of larval zebrafish and behavioral monitoring. (**a**) A tungsten wire bent at the tip is prepared for enucleation of the larvae prior to the experiment. (**b**) Illustration of enucleation process in which the muscle between the eyecup and the forebrain is cut apart in a scooping action. The procedure is performed at least 1 day before the experiment. (**c**) Fitting the 3D-printed mounting claw to the dish to eliminate motion of the agarose gel during the experiment. (**d**) Suggested placement of the infrared light emitting device (LED) to ensure good illumination of the larva's tail for behavioral tracking. (**e**) Tracking of the tail of a head-embedded tail-free larva and definition of the tail angle

fitted on the petri dish a 3D-printed mount claw before adding the droplet of agarose. (*see* Fig. 2c, **Note 7**). Add E3 medium to the petri dish once the agarose has solidified.

6. Free the tail of the larva at the level of the fins using a scalpel to monitor the tail moving upon optogenetic stimulation (*see* Fig. 2e). Remove the agarose on both sides of the tail from the end of the swim bladder. Insert the blade of the scalpel close to the fish and cut away from the body, perpendicular to the tail. Remove the agarose on each side of the tail by pulling on it with a forceps.

3.3 Calibration of the Holographic Illumination

Holographic illumination can be implemented either with a fully homemade optical path (*see* Fig. 1) or using a commercial module. For our experiment, we opted for a commercial setup integrated with a two-photon laser scanning microscope (Phasor and VIVO two-photon, intravital multiphoton imaging system, Intelligent Imaging). The user can define a target intensity distribution at the output of the objective, and the control software uses an iterative Fourier transform-based algorithm [18, 24] to calculate the corresponding phase patterns and displays them on the liquid crystal spatial light modulator.

3.3.1 Spatial Calibration of the Holographic Beam

Spatial registration between the field of excitation (FOE) of the SLM to the FOV of the fluorescence microscope is crucial before every experiment. This is to make sure that the patterned illumination targets the regions of interest selected on the fluorescence image. We perform this registration by bleaching fluorescent slides with a user-defined holographic pattern, reimaging the bleach spots, and aligning the two fields (*see* Fig. 3a).

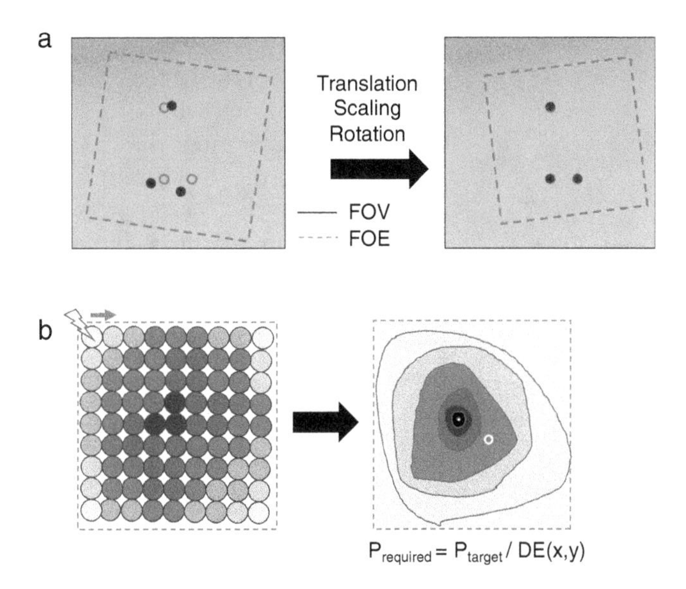

Fig. 3 Calibration of the holographic illumination. (**a**) Spatial calibration to register the field of view (FOV) of the two-photon microscope to the field of excitation (FOE) of the one-photon holography setup. (left panel) Based on the two-photon fluorescent imaging, select an asymmetric pattern in the center of the FOV. Use computer-generated holography to target the patterns to bleach the fluorescent slide (blue circles). Reimaging the fluorescent slide after bleaching. (right panel) Based on the discrepancy between the targeted patterns and the actual bleach, compute and apply the appropriate translation, scaling and rotation to match the fluorescent image to the targeted patterns. (**b**) Power calibration to correct for different diffraction efficiency across the FOE. (left panel) Measure the power at the output of the objective across the FOE and keeping the same input one-photon laser power. Circles of the same radius targeted at the peripheral of the FOE receive less power and that at the center. (right panel) By sequentially measuring laser power (P) of the same pattern at different X and Y positions, we can reconstruct a power contour map that shows the diffraction efficiency as a function of the target's lateral position. Note that there is no power delivered to the center of the FOV due to the zero-order block. To illuminate a pattern at a given power, the power required from the laser is scaled by the inverse of the diffraction efficiency (DE) at the target location

1. Place a fluorescent slide under the objective and add a droplet of distilled water between the slide and the objective. Focus the microscope on the fluorescent molecules by adjusting the Z-position to find the optimal focal plane at which the pixel count is maximized.
2. Define three asymmetrically arranged spots in the FOV and calculate the phase of the SLM to target these spots.
3. Adjust the power delivered to the spot so that it is just above the threshold required to bleach the first-order spots closest to the center of the field of excitation (FOE) (*see* **Note 8**).
4. Image the fluorescent slide again after the bleaching occurred. Calculate and apply the required transformation (i.e. scaling, translation, and rotation) to match the targeted ROIs drawn on the FOV to the actual bleached spots. From now on, this transformation will be applied on the defined patterns in order to guarantee that they will target the desired location defined on the fluorescent images.

3.3.2 Power Calibration of the Holographic Illumination

Due to the nature of the holographic illumination, a region selected near the center of the FOE will receive higher power than the one at the edge of the field. We correct for the heterogeneities of the power diffracted across the field of excitation to ensure uniform power delivery to ROIs at different locations (*see* Fig. 3b).

1. Target a circle of fixed radius (around 1/20 of the size of the FOE) with the same power and move it sequentially across the FOE to different locations.
2. Use a power meter to measure the actual power reaching the sample plane. Having measured the diffraction efficiency across the FOE, we can control the power delivered to each ROI by applying higher laser power to more peripheral positions.
3. Because the diffraction efficiency is dependent on the X,Y position of the target ROI, the power required to send through the objective to meet the target power on the sample is given as $P_{required} = P_{target}/DE(x,y)$.
4. Note that in most one-photon holography setup, there is a zero-order block at the center of the FOE to prevent the unmodulated laser light reaching the sample. It is therefore suggested to avoid creating a pattern on the center of the FOE (*see* **Note 9**).

3.4 Monitoring the Behavioral Output

1. Illuminate the larva using infrared LEDs. Place the LED at the same horizontal level as the petri dish. Keep it as close to the dish as possible, with a 45 degrees inclination (*see* Fig. 2d) to the body axis of the fish to give a good contrast all along the length of the tail (*see* **Note 10**).

2. Place the low magnification objective pointing to the petri dish. The images of the fish behavior can then be relayed to a high-speed camera through a silver mirror, an aperture, and a C-mount lens. A bandpass filter is added on the camera to filter out the imaging laser and the stimulation laser. The final distance between the petri dish and the camera chip will vary depending on the specifications of the two objectives on the light path.
3. Set the high-speed camera to start recording upon an external Transistor-to-transistor logic (TTL) trigger. The same trigger is simultaneously sent to the software controlling the stimulation sequence. The behavioral response to the optogenetic stimulation can be visually inspected at the time of the experiment at a lower frequency through the camera software.

3.5 The Optogenetic Experiment

1. Acquire a fluorescent image of the opsin expression pattern in the transgenic larva.
2. Select stimulation target region of interests (ROIs) in the field of view (FOV) using the fluorescent image as a background (*see* Fig. 4a).
3. Set the targeted power of illumination to each stimulation ROI.
4. Set the stimulation timepoints and duration for each stimulation ROI.

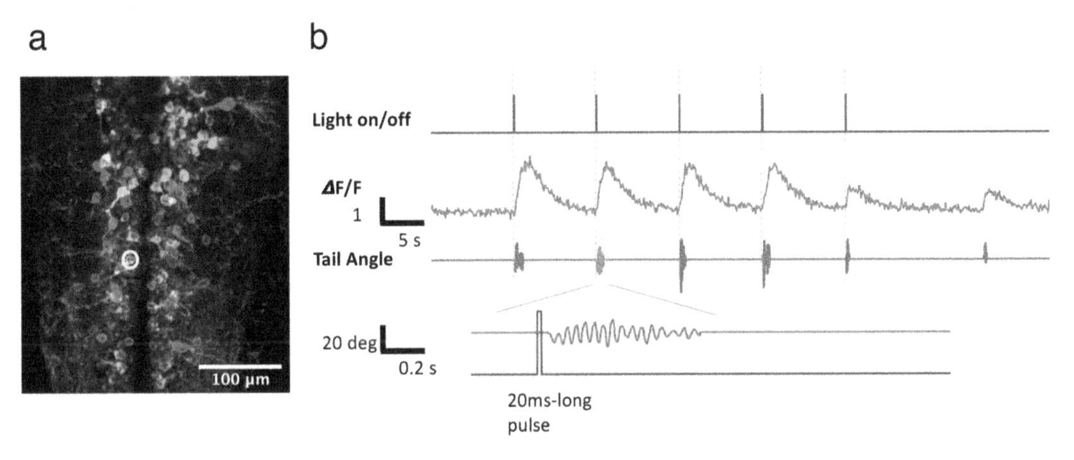

Fig. 4 A typical experiment of optogenetic stimulation and the subsequently elicited behavioral response. (**a**) A fluorescent image showing the expression of the opsin (shown in red, ChrimsonR-tdt) and the calcium indicator (GCaMP6s). A ROI is created (shown in yellow circle) to select the target neuron for stimulation. A neuron with colocalization of the opsin and sensor is selected as target. (**b**) Top trace shows the optogenetic stimulations given during a 75 s experiment where five stimulation are delivered to the fish with 10 s interval in between trials. We simultaneously recorded the calcium transient response to the stimulation to confirm activation of the target neurons. The behavioral responses of the larva were monitored using a high-speed camera as shown with the tail movement (green)

5. Set up the TTL synchronization pulses between the optogenetic stimulation and behavioral monitoring.
6. Record and monitor in real time the tail movement of the larva upon programmed illumination (*see* Fig.4b).
7. Repeat 1–7 based on behavioral and neuronal response from the larva.

In order to verify that the targeted neuron is activated by light, one can combine optogenetic activation with calcium imaging (*see* **Note 11**).

3.5.1 Choosing the Parameters of Stimulation

In principle, the intensity of photostimulation should be chosen slightly above the threshold such that a behavioral response – or a change in the readout of neuronal activity – can be observed. To identify the threshold of response for one type of neuron, adjust the stimulation pulse duration and the power density systematically and compare the behavioral response.

1. In each trial, repeat the same stimulation pulse multiple times, with at least 10 s of baseline in the beginning of the recording and typically 10 s of interval between each stimulation to allow the target cell to recover (*see* Fig. 4b).
2. Vary the stimulation pulse across trials by changing either the pulse duration or the laser irradiance, one parameter at a time. Examine the tail movement immediately following the stimulation (within a time window of typically 100 ms) and determine whether the given stimulation pulse was successful in eliciting a behavioral response.
3. For a given trial, if a behavioral response is observed for more than three pulses, we consider the stimulation used is above the threshold for behavioral response. We recommend starting with 5 ms pulse duration with varying laser irradiance from 0.2–2 mW/mm^2.

3.6 Photostimulation Control

The SLM in the holographic module can be coupled with either visible laser lines (blue, green, or amber depending on the optimal wavelength to activate the opsin) or with an infrared laser line for two-photon excitation. When using a visible laser line for excitation on the SLM, the axial confinement of the illumination should be approximately twice the diameter of the spot in the X and Y directions: so a 10 um diameter spot in X,Y, will typically extend over 20 um in Z. Our setup operates with a set of visible laser lines and we used a 594 nm laser for the activation of ChrimsonR (*see* Fig. 4).

In order to establish a causal link between the manipulations of neuronal activity and the behavioral output, it is essential that we ensure the behavioral response results from the optogenetic stimulation and not from any thermal, visual stimulation or out-of-target stimulation. Here, we proposed several positive or negative control experiment ideas to run alongside the main experiment while also suggesting some key behavioral statistics to compare on.

3.6.1 Controls in Non-opsin-Expressing Larvae

Nonspecific effects can include thermal, visual responses and others (*see* **Note 12**). For the purpose of our experiment, the eyes of the zebrafish larvae have been removed to eliminate any visual response to stimulation. It is important to perform this control with non-opsin-expressing siblings using the same stimulation protocol. Examine the possible effects of heating and visible light stimulation on the behavior. The occurrences of large tail movement such as struggle and escape are higher during heating and visual stimulations.

3.6.2 Out-of-Target Optogenetic Stimulation in Larvae Expressing the Opsin

In order to conclude our target neurons of interest are responsible for eliciting behaviors, we examine the effect from stimulating the possible axonal tracts or dendritic trees above or below our plane of interest (*see* **Note 13**)

1. Acquire an anatomical Z-stack of the opsin expression typically over ~30 um above and below the cells of interest.
2. Target the neuropils within this volume with the same stimulation protocol and monitor the behavioral response for comparison.
3. Examine the overall stimulation response success rate and the trials' latency to response between the out-of-target control and the main experiment.

Note that excess power from the one-photon laser could also lead to out-of-target stimulation of opsins expressed far from the target (*see* **Note 14**).

3.6.3 Ablation Control to Inspect the Contribution of the Targeted Neuron(s)

1. Conduct main optogenetics experiment on target neurons with stimulation protocol at calibrated power intensity.
2. Perform two-photon laser ablation of the target cell. Image after ablation to confirm cell death.
3. Keeping the position of the stimulation, repeat the same experiment on the rest of the population as a control.
4. Look for the difference in stimulation response rate, latency, and tail movement kinematics between the behavioral responses in the main experiment and the ablated control to identify the contribution from the ablated neurons.

4 Notes

1. When deciding which opsin to use, consider the excitability of the target neuron cell types and typical membrane resistance. In the cells with high membrane resistance, for example, the cerebrospinal fluid-contacting neurons which have giga ohms membrane resistance, the channelrhodopsin-2 is effective in activating these types of cells. However, more efficient opsins are necessary for larger neurons, such as the primary motor neurons, that have smaller membrane resistance. Consequently, the threshold of optogenetic stimulation for behavioral response can also differ. It is important to calibrate the behavioral response to identify the optimal stimulation parameters for each cell type individually.
2. There is a huge diversity of opsins to control the activity of neurons [25]. Over the last 15 years, there are recurring concerns of variegation, in particular but not only via the GAL4/UAS combinatorial expression system. A recent optogenetic toolbox provides an electrophysiological calibration of nine opsins expressed as transgenic lines using the lines. However, the variegation of opsin-expressing transgenic lines nonetheless imposes a very strict screening and sorting of the best expressers in order to avoid a renewal of transgenic lines.
3. The fluorescent slide for spatial calibration can be easily prepared at the bench. Take a green fluorescent marker and apply a generous amount of fluorescent ink to draw a square on a mount slide. Wait for 5–10 min and cover the fluorescent square with a coverslip. Use transparent nail polish to seal the edge.
4. For one-photon activation, it is important that the wavelength of the laser line used is within the optimal range of the opsin action spectrum [19]. The range of the usable wavelength for two-photon activation gets wider [26, 27] and the choices available for high-power lasers sufficient to trigger multiple spots becomes the limiting factors [28].
5. Because our experiment investigates the effect in behaviors of activating subsets of reticulospinal neurons, we choose to perform experiments on 5–6 dpf larvae to limit the occurrence of spontaneous swimming.
6. The optimal sorting windows depend on the GAL4/UAS combinations. For a combination that starts expression early on, we screen at 1 dpf to look for lasting coiling inside the chorion upon illumination. For late expression, screen around 3 dpf when the larvae have hatched but still have minimal spontaneous swimming. This prevents the use of tricaine

during sorting and allows us to look for behavioral response to opsin stimulation.

7. We use 3% low melting point agarose for mounting during behavioral experiments to minimize the effects of tail movement during calcium imaging and reduce the motion artefacts. The tail freeing procedure decreases the area of contact between the agarose sheet and the petri dish. We use the 3D printed mount claw to prevent the agarose sheet and the larva from floating in the dish during the experiment.
8. For spatial calibration, it is preferable to perform a few "bleach" tests to identify the optimal laser power for bleaching on your system and laser wavelength. The goal is to avoid higher-order bleaching on the slide that causes confusion for calibration. Increase the laser power if the contrast of the bleach spot is low. Reduce the laser power if the bleached pattern repeats itself close to the edge of the FOE.
9. The zeroth order light arises from the mirror grid between the pixels of the SLM where the incoming light is simply reflected and converges on the center of the FOE without any phase modulation. It is typically removed by placing a zero-order block in an intermediary image plane on an one-photon holography setup (*see* Fig. 1). It can be elegantly suppressed in a two-photon holography setup where a cylindrical lens is used to stretch the zero-order spot into a line, and using the SLM to correct for the added optical aberration [29].
10. Optimizing the infrared illumination for the behavioral monitoring can help the tracking software to better localize the tail in motion, which simplifies the tail angle extraction process.
11. In order to confirm that the targeted neuron is activated and to monitor the other cells that were recruited, it is possible to use transgenic larvae expressing a calcium sensor together with an opsin in the same neurons [24]. In this case, synchronize optogenetic stimulation with calcium imaging before and after the light pulse to monitor the response of the targeted neurons and other neurons in the vicinity.

    When performing two-photon calcium imaging alongside an optogenetic experiment, it is important to optimize the choice of opsins and calcium indicators, as well as the imaging wavelength to minimize spectral crosstalk [30]. Many opsins have a very large absorption spectrum [19] and can therefore be activated by the laser line used for imaging green or red calcium sensors. Although Chronos (and more recent variants [31, 32]) offered many opportunities as a fast red-shifted opsin in mammals, it has not been possible to effectively express this opsin in zebrafish [19]. A typical sensor-actuator choice would be combining green calcium sensors such as GCaMP with the red-shifted opsin ChrimsonR [33].

12. Be aware of endogenous opsins expressed in the zebrafish nervous system: Multiple opsins are endogenously expressed in the central nervous system of larval zebrafish in the brain [34] and embryonic spinal cord [35].
13. Soma-targeted opsin can be very effective in reducing out-of-target stimulation of nearby processes [31, 36]. However, it is important to verify the expression in dendrites and axons especially in the axonal initial segment by imaging before the experiment. We observe that opsins referred to as soma-targeted in mammals can be expressed widely in the neuropils (Jia and Wyart, *unpublished observations*).
14. Caution when the output power of the stimulation laser is too high, the overall distribution of the laser intensity at the sample increases accordingly. Be careful that in this case the "stray light" within the FOE can be sufficient to activate the opsin to elicit behavioral response.

Acknowledgments

The authors would like to acknowledge Martin Carbó-Tano for the development of the behavioral monitoring protocol including the 3D printed claw and the enucleation protocol that leads to Fig. 2. in this chapter and Mathilde Lapoix for her advice in the implementation of the behaviour protocol and the members of the Wyart Lab for feedback and discussion. We also thank Dimitrii Tanese, Nelson Rebola and Joana Guedes for proofreading of the manuscript. We thank Ben Shababo and Osnath Assayag from Intelligent Imaging for the technical support on using Phasor. This project has received funding from the European Union's Horizon 2020 research and innovation program under the Marie Skłodowska-Curie grant agreement #813457 as well as Fondation pour la Recherche Médicale (FRM, grant #EQU202003010612) and the Fondation Bettencourt-Schueller (FBS, grant #Don0063).

References

1. Grillner S, El Manira A (2020) Current principles of motor control, with special reference to vertebrate locomotion. Physiol Rev 100:271–320. https://doi.org/10.1152/physrev.00015.2019
2. Grätsch S, Büschges A, Dubuc R (2019) Descending control of locomotor circuits. Curr Opin Physio 8:94–98. https://doi.org/10.1016/j.cophys.2019.01.007
3. Wyart C, Del Bene F (2011) Let there be light: zebrafish neurobiology and the optogenetic revolution. Rev Neurosci 22:121. https://doi.org/10.1515/rns.2011.013
4. Boyden ES, Zhang F, Bamberg E, Nagel G, Deisseroth K (2005) Millisecond-timescale, genetically targeted optical control of neural activity. Nat Neurosci 8:1263–1268. https://doi.org/10.1038/nn1525

5. Deisseroth K (2015) Optogenetics: 10 years of microbial opsins in neuroscience. Nat Neurosci 18:1213–1225. https://doi.org/10.1038/nn.4091
6. Kawakami K, Patton EE, Orger M (2016) Zebrafish: methods and protocols. Springer, New York
7. Arrenberg AB, Del Bene F, Baier H (2009) Optical control of zebrafish behaviour with halorhodopsin. Proc Natl Acad Sci 106: 17968–17973. https://doi.org/10.1073/pnas.0906252106
8. Douglass AD, Kraves S, Deisseroth K, Schier AF, Engert F (2008) Escape behaviour elicited by single, channelrhodopsin-2-evoked spikes in zebrafish somatosensory neurons. Curr Biol 18:1133–1137. https://doi.org/10.1016/j.cub.2008.06.077
9. Wyart C, Bene FD, Warp E, Scott EK, Trauner D, Baier H, Isacoff EY (2009) Optogenetic dissection of a behavioural module in the vertebrate spinal cord. Nature 461:407–410. https://doi.org/10.1038/nature08323
10. Warp E, Agarwal G, Wyart C, Friedmann D, Oldfield CS, Conner A, Del Bene F, Arrenberg AB, Baier H, Isacoff EY (2012) Emergence of patterned activity in the developing zebrafish spinal cord. Curr Biol 22:93–102. https://doi.org/10.1016/j.cub.2011.12.002
11. Hubbard JM, Böhm UL, Prendergast A, Tseng P-EB, Newman M, Stokes C, Wyart C (2016) Intraspinal sensory neurons provide powerful inhibition to motor circuits ensuring postural control during locomotion. Curr Biol 26: 2841–2853. https://doi.org/10.1016/j.cub.2016.08.026
12. Rickgauer JP, Tank DW (2009) Two-photon excitation of channelrhodopsin-2 at saturation. Proc Natl Acad Sci 106:15025–15030. https://doi.org/10.1073/pnas.0907084106
13. Ronzitti E, Ventalon C, Canepari M, Forget BC, Papagiakoumou E, Emiliani V (2017) Recent advances in patterned photostimulation for optogenetics. J Opt 19:113001. https://doi.org/10.1088/2040-8986/aa8299
14. Papagiakoumou E, Ronzitti E, Emiliani V (2020) Scanless two-photon excitation with temporal focusing. Nat Methods 17:571–581. https://doi.org/10.1038/s41592-020-0795-y
15. Accanto N, Molinier C, Tanese D, Ronzitti E, Newman ZL, Wyart C, Isacoff E, Papagiakoumou E, Emiliani V (2018) Multiplexed temporally focused light shaping for high-resolution multi-cell targeting. Optica 5: 1478. https://doi.org/10.1364/OPTICA.5.001478
16. Chen I-W, Papagiakoumou E, Emiliani V (2018) Towards circuit optogenetics. Curr Opin Neurobiol 50:179–189. https://doi.org/10.1016/j.conb.2018.03.008
17. Lutz C, Otis TS, DeSars V, Charpak S, DiGregorio DA, Emiliani V (2008) Holographic photolysis of caged neurotransmitters. Nat Methods 5:821–827. https://doi.org/10.1038/nmeth.1241
18. Stroh A (2018) Optogenetics: a roadmap. Springer, New York
19. Antinucci P, Dumitrescu A, Deleuze C, Morley HJ, Leung K, Hagley T, Kubo F, Baier H, Bianco IH, Wyart C (2020) A calibrated optogenetic toolbox of stable zebrafish opsin lines. eLife 9:e54937. https://doi.org/10.7554/eLife.54937
20. Vladimirov N, Wang C, Höckendorf B, Pujala A, Tanimoto M, Mu Y, Yang C-T, Wittenbach JD, Freeman J, Preibisch S, Koyama M, Keller PJ, Ahrens MB (2018) Brain-wide circuit interrogation at the cellular level guided by online analysis of neuronal function. Nat Methods 15:1117–1125. https://doi.org/10.1038/s41592-018-0221-x
21. Scott EK (2009) The Gal4/UAS toolbox in zebrafish: new approaches for defining behavioural circuits. J Neurochem 110:441–456. https://doi.org/10.1111/j.1471-4159.2009.06161.x
22. Goll MG, Anderson R, Stainier DYR, Spradling AC, Halpern ME (2009) Transcriptional silencing and reactivation in transgenic zebrafish. Genetics 182:747–755. https://doi.org/10.1534/genetics.109.102079
23. Goll MG, Halpern ME (2011) DNA Methylation in Zebrafish. In: Progress in molecular biology and translational science. Elsevier, pp 193–218
24. Hernandez O, Papagiakoumou E, Tanese D, Fidelin K, Wyart C, Emiliani V (2016) Three-dimensional spatiotemporal focusing of holographic patterns. Nat Commun 7:11928. https://doi.org/10.1038/ncomms11928
25. Friedman JM (2021) How the discovery of microbial opsins led to the development of optogenetics. Cell 184:5687–5689. https://doi.org/10.1016/j.cell.2021.10.008
26. Sridharan S, Gajowa M, Ogando MB, Jagadisan U, Abdeladim L, Sadahiro M, Bounds H, Hendricks WD, Tayler I, Gopakumar K, Antón Oldenburg I, Brohawn SG, Adesnik H (2021) High performance microbial opsins for spatially and temporally precise perturbations of large neuronal networks. Neuroscience 110(7):1139–1155

27. Drobizhev M, Makarov NS, Tillo SE, Hughes TE, Rebane A (2011) Two-photon absorption properties of fluorescent proteins. Nat Methods 8:393–399. https://doi.org/10.1038/nmeth.1596
28. Chen I-W, Ronzitti E, Lee BR, Daigle TL, Dalkara D, Zeng H, Emiliani V, Papagiakoumou E (2019) *In vivo* sub-millisecond two--photon optogenetics with temporally focused patterned light. J Neurosci:1785–1718. https://doi.org/10.1523/JNEUROSCI.1785-18.2018
29. Hernandez O, Guillon M, Papagiakoumou E, Emiliani V (2014) Zero-order suppression for two-photon holographic excitation. Opt Lett 39:5953. https://doi.org/10.1364/OL.39.005953
30. McRaven C, Zhang L, Yang C-T, Ahrens MB, Emiliani V, Koyama M (2020) High-throughput cellular-resolution synaptic connectivity mapping in vivo with concurrent two-photon optogenetics and volumetric Ca2 imaging. Neuroscience. https://doi.org/10.1101/2020.02.21.959650
31. Mardinly AR, Oldenburg IA, Pégard NC, Sridharan S, Lyall EH, Chesnov K, Brohawn SG, Waller L, Adesnik H (2018) Precise multimodal optical control of neural ensemble activity. Nat Neurosci 21:881–893. https://doi.org/10.1038/s41593-018-0139-8
32. Ronzitti E, Conti R, Zampini V, Tanese D, Foust AJ, Klapoetke N, Boyden ES, Papagiakoumou E, Emiliani V (2017) Submillisecond optogenetic control of neuronal firing with two-photon holographic photoactivation of chronos. J Neurosci 37:10679–10689. https://doi.org/10.1523/JNEUROSCI.1246-17.2017
33. Förster D, Dal Maschio M, Laurell E, Baier H (2017) An optogenetic toolbox for unbiased discovery of functionally connected cells in neural circuits. Nat Commun 8:116. https://doi.org/10.1038/s41467-017-00160-z
34. Frøland Steindal IA, Beale AD, Yamamoto Y, Whitmore D (2018) Development of the Astyanax mexicanus circadian clock and non-visual light responses. Dev Biol 441:345–354. https://doi.org/10.1016/j.ydbio.2018.06.008
35. Friedmann D, Hoagland A, Berlin S, Isacoff EY (2015) A spinal opsin controls early neural activity and drives a behavioural light response. Curr Biol 25:69–74. https://doi.org/10.1016/j.cub.2014.10.055
36. Shemesh OA, Tanese D, Zampini V, Linghu C, Piatkevich K, Ronzitti E, Papagiakoumou E, Boyden ES, Emiliani V (2017) Temporally precise single-cell-resolution optogenetics. Nat Neurosci 20:1796–1806. https://doi.org/10.1038/s41593-017-0018-8

Chapter 9

Brain Imaging and Registration in Larval Zebrafish

Ashwin A. Bhandiwad, Tripti Gupta, Abhignya Subedi, Victoria Heigh, George A. Holmes, and Harold A. Burgess

Abstract

Registration of larval zebrafish brain scans to a common reference brain enables comparison of transgene and gene expression patterns, neuroanatomy, and morphometry. Here we describe methods for staining and mounting larval zebrafish to facilitate whole-brain fluorescence imaging. Following image acquisition, we provide a template for aligning brain images to a reference atlas using nonlinear registration with the ANTs software package.

Key words Brain imaging, Zebrafish, Confocal, Registration, ANTs, LysoTracker

1 Introduction

Three-dimensional image registration is widely used in human brain imaging studies so that samples can easily be compared or subjected to quantitative statistical analysis. Recently, brain registration methods have also been used in zebrafish, where the entire larval brain can be imaged at cellular resolution using standard confocal or light-sheet microscopy. Following acquisition, images may be registered to a digital brain atlas, allowing samples to be automatically annotated with neuroanatomical information, compared to results from previous experiments, or used for morphometric studies. Brain registration accuracy in zebrafish is primarily limited by intrinsic biological variability, resulting in alignment errors that are approximately the diameter of a single neuron [1]. Registration algorithms require one or more reference brain images generated from fluorescent markers that are present in the target brain atlas. Suitable references include transgenic lines with broad expression of a fluorescent protein (e.g., *Tg(vGlut:dsRed)*), immunohistochemical stains against common proteins (e.g.,

Ashwin A. Bhandiwad and Tripti Gupta contributed equally to this work

James F. Amatruda et al. (eds.), *Zebrafish: Methods and Protocols*, Methods in Molecular Biology, vol. 2707,
https://doi.org/10.1007/978-1-0716-3401-1_9,

tERK), or fluorescent dyes (e.g., LysoTracker). Images acquired from either live or fixed samples may be used; in fact, when fixed samples are registered to a live brain reference, distortions introduced by fixation may be corrected by local elastic alignment [1]. Several open-source packages have been successfully used to align zebrafish brain images to a reference, including CMTK [2, 3], minctracc [4], and Advanced Normalization Tools (ANTs) [1, 5]. These algorithms each first apply linear transformations (such as rotation and scaling) to loosely align images to the reference brain, then achieve a precise match through local nonlinear deformations. Here we describe a protocol for whole-brain imaging of larval zebrafish brains, followed by computational registration to a brain atlas using Advanced Normalization Tools (ANTs).

2 Materials

Sample files and other resources referenced below may be downloaded from the Zenodo Registration Utility archive (https://doi.org/10.5281/zenodo.5847085) associated with this chapter.

2.1 Reagents and Equipment for Mounting and Imaging Embryos

1. E3 embryo medium (5 mM NaCl, 0.17 mM KCl, 0.33 mM $CaCl_2$, 0.33 mM $MgSO_4$ in ddH_2O).
2. N-Phenylthiourea (PTU, Sigma, Catalog #222909), prepared as 300 μM solution in E3 embryo medium.
3. Tricaine methanesulfonate (Sigma, Catalog #E10521). Stock solution prepared as 0.2% solution in 1 L MilliQ H_2O, pH 7.2.
4. Low melting temperature agarose: NuSieve™ GTG™ Agarose (Lonza, Catalog #50080), prepared as 2.5% solution in H_2O.
5. Thermo Scientific Lab-Tek II Chambered Coverglass, 2-well (Fisher, Catalog #12565336).
6. 3D-printed chambered coverglass insert (Zenodo Registration Utility archive: **mounting_insert.stl**).
7. Laser scanning confocal microscope with 488, 561, and 633 laser lines, 20× or 25× water objective, and software with tile scanning ability.
8. Fiji (i.e., ImageJ) software (https://fiji.sc/) for image processing. To load and save images in the NIfTI format, download and install the NIfTI Input/Output plugin from https://imagej.nih.gov/ij/plugins/nifti.html.
9. If staining brains for live imaging, LysoTracker™ Deep Red (Invitrogen, Catalog #L12492).

2.2 Brain Registration

1. Hardware requirements. ANTs requires around 12 Gb of memory to register a 2 × 2 × 2 μm resolution image. For 1 μm isotropic resolution, around 80 Gb memory is needed.
2. Download and install ANTs (http://stnava.github.io/ANTs/). ANTs can be run on several operating systems. The instructions below are for a linux system (*see* **Note 1**).
3. Reference brain image. Several reference brains are available in the Zenodo Reference Brain archive: https://doi.org/10.5281/zenodo.3367708
4. Digital brain atlas file (Zenodo Registration Utility archive: **zbb-region-atlas.nii.gz**) for brain region annotation.

3 Methods

3.1 Select a Registration Channel

1. Select a transgene, dye, or antibody stain that will be used to drive the brain registration, based on your experimental design and the spectral lines that are available on your microscope. Recommended registration channels are described below and summarized in Table 1. Live imaging is preferred where possible, as tissue fixation distorts brain structure, although these changes can be partially corrected by the brain registration procedure.
2. For imaging the expression pattern in a transgenic line, cross the line to *vGlut2a:GFP*, *vGlut2a:DsRed*, or *tuba:mCardinal* (*see* **Note 2**), as these expression patterns can be directly aligned to the **vglut-ref-01** and **tuba-mcar-ref-01** reference

Table 1
Suggested fluorescent labels and matching reference brains for common experimental conditions

Experimental condition	Fluorescent label in sample	Reference brain for registration
Live imaging red fluorescence	*Tg(vGlut2a:DsRed)nns14*	vglut-ref-01.nii.gz[a]
Live imaging far red fluorescence	*Tg(tuba:mCardinal)y516*	tuba-mcar-ref-01.nii.gz[a]
Live imaging green fluorescence	*Tg(vGlut2a:GFP)nns14*	vglut-ref-01.nii.gz[a]
Whole-brain calcium imaging	*Tg(elavl3:GCaMP)*[b]	huc-gcamp5g-ref-01.nii.gz
Structural brain imaging	LysoTracker vital dye	lyso-ref-01.nii.gz[a]
Immunofluorescence	anti-tERK	terk-ants-ref-02.nii.gz
Hybridization chain reaction in situ	*Tg(vGlut2a:GFP)nns14*	vglut-ref-01.nii.gz[a]
Hybridization chain reaction in situ	HCR using *elavl3* probe	HCR-huc-ref-01.nii.gz

[a]Reference brains where a downsampled 2 × 2 × 2 μm reference brain is also available in the Zenodo Reference Brain archive (indicated by "R2" in the cognate filename)
[b]Various *elavl3:GCaMP* lines are available and have sufficiently similar patterns to match the suggested reference brain

brain images, respectively. Whole-brain GCaMP experiments can be registered to the **huc-gcamp5g-ref-01** reference brain using the GCaMP expression pattern.

3. If transgenic lines are not available or are inconvenient (e.g., when performing morphometric analysis of brain structure in mutant animals), live imaging can be performed using LysoTracker vital dye staining (Subheading 3.3) and registered using the **lyso-ref-01** reference brain image. Morphometry can also be achieved by labeling brain tissue with antibody staining [6].
4. For immunohistochemical markers, co-staining with antibody labeling against MAP Kinase (tERK) is recommended for registration to the **terk-ref-01** reference image.
5. For visualizing RNA expression using hybridization chain reaction (HCR) in situ labeling, we recommend performing the experiment in transgenic *vGlut2a:GFP* expressing embryos, as GFP fluorescence survives the procedure and can be aligned to the **vglut-ref-01** reference brain image. Alternatively, if this is not possible, we recommend simultaneously using a second HCR probe set against *elavl3*, which is commonly used in HCR protocols as a positive control. The **HCR-huc-ref-01** reference image should be used for registration with *elavl3* HCR staining.

3.2 Raise Larvae

1. Raise embryos at 28 °C in 300 μM PTU dissolved in E3 embryo medium until 6 days postfertilization (dpf).
2. If imaging transgenes, sort larvae for transgene expression. Larvae are easiest to sort before the swim bladder inflates at 4 dpf (*see* **Note 3**).

3.3 Label Brain Tissue with a Fluorescent Marker

1. For tERK antibody labeling or *elavl3* HCR, refer to published protocols [7, 8].
2. For LysoTracker staining [9], follow **steps 3–4**.
3. At 5 dpf, transfer larvae raised in 300 μM PTU to 300 μM PTU containing 1% DMSO and 1 μM LysoTracker.
4. Incubate 12–16 h at 28 °C in dim light, then transfer from LysoTracker solution to fresh PTU just prior to imaging.

3.4 Mount Larvae for Imaging

The steps below describe mounting larvae for imaging using an inverted confocal microscope. Larvae can also be imaged using an upright microscope (*see* **Note 4**)

1. Boil 2.5% low melt agarose and cool to 55 °C in an incubator or water bath.
2. Place 3D printed chamber inserts into cover glass chambers.

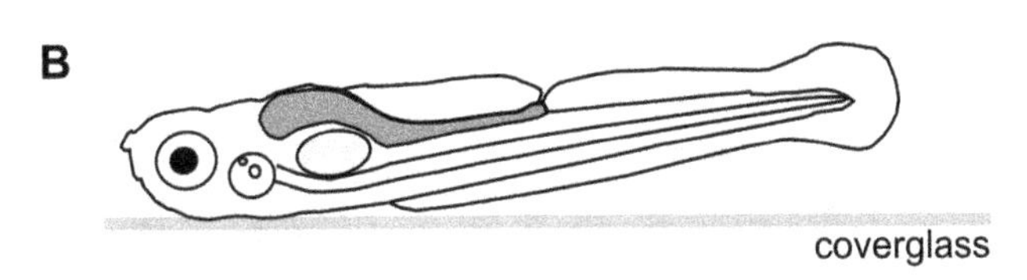

Fig. 1 Mounting larval zebrafish for confocal imaging. (**a**) Schematic of mounting 6 dpf zebrafish using 3D-printed chamber insert. (**b**) Orientation of a well-mounted zebrafish larva with dorsal surface of the head in contact with coverslip (blue) for imaging on an inverted confocal microscope

3. Anesthetize larvae in 10 mL of E3 with 0.5 mL Tricaine stock solution.
4. Draw a larva into the tip of a glass Pasteur pipette with ~5 μL E3.
5. Fill a well of the chamber insert with the low melt agarose.
6. Pipette the larva into the agarose (*see* Fig. 1a).
7. Immediately push the larva to the bottom of the chamber by applying gentle pressure at the swim bladder with a pipette tip or forceps tips. The larva should be oriented ventral side up. The dorsal surface of the head should be flush against the coverslip with the tail sticking up (*see* Fig. 1b). Hold the larva down until the agarose starts to set, making sure that the larva does not float up from the coverslip.
8. Mount remaining larvae in other chambers and cover with a few drops of E3 with tricaine.

3.5 Confocal Imaging

Any conventional confocal microscope can be used for imaging. To achieve the required resolution for registration, a 20× or 25× objective with a high numerical aperture (~0.95 or higher) and a working distance of at least 350 μm should be used. In most cases, tile scanning will be required to image the entirety of the brain from the olfactory epithelia in the forebrain to the caudal-most region of the hindbrain (*see* Fig. 2, **Note 5**).

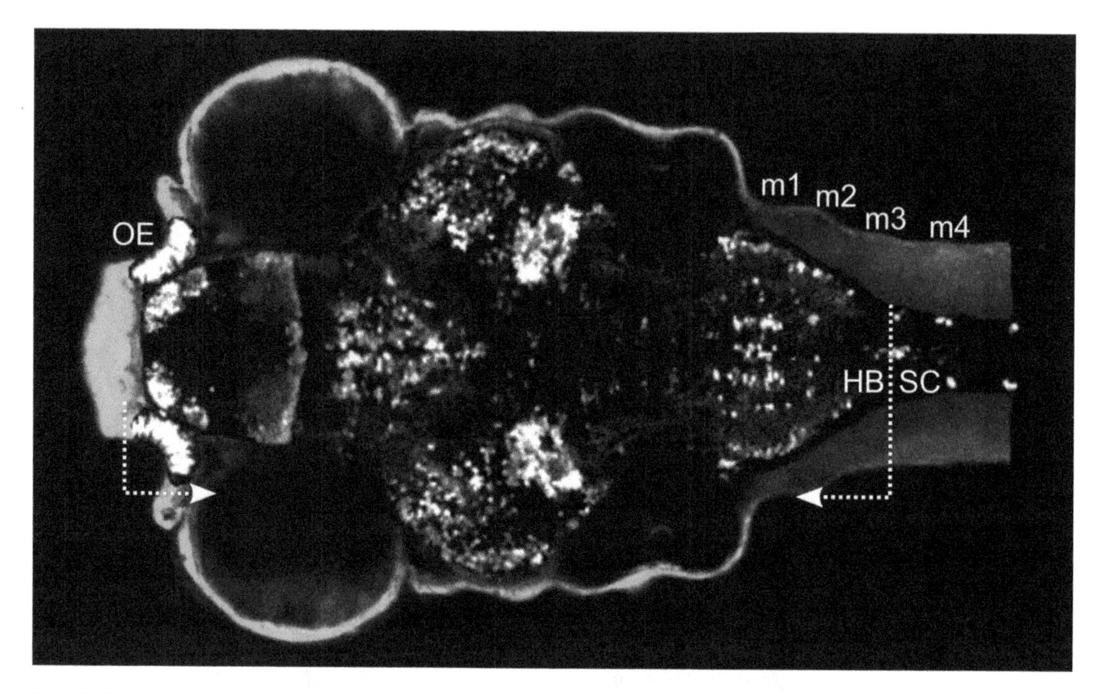

Fig. 2 Sample whole-brain image. Single slice image of *vGlut2a:DsRed* reference brain (grey) showing olfactory epithelia (OE) at anterior boundary and posterior hindbrain boundary between myomeres 3 and 4 (visualized using *bActin:GFP*). Image stacks should comprise the area between the dotted lines

1. Set up tile scanning so that the entire brain will be included in the image once tiled.
2. Set voxel size to 1 × 1 × 1 or 2 × 2 × 2 μm depending on the resolution needed for your experiment (*see* **Note 6**).
3. Make sure no image processing features are set that can affect pixel intensity or noise filtering during the acquisition.
4. Set up z-stack from the dorsal surface of the brain to the ventral hypothalamus (*see* Fig. 3). Set step size to 1 or 2 μm resolution.
5. Display saturated pixels. Set up z-compensation by scanning every 50 μm and adjusting laser power so that the brightest pixels in the image have reached saturation. You will likely have to increase the laser power for all channels as you image more ventrally.
6. Acquire image stacks for all larvae.

3.6 Post-imaging Processing

1. If image stacks were acquired using tiling, stitch tiles together into a single image stack comprising the entire brain, using either Fiji or software included with the microscope.
2. Use Fiji to rotate image stacks so that they are oriented in the same direction as the reference image. For the sample images here, they should be rotated so that the anterior end of the

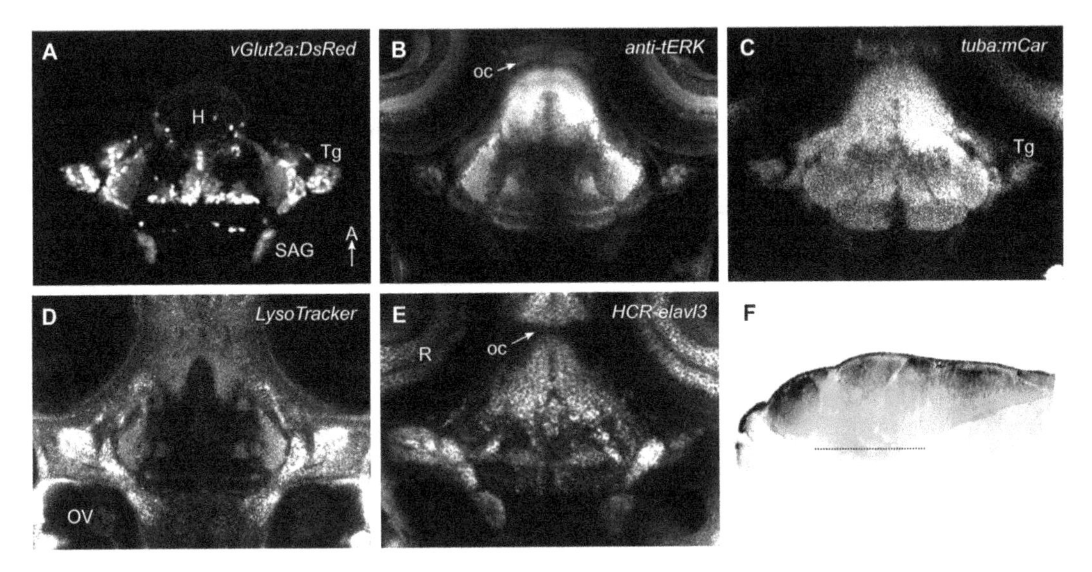

Fig. 3 Imaging depth required to resolve ventral brain structures. (**a**–**e**). Examples of single slices through confocal images of 6 dpf brains using fluorescent labels in Table 1, showing desirable image quality 280 μm from dorsal surface. (**f**) Schematic indicating depth of slices shown (dotted line). h, Hypothalamus; oc, optic chiasm; OV, otic vesicle; R, retina; SAG, statoacoustic ganglion; Tg, Trigeminal ganglion. Arrow indicates anterior

head faces left. Similarly, ensure that the image stacks have the same dorsoventral direction as the reference.

3. The image dimension units (i.e., the units of pixel length/ width and depth) of your stacks should match those of the reference brain. Typically, this will be in either mm or μm (*see* **Note 7**).
4. If using multiple fluorophores, split stacks into separate channels using Fiji (*see* **Note 8**).
5. Save stacks as 16-bit NIfTI files (*see* **Note 9**).
6. Following processing, images should match the reference brain dimension units and orientation. For processing large numbers of images, a sample Fiji macro (**stitch-3channel-image-stacks.ijm**) that performs all of these steps is available in the Zenodo Registration Utility archive.

3.7 Brain Registration

Two ANTs programs are run to generate a brain image that is registered to the reference brain: *antsRegistration* and *antsApplyTransforms*. *antsRegistration* computes and saves a transformation matrix by comparing the brain image to the reference brain. *antsApplyTransforms* applies the transformation matrix to the brain image and saves a registered image. The sample files referenced below can be downloaded from the Zenodo Registration Utility archive and used to troubleshoot the image registration process (*see* **Note 10**).

1. Save the reference brain file (**reference.nii.gz**), and the brain image files to the same folder. If the brain image (**brain1**) has multiple fluorescence channels, then name the file that has the same fluorescence marker as the reference **brain1-01.nii.gz**, and name the other channel **brain1-02.nii.gz**.
2. Run the *antsRegistration* command (*see* **Note 11**). Adapt the command from Step 1 in Table 2 substituting filenames as needed. Images acquired from fixed brain tissue typically require greater deformation to achieve a close match than is needed for live brain scans. Therefore, depending on the samples, select either the live (Step 1a) or the fixed tissue (Step 1b) command from Table 2 (see **Note 12**). In some cases you may have acquired multiple fluorescence channels that are also present in the target image. Multiple channels can be used simultaneously during registration to improve accuracy.
3. For a 2 μm isotropic resolution image, *antsRegistration* may take 1–2 h to run (*see* **Note 13**), and on completion will output three transformation matrices. If using the sample files, these will be designated **brain1_0GenericAffine.mat**, **brain1_1-Warp.nii.gz** and **brain1_1InverseWarp.nii.gz.**
4. Run the *antsApplyTransforms* command (adapt Step 2a in Table 2) to apply the transformation matrix to the first channel. This command should typically take around 1 min to complete and will save the registered brain image as **brain1-01_registered.nii** (*see* **Note 14**).
5. If your brain image has multiple fluorescent channels, you may also run the *antsApplyTransforms* command to apply the same transformation matrix to additional channels (Step 2b in Table 2). The registered second channel will be saved as **brain1-02_registered.nii**. If you have more than two channels, modify the command in Step 2b accordingly (change brain1-02 to brain1-03, brain1-04, etc.).
6. The registered brain images can now be compared with a digital brain atlas file, for example, by loading both files into Fiji and synchronously scrolling through to check annotations, or by loading the registered file into the Zebrafish Brain Browser (*see* **Note 15**) .
7. To measure the size of brain regions in the original images, it may be useful to inverse-transform an atlas file back onto the original brain scans. To do so, run the *antsApplyTransforms* command with the inverse transformation matrix (Step 3a in Table 2). When complete, the inverse-transformed atlas will be saved in your folder as **brain1_regions.nii.gz**. You may then estimate the size of a brain region by summing the number of voxels labeled with the matching index (*see* **Note 16**). Alternatively, you may inverse-transform an atlas with larger brain

Table 2
ANTs commands for brain registration using sample images provided in the Zenodo Registration Utility archive

Step	Purpose	Command
1a	Produce matrix that transforms brain scan from live brain scan to reference image	antsRegistration -d 3 --float 1 -o [brain1_] --interpolation WelchWindowedSinc --use-histogram-matching 0 -r [reference.nii.gz, brain1-01.nii.gz,1] -t rigid[0.1] -m MI [reference.nii.gz,brain1-01.nii.gz,1,32, Regular,0.25] -c [200 × 200 × 200 × 0,1e-8,10] --shrink-factors 12 × 8 × 4 × 2 --smoothing-sigmas 4 × 3 × 2 × 1vox -t Affine[0.1] -m MI [reference.nii.gz,brain1–01.nii.gz,1,32, Regular,0.25] -c [200 × 200 × 200 × 0,1e-8,10] --shrink-factors 12 × 8 × 4 × 2 --smoothing-sigmas 4 × 3 × 2 × 1vox -t SyN[0.05,6,0.5] -m CC[reference.nii.gz,brain1-01.nii.gz,1,2] -c [200 × 200 × 200 × 200 × 10,1e-7,10] --shrink-factors 12 × 8 × 4 × 2 × 1 --smoothing-sigmas 4 × 3 × 2 × 1 × 0vox
1b	Produce matrix that transforms brain scan from fixed tissue brain scan to reference image	antsRegistration -d 3 --float 1 -o [brain1_] --interpolation WelchWindowedSinc --use-histogram-matching 0 -r [reference.nii.gz, brain1-01.nii.gz,1] -t rigid[0.1] -m MI [reference.nii.gz,brain1-01.nii.gz,1,32, Regular,0.25] -c [200x200x200x0,1e-8,10] --shrink-factors 12x8x4x2 --smoothing-sigmas 4x3x2x1vox -t Affine[0.1] -m MI[reference.nii.gz,brain1-01.nii.gz,1,32,Regular,0.25] -c [200 × 200 × 200 × 0,1e-8,10] --shrink-factors 12 × 8 × 4 × 2 -- smoothing-sigmas 4 × 3 × 2 × 1vox -t SyN[0.1,6,0] -m CC [reference.nii.gz,brain1-01.nii.gz,1,2] -c [200 × 200 × 200 × 200 × 10,1e-7,10] --shrink-factors 12 × 8 × 4 × 2 × 1 --smoothing-sigmas 4 × 3 × 2 × 1 × 0vox
2a	Produce registered brain image for first channel	antsApplyTransforms -d 3 -v 0 --float -n WelchWindowedSinc -i brain1-01.nii.gz -r reference.nii.gz -o brain1-01_registered.nii.gz -t brain1_1Warp.nii.gz -t brain1_0GenericAffine.mat
2b	Produce registered brain image for second channel	antsApplyTransforms -d 3 -v 0 --float -n WelchWindowedSinc -i brain1-02.nii.gz -r reference.nii.gz -o brain1-02_registered.nii.gz -t brain1_1Warp.nii.gz -t brain1_0GenericAffine.mat
3a	Annotate original brain scan with atlas of brain regions	antsApplyTransforms -d 3 -v 0 --float -n MultiLabel -i zbb-region-atlas.nii.gz -r brain1-01.nii.gz -o brain1_regions.nii.gz -t [brain1_0GenericAffine.mat,1] -t brain1_1InverseWarp.nii.gz

(continued)

Table 2
(continued)

Step	Purpose	Command
3b	Annotate original brain scan with atlas of brain divisions	antsApplyTransforms -d 3 -v 0 --float -n MultiLabel -i zbb-division-atlas.nii.gz -r brain1-01.nii.gz -o brain1_divisions.nii.gz -t [brain1_0GenericAffine.mat,1] -t brain1_1InverseWarp.nii.gz
4	Visualize the transformation matrix	CreateJacobianDeterminantImage 3 brain1_1InverseWarp.nii.gz brain1_jac.nii.gz 1
5	Register neuron trace to reference image	antsApplyTransformsToPoints -d 3 -i brain1-neuron1.csv -o brain1-neuron1_registered_points.csv -t [brain1_0GenericAffine.mat,1] -t brain1_1InverseWarp.nii.gz

divisions (e.g., pallium, cerebellum) annotated (Step 3b in Table 2; *see* **Note 17**).

8. When comparing wild-type and mutant brain structure, it may be instructive to examine which parts of the brain were locally dilated or contracted to fit the reference brain. To create an image stack that contains this information, run the command in Step 4 of Table 2. This will produce a new file called **brain1_jac.nii.gz** in which the value of each pixel indicates if the pixel in the original image was expanded or contracted to match the reference (*see* **Note 18**).
9. Brain scans may be used to reconstruct neuronal projections. To register a set of x,y,z coordinates that describe a neuron's morphology into the coordinate system of the reference brain, first save a comma separated text file **brain1-01.csv** in the directory above. The first line (header) of this file should be "x,y,z,t,label,comment". The first three columns are the x,y,z coordinates (in millimeters, not microns); t is 0; label is the point number; comment is the point to which it connects. A sample file is available in the Zenodo Registration Utility archive. A python script that converts standard swc format files to csv files is available the same archive (**swc-to-ants.py**). Finally, run the *antsApplyTransformsToPoints* command (Step 5 in Table 2, *see* **Note 19**).
10. After registration, brain images may be compared to determine if genes or transgenes express in similarly located cells (*see* **Note 20**).

4 Notes

1. We recommend installing ANTs using the script **installANTs.sh**, which can be downloaded from https://github.com/cookpa/antsInstallExample. You will need *cmake* and *git* already installed on your system.
2. *vGlut2a:DsRed* is *TgBAC(slc17a6b[vglut2a]:loxP-DsRed-loxP-GFP)nns14)* [10]. This line can be readily converted to a stable GFP expressing line (referred to here as *vGlut2a:GFP*) by injection of Cre mRNA. *tuba:mCar* is *Tg(Cau.Tuba1a:MCardinal)y516* [11], which is available from the Zebrafish International Resource Center (http://zebrafish.org).
3. *vglut2a:GFP/DsRed* and *tuba:mCardinal* transgenes begin expressing by 1 dpf and can be conveniently sorted at this stage.
4. For imaging using an upright microscope, larvae need to be mounted dorsal side up. Dorsal mounting can be facilitated using a specialized mold [12], or by mounting ventral side up, then removing the block of agarose containing the larva from the mounting insert, flipping it over, and replacing it in the mounting insert.
5. The caudal boundary of the hindbrain aligns with the end of the third myomere but otherwise does not have a salient morphological feature [13].
6. We suggest using 1 μm resolution for imaging transgene expression patterns, tERK staining, HCR, and reconstructing neuronal projections. Brain morphometry may be performed at 2 μm resolution, as in any case, images will likely be smoothed and downsampled to facilitate downstream computation. If using 2 μm resolution images, then it will be most efficient to register images to the 2 μm resolution reference brains in the Zenodo Reference Brain archive, which have the "R2" designation in the filename.
7. Set image dimension units to microns in Fiji by entering the word "micron" in the image property "Unit of length" field.
8. If using simultaneous acquisition for multiple channels, you may need to perform spectral unmixing to eliminate bleed-through. This can be done using confocal-specific software or Fiji.
9. Although these are 16-bit files, do not include intensity values greater than 32,767 as this will result in negative pixel values in the output image. If necessary, use Fiji to rescale intensity values to remain below this limit. The sample files are gzipped NIfTI files (.nii.gz); however, compressing input files is not essential.

10. The sample reference brain image (**reference.nii.gz**) is the **vglut-ref-R2-01** reference brain. In the brain that will be registered to this reference, channel 1 (**brain1-01.nii.gz**) is DsRed fluorescence from the *vglut2a:DsRed* transgene, and channel 2 (**brain1-02.nii.gz**) is Kaede fluorescence from *Et (SCP1:GAL4FF)y265, Tg(UAS-E1b:Kaede)s1999t* [14].
11. Before running, you may limit the number of threads that ANTs uses with the command *export ITK_GLOBAL_DEFAULT_NUMBER_OF_THREADS = 8* (or other suitable value). This is especially important on a cluster or shared system to avoid overusing resources.
12. The major difference between the live and fixed parameter sets is that the fixed parameter set allows greater local elastic registration. While this is useful for correcting fixation artifacts, it can also introduce potentially severe local distortions into the sample image, for example, causing axon tracts to appear bent, or cell bodies to elongate.
13. Using 12 threads, on Intel Xeon 6140 processors.
14. After registration, the image (in this case **brain1-01_registered.nii**) should have the same width, height, and depth (in pixels) as the reference brain. However, it will be a 32-bit image and the dimensions will be specified in millimeters instead of microns.
15. To load stacks into the web-based version of the Zebrafish Brain Browser (ZBB), stacks should be converted to 2D montage files, as described at http://vis.arc.vt.edu/projects/zbb/help/. NIfTI files can be directly loaded into the desktop version of ZBB.
16. After transforming the ZBB atlas back onto the original brain scans, the output file should have the same dimensions (in x,y, z) as the scan. The boundaries of the inverse-transformed atlas should closely match the contour of the brain in the image stack.
17. The larger brain divisions in **zbb-division-atlas.nii.gz** are agglomerations of the small computationally derived regions in the **zbb-region-atlas.nii.gz** file.
18. With these parameters, this command calculates the log of the Jacobian determinant of the image. The output file has the same dimensions as the reference brain: pixels that were inflated to match the reference brain have positive values, and pixels that were compressed to match the reference have negative values.
19. The dimension units in the output file will be in mm, even if all the input files and reference are in microns.

20. Due to registration errors and biological variability, essential conclusions about colocalization should be empirically confirmed using multiple probes in the same sample.

Acknowledgments

This work was supported by the Intramural Research Program of the *Eunice Kennedy Shriver* National Institute for Child Health and Human Development (NICHD) and utilized the high-performance computational capabilities of the Biowulf Linux cluster at the National Institutes of Health, Bethesda, MD.

References

1. Marquart GD, Tabor KM, Horstick EJ et al (2017) High precision registration between zebrafish brain atlases using symmetric diffeomorphic normalization. Gigascience 6. https://doi.org/10.1093/gigascience/gix056
2. Rohlfing T, Maurer CR Jr (2003) Nonrigid image registration in shared-memory multiprocessor environments with application to brains, breasts, and bees. IEEE Trans Inf Technol Biomed 7:16–25
3. Portugues R, Feierstein CE, Engert F, Orger MB (2014) Whole-brain activity maps reveal stereotyped, distributed networks for visuomotor behavior. Neuron 81:1328–1343. https://doi.org/10.1016/j.neuron.2014.01.019
4. Kenney JW, Steadman PE, Young O et al (2021) A 3D adult zebrafish brain atlas (AZBA) for the digital age. elife 10:e69988. https://doi.org/10.7554/eLife.69988
5. Avants BB, Epstein CL, Grossman M, Gee JC (2008) Symmetric diffeomorphic image registration with cross-correlation: evaluating automated labeling of elderly and neurodegenerative brain. Med Image Anal 12: 26–41. https://doi.org/10.1016/j.media.2007.06.004
6. Heffer A, Marquart GD, Aquilina-Beck A et al (2017) Generation and characterization of Kctd15 mutations in zebrafish. PLoS One 12: e0189162. https://doi.org/10.1371/journal.pone.0189162
7. Randlett O, Wee CL, Naumann EA et al (2015) Whole-brain activity mapping onto a zebrafish brain atlas. Nat Methods 12:1039–1046
8. Choi HMT, Schwarzkopf M, Fornace ME et al (2018) Third-generation in situ hybridization chain reaction: multiplexed, quantitative, sensitive, versatile, robust. Development 145: dev165753. https://doi.org/10.1242/dev.165753
9. Trivellin G, Tirosh A, Hernández-Ramírez LC et al (2021) The X-linked acrogigantism-associated gene gpr101 is a regulator of early embryonic development and growth in zebrafish. Mol Cell Endocrinol 520:111091. https://doi.org/10.1016/j.mce.2020.111091
10. Satou C, Kimura Y, Higashijima S (2012) Generation of multiple classes of V0 neurons in zebrafish spinal cord: progenitor heterogeneity and temporal control of neuronal diversity. J Neurosci 32:1771–1783. https://doi.org/10.1523/JNEUROSCI.5500-11.2012
11. Gupta T, Marquart GD, Horstick EJ et al (2018) Morphometric analysis and neuroanatomical mapping of the zebrafish brain. Methods 150:49–62. https://doi.org/10.1016/j.ymeth.2018.06.008
12. Geng Y, Peterson RT (2021) Rapid mounting of zebrafish larvae for brain imaging. Zebrafish 18:376–379. https://doi.org/10.1089/zeb.2021.0062
13. Ma L-H, Gilland E, Bass AH, Baker R (2010) Ancestry of motor innervation to pectoral fin and forelimb. Nat Commun 1:49. https://doi.org/10.1038/ncomms1045
14. Marquart GD, Tabor KM, Brown M et al (2015) A 3D searchable database of transgenic zebrafish Gal4 and Cre Lines for functional neuroanatomy studies. Front Neural Circ 9: 78. https://doi.org/10.3389/fncir.2015.00078

Chapter 10

Simultaneous Behavioral and Neuronal Imaging by Tracking Microscopy

Drew N. Robson and Jennifer M. Li

Abstract

Tracking microscopy enables whole-brain cellular resolution imaging in freely swimming animals. This technique enables both structural and functional imaging without immobilizing the animal, and greatly expands the range of the behaviors accessible to neuroscientists. We use infrared imaging to track the target animal in a behavioral arena. Based on the predicted trajectory of the brain, we apply optimal control theory to a motorized stage system to cancel brain motion in three dimensions. We have combined this motion cancellation system with Differential Illumination Focal Filtering (DIFF), a form of structured illumination microscopy, which enables us to image the brain of a freely swimming larval zebrafish for over an hour. Here we describe the typical experimental procedure for data acquisition and processing using the tracking microscope.

Key words Tracking microscopy, neuronal imaging, animal behavior, whole brain activity, applied control theory, Differential Illumination Focal Filtering

1 Introduction

Among vertebrate model systems, the larval zebrafish is an ideal system for brain-wide cellular resolution calcium imaging due to its small nervous system and high optical transparency [1–4]. However, imaging methods that can achieve cellular resolution whole-brain neural imaging typically require this animal to be tethered under a microscope, which not only inhibits spontaneous movement [4, 5], but also restricts the range of behaviors the animal can perform.

Tracking microscopes are imaging systems designed to keep the brain of a moving animal within the Field of View (FOV) of a microscope, allowing for continuous neural imaging with both high spatial and temporal resolution. Tracking microscopes have

---

Drew N. Robson and Jennifer M. Li contributed equally to this work

James F. Amatruda et al. (eds.), *Zebrafish: Methods and Protocols*, Methods in Molecular Biology, vol. 2707, https://doi.org/10.1007/978-1-0716-3401-1_10,

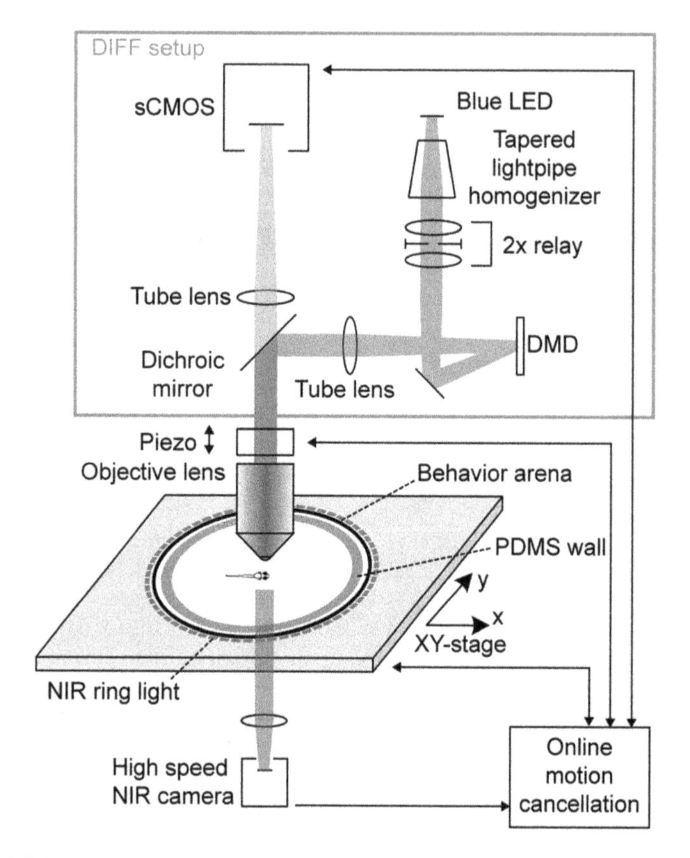

Fig. 1 Schematic of the tracking microscope. The microscope consists of a high-speed NIR camera for behavioral imaging, a motion cancellation system, and a DIFF microscope for neural imaging. As a larval zebrafish swims within an enclosed behavioral arena, brain motion is tracked and cancelled in XY by a near-infrared (NIR) camera and motorized stages using optimal control theory. A DIFF fluorescence microscope optically sections the brain, while a piezo Z stage sweeps the objective for volumetric imaging. A live Z registration pipeline analyzes each structured fluorescence brain image immediately after collection to implement closed loop axial scanning

been successfully implemented for *Escherichia coli* [6], *Caenorhabditis elegans* [7–10], *Drosophila* [7, 11], and zebrafish [4, 12]. For fast-moving animals such as zebrafish, two criteria for a tracking microscope are essential. First, tracking and motion cancellation must be performed at high speed. Second, fluorescence image capture must be robust to variable animal posture, as well as motion blur induced by high-velocity movement.

Tracking microscopy in zebrafish combines a motorized stage system that utilizes optimal control theory [13] to cancel brain motion in three dimensions (*see* Fig. 1) with a variant of HiLo microscopy [14] which we call Differential Illumination Focal Filtering (DIFF). DIFF microscopy performs cellular resolution imaging at 100 optical sections/s and two brain volumes/s. Over

98% of the time, a majority of the brain is within the microscope's FOV, with an average tracking error of <50 μm.

In addition to DIFF, the high-speed motion cancellation system can be easily combined with other fluorescent microscopy techniques such as selective plane illumination and light field. The experimental procedure for data acquisition and processing can be easily adapted for each fluorescent imaging strategy.

We describe below the experimental procedure for data acquisition and post-processing.

2 Materials

For the imaging, the user should prepare (1) a gas permeable behavior chamber with a freely swimming larval zebrafish, and (2) a thin fluorescent sheet for calibration of the DIFF microscope. Preparations can be done at room temperature.

2.1 Materials for Chamber Assembly

1. Glass bottom (#076-015P; GM Associates): 100 mm diameter, 0.381 mm thick. Dimensions are adjustable by the user.
2. The chamber walls are molded polydimethylsiloxane (PDMS; Sylgard 184, Dow Corning), which is gas-permeable for specimen health and optically transparent for side illumination.
3. Top glass is a standard #1 coverslip (#260462; Ted Pella): 100 mm in diameter.
4. E3 buffer solution [15].

2.2 Material for Thin Fluorescent Sheet

A sub-μm thin fluorescent sheet is used to fit the phase gratings for DIFF and calibrate brightness variations across the field of view. A 0.01 μm thin fluorescent sheet can be generated by spin-coating a clean coverslip with a film of fluorescent $CsPbBr_3$ perovskite nanocrystals. Alternative methods can be determined by the user.

3 Methods

3.1 Chamber Assembly

1. Clean the surface of the top and bottom glass with 70% isopropanol, then rinse with distilled water. Avoid touching the cleaned surfaces.
2. Prepare PDMS walls. These can be manually cut from preformed PDMS sheets. Alternatively, freshly prepared PDMS can be poured into a 3D mold and cured.
   (a) The PDMS contact surface with the top and bottom glass must be smooth for proper adhesion.
   (b) The PDMS inner walls facing the animal must be sloped at a 45° angle. The sloped walls can be created by PDMS

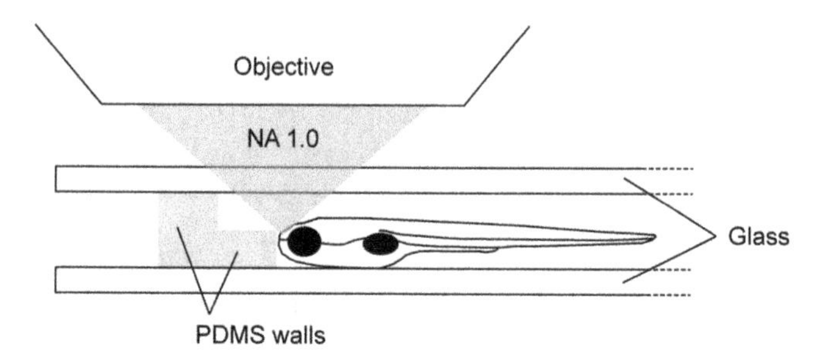

Fig. 2 Design of behavior chamber wall with terraced layers of PDMS. The terraced design allows fluorescent light from the brain to reach the imaging objective (NA = 1.0) without obstruction or aberration

molding or through a terraced design using PDMS sheets of differing thicknesses (*see* Fig. 2).

(i) The angle of the wall is determined by the NA of the fluorescence objective. A 90-degree chamber wall perturbs the optical path length of a substantial portion of the emitted fluorescent light from the brain that is collected by a high-NA objective. This degrades both the brightness and resolution of the fluorescent image. A sloped chamber wall resolves this problem, and is thus critical for any fluorescent method using a high-NA objective with a free-swimming animal.

3. Clean the surface of PDMS with 70% isopropanol, then rise with distilled water. Avoid touching the cleaned surfaces.
4. Attach the PDMS to the bottom glass. For complete adhesion, perform the procedure in a clean environment with a plasma cleaner.
5. Fill the chamber with E3 buffer.
6. Add larval zebrafish to chamber with a pipette.
7. Seal the chamber with the top cover glass.

3.2 Collection of Calibration Data with Thin Fluorescent Sheet

1. Place thin fluorescent sheet (*see* Subheading 2.2) under the DIFF microscope.
   (a) Ensure that the fluorescent sheet covers the entire field of view (FOV) of the microscope.
   (b) Ensure that the fluorescent sheet is uniform over the entire FOV of the microscope.
2. Using DIFF imaging, collect and save an image stack with 1 um spacing that encompasses the thin fluorescent sheet.

3.3 Collection of Reference Brain Stack for Online and Offline Registration

1. Placed the assembled chamber with larval zebrafish under the tracking microscope.
2. Enable tracking with the NIR behavioral camera (*see* **Note 1**).
3. Load structured illumination patterns onto the Digital Mirror Device (DMD).
   (a) Configure the DMD to generate two complementary vertical grating patterns. The patterns are strictly complementary, that is, each on pixel in the A pattern is off in the B pattern and vice versa. Grating period should be determined by the user based on the desired axial imaging resolution.
4. Turn on fluorescent LED for DIFF imaging.
   (a) Total illumination time should be minimized (e.g., < 200 μs) to prevent blur during motion (*see* Fig. 3).
   (b) For each complementary image pair, time between LED pulses should be kept to a minimum (e.g., 60 μs to allow for charge transfer time of global shutter image sensor).
5. Collect initial fluorescent brain volumes across a 400 μm axial extent.
   (a) Imaging should continue until a full sweep of the brain is collected while the fish is stationary.
   (b) Examine the collected imaging and behavioral data in real time to determine whether a satisfactory reference stack has been saved.
6. Identify a satisfactory reference sweep of the brain. In software, indicate the top, bottom, and center of the brain in three dimensions. This reference sweep will the used for two purposes:

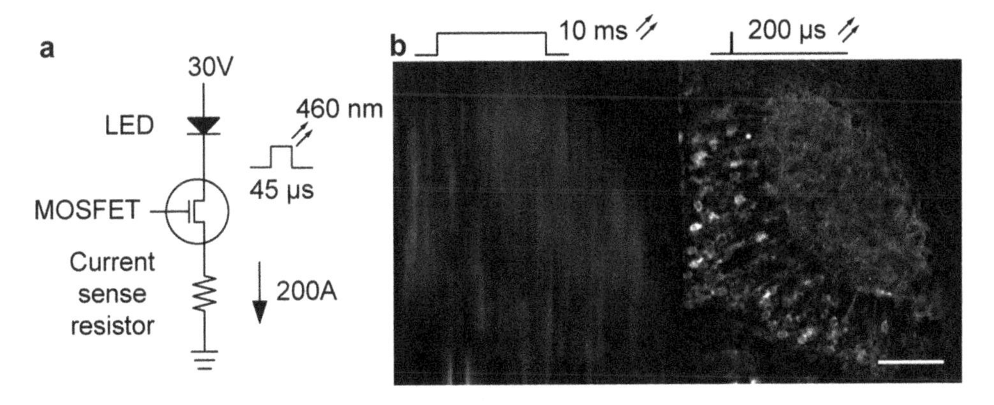

Fig. 3 Pulsed illumination is needed to minimize motion blur. To minimize motion blur, the fluorescence illumination is pulsed for 45 μs per image using a pulse-gated high-current drive circuit

(a) Closed-loop axial motion cancellation during experimental data collection.

(b) Post-imaging registration to recover cellular-resolution neural signals.

3.4 Data Acquisition During Experiment

For technical explanation of DIFF imaging, see **Note 2**.

1. Once a reference brain volume has been defined, load the reference brain volume into the imaging software.
2. Enable closed-loop axial motion cancellation. This will adjust the axial position of the objective to be centered on the brain the animal (*see* Fig. 4, **Note 3** for detailed explanation of the axial motion cancellation process).
3. Begin experimental data collection.
4. The NIR tracking system stably locks onto the fish for duration of the experiment, as determined by the user.

3.5 DIFF Solve Post-imaging

1. Use imaging data from thin fluorescent sheet (*see* Subheading 3.2) to fit two complementary DIFF grating patterns (A and B). These two grating patterns will be used in the DIFF algorithm to obtain optically sectioned images from the thin fluorescent calibration sheet, the reference brain volume, and each brain volume collected during the experiment.
   (a) For a detailed description of the DIFF solver, see Fig. 5 and **Note 2**.
2. DIFF-sectioned fluorescent images from the thin fluorescent calibration sheet will be used to correct for brightness variations across the FOV.

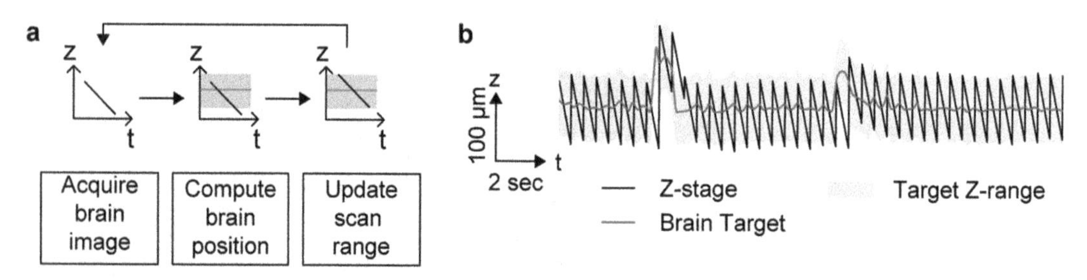

Fig. 4 Schematic of axial motion cancellation. (**a**) During each piezo Z sweep (black line), each structured fluorescence image is analyzed in real time to compute an estimate for its axial location (red line) within the targeted brain volume (red shaded region). Near the end of each sweep, the estimated axial brain position is used to adjust the range of the next piezo Z sweep to recenter the brain. (**b**) Example of closed-loop live Z tracking. As the targeted brain volume shifts axially (red line and shaded region), the piezo Z sweeps are adjusted to center the brain within the next axial sweep

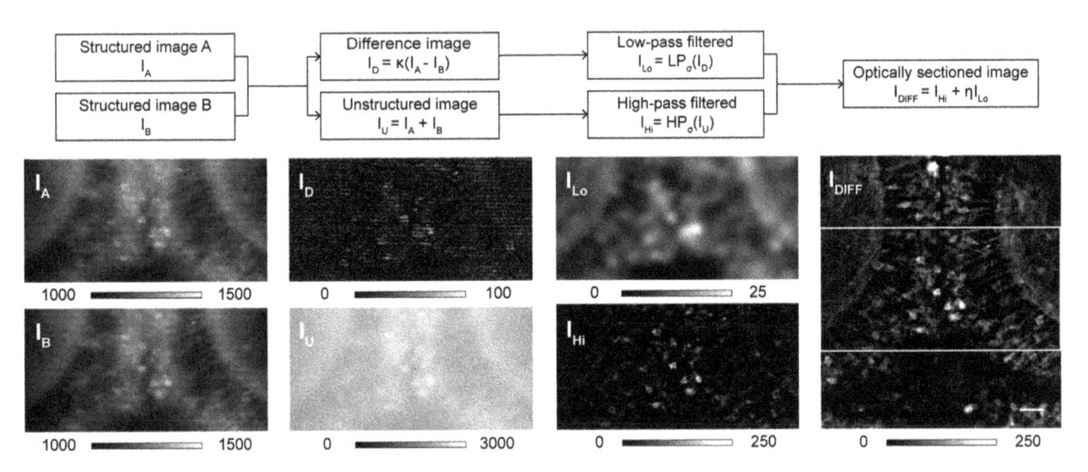

Fig. 5 Optical sectioning algorithm of DIFF microscopy. Top, schematic of the DIFF optical sectioning algorithm. Bottom, corresponding images at each step of the DIFF processing pipeline. Scale bar: 20 μm

3. DIFF-sectioned fluorescent images from the reference brain volume will be used for offline image registration (*see* Subheading 3.6).
4. Each DIFF-sectioned fluorescence image from the experimental brain volume will undergo offline registration against the reference brain volume (see Subheading 3.6). Post-registration, neural activity will be extracted from each neuron within the brain.

3.6 Registration Post-imaging

Each DIFF-sectioned fluorescence image (the "moving image") is registered to a high-resolution reference brain volume collected from the same animal (described above).

1. For each moving image, we perform offline rigid registration with six degrees of freedom against the saved reference stack. This initial registration is obtained by optimizing a 3D rigid transformation mapping the moving image to a (possibly tilted) plane within the reference brain volume (*see* **Note 4** for detailed implementation).
2. After rigid registration, each moving image undergoes non-rigid registration with a piecewise affine transformation against the reference brain volume. The planar surface of each moving image is then finely subdivided into a deformable surface that is locally adjusted within the reference volume using a regularized piecewise affine transform (*see* **Note 5** for detailed implementation).
3. The resulting registered image volumes across time are saved and ready for analysis of neural activity. Extraction of neural activity can be performed with a variety of algorithms (e.g., nonnegative matrix factorization, independent component analysis, etc.) depending on the needs of the user.

4 Notes

1. Model Predictive Control (MPC) of X-Y motorized stages. The MPC implementation uses a predictive model of the moving target and a predictive model of the motorized stages (*see* Fig. 6). To predict the future trajectory of the brain, we use position and heading information from the past six timepoints (−20 ms to 0 ms). We extrapolate a future trajectory based on the current velocity and heading. However, the oscillatory nature of fish motion creates a large perpendicular velocity component that impairs a naive extrapolation. Therefore, we use the past six timepoints to calculate a median heading vector, project the last six positions onto a line defined by this heading vector, and then predict the next seven brain positions (+4 ms to +28 ms) based on the estimated velocity of this projection. We use a dampening factor at the beginning of each movement to avoid overextrapolation of large angle turns.

   To build a predictive model of stage motion, we measured stage velocities in response to white noise inputs. Treating each stage as a linear time invariant (LTI) system, we estimate the impulse response function of the system as the least squares solution of a linear system that transforms a sliding 100 ms history of white noise input to the measured stage output.

   MPC then uses both the predicted fish trajectory and the motion model of the motorized stages to generate an optimal series of stage inputs that minimizes the sum of squares tracking error over seven time steps into the future, which constitutes the prediction horizon of our system. This optimization is performed every 4 ms, as soon as image processing is complete, so that the calculated optimal stage input can be provided to the stage controller by an update that is synchronized to the start of the next frame.

   We implement two additional safeguards to counteract instability in the controller. Firstly, to prevent the stages from exceeding their thermal limit, we apply an L2 norm penalty to the planned future acceleration vector. Secondly, any gradual accumulation of errors due to inaccuracies in the stage model

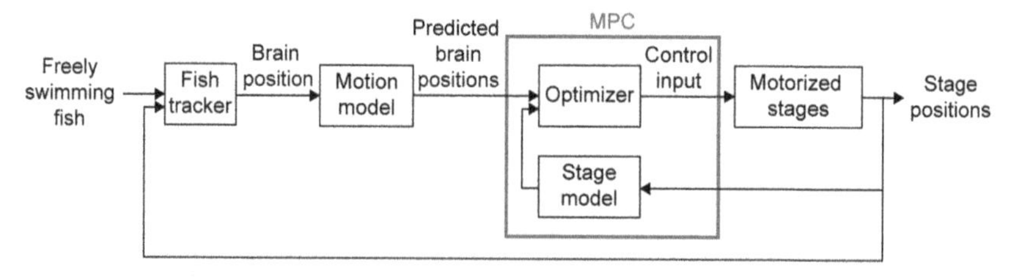

Fig. 6 A custom Model Predictive Control (MPC) implementation is used to keep the brain within the field of view of the DIFF microscope while the animal moves freely in the behavior arena

are removed by a slow feedback loop. This compensation accounts for ~1% of the input signal to the motorized stage system.

2. DIFF algorithm.

We perform optical sectioning with DIFF as follows:

$$I_D = \kappa(I_A - I_B)$$

$$I_U = I_A + I_B$$

$$I_{Lo} = \mathrm{LP}_\sigma(I_D)$$

$$I_{Hi} = \mathrm{HP}_\sigma(I_U)$$

$$I_{DIFF} = I_{Hi} + \eta I_{Lo}$$

Each step of this optical sectioning pipeline is explained below.

$I_A(i,j)$ and $I_B(i,j)$ are the raw structured fluorescence images collected in DIFF microscopy. This image pair is obtained by sequentially illuminating the sample with a pair of complementary grating patterns, preferably using a global shutter sensor so the images can be illuminated by a "double shot" strategy with minimal time delay.

In DIFF, $\kappa\left(\vec{p}\right)$ simply flips the sign of I_D for any pixel $\vec{p} = (i,j)$ where grating pattern B is brighter than grating pattern A. That is, we define

$$\kappa\left(\vec{p}\right) = \begin{cases} 1 & \text{grating A}\left(\vec{p}\right) \geq \text{grating B}\left(\vec{p}\right) \\ -1 & \text{grating A}\left(\vec{p}\right) < \text{grating B}\left(\vec{p}\right) \end{cases}$$

LP_σ and HP_σ denote 2D low-pass and high-pass filters with a Gaussian kernel with standard deviation σ. The parameter σ should be at least half the grid period p. We use $\sigma = 0.56p$.

I_{Lo} and I_{Hi} contain the low-frequency in-focus and high-frequency in-focus components of the optically sectioned image. I_{DIFF} contains the final optically sectioned image, and η controls the relative contribution of I_{Lo} and I_{Hi}.

For images collected with camera binning enabled, we use $\eta = 3$, which can be determined by measuring the modulation depth with a thin fluorescent sheet as a calibration sample. As M decreases as a function of tissue depth, η should ideally be adjusted to maintain a consistent ratio of high- and low-frequency information. Thus, in the current implementation, low-frequency information may be underestimated in the ventral parts of the brain.

The optical sectioning thickness of DIFF is related to the illumination grating period. A finer grating period leads to thinner sectioning because the patterns quickly become defocused away from the focal plane. Conversely, larger patterns lead to thicker axial sectioning because the grating pattern retains significant contrast over a larger axial range before becoming defocused. However, there is a trade-off in that the modulation depth tends to decrease as the grating period is reduced to near the diffraction limit.

3. Online Z tracking by GPU-accelerated live registration. At the beginning of an experiment, a series of sweeps over the entire piezo Z adjustment range are performed to facilitate both online and offline registration of moving images to a reference volume acquired from the same animal.

   To implement closed-loop Z scanning adjustment, each moving image is analyzed in real time to determine its approximate location in the brain. To ensure low latency, a downsampled registration pipeline is implemented on a dedicated GPU (distinct from the NIR tracking GPU). We implement an established rotation invariant form of the phase correlation technique[49] to rapidly detect the best matching reference plane regardless of the heading of the animal in the fluorescence image. Briefly, we downsample the moving image by eightfold on each axis to obtain a 128 × 128 (padded) downsampled image. We then high-pass with an 11 × 11 pixel Gaussian filter (NPP, NVIDIA), calculate the Fourier transform of the image (cuFFT, NVIDIA), and then calculate the magnitude of each complex Fourier coefficient. We then extract a semicircular subset of the resulting real-valued image by converting to polar coordinates, with r ranging from 1.0 to 13.4 in increments of 0.1 and θ ranging from $-\frac{\pi}{2}$ to $\frac{\pi}{2}$ in increments of 0.025 (1.4°). Finally, we use FFT-based convolution to compute the dot product between the polar image and the corresponding precomputed polar image for each reference plane, for all circular permutations $\Delta\theta$ on the θ axis of the polar image. The best matching Z plane is obtained from

$$\underset{\Delta\theta, z}{\operatorname{argmax}} \sum_{r,\theta} P_{\mathrm{mov}}(r, \theta + \Delta\theta) P_{\mathrm{ref}}^{z}(r, \theta)$$

where $P_{\mathrm{mov}}(r, \theta)$ and $P_{\mathrm{ref}}^{z}(r, \theta)$ are the polar images of the moving image and reference image z, respectively. This procedure also provides the yaw (modulo π), but this information is not used during live tracking.

4. High-resolution offline registration: 6 Degree-of-Freedom (6-DoF) rigid transform.

   For each pixel location $\vec{p}_{\mathrm{mov}}^{\,ij} = [i, j]^T$ in the 2D moving image, we associate a 3D point in the reference volume $\vec{p}_{\mathrm{ref}}^{\,ij} = \left[p_x, p_y, p_z\right]^T$ according to

$$\vec{p}^{\,ij}_{\text{ref}}\left(\vec{x}\right) = \vec{p}^{\,ij}_{\text{ref}}\left(t_x, t_y, t_z, \phi, \theta, \psi\right) = R_\psi^{-1} R_\theta^{-1} R_\phi^{-1} \begin{bmatrix} i \\ j \\ 0 \end{bmatrix} + \begin{bmatrix} t_x \\ t_y \\ t_z \end{bmatrix}$$

where $\vec{x} = \left[t_x, t_y, t_z, \phi, \theta, \psi\right]^T$ are the parameters of the 6-DoF transform. The parameters t_x, t_y, and t_z represent translations, and R_ϕ, R_θ, and R_ψ represent rotation matrices for roll, pitch, and yaw (heading), respectively.

We optimize the 6-DoF correlation function

$$f\left(\vec{x}\right) = \underset{\text{i,j}}{\text{cor}}\left(I_{\text{mov}}\left(\vec{p}^{\,ij}_{\text{mov}}\right), I_{\text{ref}}\left(\vec{p}^{\,ij}_{\text{ref}}\left(\vec{x}\right)\right)\right)$$

where $I_{mov} : \mathbb{Z}^2 \to \mathbb{R}$ is the moving image, $I_{ref} : \mathbb{R}^3 \to \mathbb{R}$ is the reference volume extended to $\mathbb{R}^3$ using trilinear interpolation, and $\underset{\text{i,j}}{\text{cor}}(\vec{n})$ is the Pearson correlation taken over all pixel coordinates $[i, j]^T$ in the 2D moving image. We use a three-step procedure to locate the global solution $\vec{x}^{\,*} = \left[t_x^*, t_y^*, t_z^*, \phi^*, \theta^*, \psi^*\right]^T$ of the 6-DoF registration problem $\text{argmax}\, f(\vec{x})$.

First, we solve $\underset{[t_x, t_y, \psi]^T \in \Omega_1}{\text{argmax}} f\left(t_x, t_y, \psi\right)$ over a gridded domain $\Omega_1 \subset \mathbb{R}^3$, taking advantage of the fact that this simpler objective function can be optimized by FFT-based cross correlation.

Second, we solve

$$\underset{\vec{x} \in \Omega_2}{\text{argmax}}\, f\left(\vec{x}\right)$$

by block coordinate descent over a gridded domain $\Omega_2 \subset \mathbb{R}^6$.

Finally, we iteratively solve

$$\underset{\vec{x}}{\text{argmax}}\, f\left(\vec{x}\right)$$

for $\vec{x} \in \mathbb{R}^6$ using the Broyden–Fletcher–Goldfarb–Shanno (BFGS) algorithm.

5. High-resolution offline registration: Nonrigid registration with piecewise affine transform. To account for nonrigid motion within the brain, we subdivide the plane defined by the 6-DoF rigid transform into a regularly spaced grid of points. These points induce a piecewise affine transformation linking the pixels of the moving image to a continuous, deformable 3D surface in the reference brain volume. The locations of these points are adjusted by iterative optimization using the limited memory Broyden-Fletcher-Goldfarb-Shanno (L-BFGS) algorithm.

   A 2D grid of control points P_{mov} are defined in the moving image, spaced by 64 pixels in x and y. These control points are fixed. A corresponding set of control points P_{ref} are defined in the reference brain volume, with initial values set by the optimal solution x^* of the 6-DoF problem (described above). If $\vec{p}_{\text{mov}} = [i, j]^T$ is a control point in the moving image, then the corresponding control point in the reference brain volume is initialized as

$$\vec{p}^{\,ij}_{\text{ref}}\left(\vec{x}^{\,*}\right) = \vec{p}^{\,ij}_{\text{ref}}\left(t^*_x, t^*_y, t^*_z, \phi^*, \theta^*, \psi^*\right) = R^{-1}_{\psi^*} R^{-1}_{\theta^*} R^{-1}_{\phi^*} \begin{bmatrix} i \\ j \\ 0 \end{bmatrix} + \begin{bmatrix} t^*_x \\ t^*_y \\ t^*_z \end{bmatrix}$$

We define a triangulation over the control points by connecting each control point to its four immediate neighbors and connecting $\vec{p}^{\,ij}_{\text{mov}}$ to $\vec{p}^{\,i+1,j+1}_{\text{mov}}$ for all i, j. For a moving image of 1024×772 pixels, we generate 221 control points and 384 triangles. Since each moving image pixel belongs to a triangle $\left(\vec{p}^{\,1}_{\text{mov}}, \vec{p}^{\,2}_{\text{mov}}, \vec{p}^{\,3}_{\text{mov}}\right)$, it can be mapped into the reference brain volume by interpolating its location within the plane defined by the corresponding points $\left(\vec{p}^{\,1}_{\text{ref}}, \vec{p}^{\,2}_{\text{ref}}, \vec{p}^{\,3}_{\text{ref}}\right)$. Each triangle defines an affine transform, and collectively the triangles define a piecewise affine transform.

Given the locations of the control points $P_{\text{ref}} = \left\{\vec{p}_{\text{ref}}\right\}$, we compute the similarity of the moving image and the piecewise affine interpolation of the reference image by their dot product:

$$f(P_{\text{ref}}) = \sum\nolimits_{i,j} \text{HP}_\sigma(I_{\text{mov}}(i,j))\text{HP}_\sigma\left(I_{\text{ref}}\left(\vec{p}^{\,ij}_{\text{ref}}\right)\right)$$

where $\text{HP}_\sigma(\cdot)$ denotes a Gaussian high-pass filter with $\sigma = 20$.

We solve

$$P^*_{\text{ref}} = \underset{P_{\text{ref}}}{\text{argmax}} f(P_{\text{ref}}) - \lambda \sum_{i,j}\left(\left\|D_{ii}\vec{p}^{\,ij}_{\text{ref}}\right\|^2_2 + \left\|D_{jj}\vec{p}^{\,ij}_{\text{ref}}\right\|^2_2\right)$$

where D_{ii} and D_{jj} represent second-order difference operators, and $\lambda = 1$ is a regularization penalty. We use L-BFGS, storing only the past $m = 10$ steps.

References

1. Ahrens MB, Li JM, Orger MB, Robson DN, Schier AF, Engert F, Portugues R (2012) Brain-wide neuronal dynamics during motor adaptation in zebrafish. Nature 485:471–477
2. Ahrens MB, Orger MB, Robson DN, Li JM, Keller PJ (2013) Whole-brain functional imaging at cellular resolution using light-sheet microscopy. Nat Methods 10:10
3. Portugues R, Feierstein CE, Engert F, Orger MB (2014) Whole-brain activity maps reveal stereotyped, distributed networks for visuomotor behavior. Neuron 81:1328–1343
4. Kim DH, Kim J, Marques JC, Grama A, Hildebrand DGC, Gu W, Li JM, Robson DN (2017) Pan-neuronal calcium imaging with cellular resolution in freely swimming zebrafish. Nat Methods 14:1107–1114

5. Severi KE, Portugues R, Marques JC, O'Malley DM, Orger MB, Engert F (2014) Neural control and modulation of swimming speed in the larval zebrafish. Neuron 83:692–707
6. Berg HC, Brown DA (1972) Chemotaxis in escherichia coli analysed by three-dimensional tracking. Nature 239:500–504
7. Venkatachalam V, Ji N, Wang X, Clark C, Mitchell JK, Klein M, Tabone CJ, Florman J, Ji H, Greenwood J et al (2016) Pan-neuronal imaging in roaming caenorhabditis elegans. Proc Natl Acad Sci 113:E1082–E1088
8. Nguyen JP, Shipley FB, Linder AN, Plummer GS, Liu M, Setru SU, Shaevitz JW, Leifer AM (2016) Whole-brain calcium imaging with cellular resolution in freely behaving caenorhabditis elegans. Proc Natl Acad Sci 113:E1074–E1081
9. Ben Arous J, Tanizawa Y, Rabinowitch I, Chatenay D, Schafer WR (2010) Automated imaging of neuronal activity in freely behaving caenorhabditis elegans. J Neurosci Methods 187:229–234
10. Leifer AM, Fang-Yen C, Gershow M, Alkema MJ, Samuel ADT (2011) Optogenetic manipulation of neural activity in freely moving caenorhabditis elegans. Nat Methods 8:147–152
11. Grover D, Katsuki T, Greenspan RJ Flyception: imaging brain activity in freely walking fruit flies. Nat Methods 13:569–572
12. Cong L, Wang Z, Chai Y, Hang W, Shang C, Yang W, Bai L, Du J, Wang K, Wen Q (2017) Rapid whole brain imaging of neural activity in freely behaving larval zebrafish (Danio rerio). elife 6:6
13. García CE, Prett DM, Morari M (1989) Model predictive control: theory and practice – a survey. Automatica 25:335–348
14. Mertz J (2011) Optical sectioning microscopy with planar or structured illumination. Nat Methods 8:811–819
15. E3 medium (for zebrafish embryos) (2011) Cold Spring Harb Protoc, 2011:pdb.rec66449

Chapter 11

Genetic Identification of Neural Circuits Essential for Active Avoidance Fear Conditioning in Adult Zebrafish

Pradeep Lal, Hideyuki Tanabe, and Koichi Kawakami

Abstract

Inhibition or ablation of neuronal activity combined with behavioral assessment is crucial in identifying neural circuits or populations essential for specific behaviors and to understand brain function. In the model vertebrate zebrafish, the development of genetic methods has allowed not only visualization but also targeted manipulation of neuronal activity, and quantitative behavioral assays allow precise measurement of animal behavior. Here, we describe a method to inhibit a specific neuronal population in adult zebrafish brain and assess their role in a learning behavior. We employed the Gal4–UAS system, gene trap and enhancer trap methods, and isolated transgenic zebrafish lines expressing Gal4FF transactivator in specific populations of neurons in the adult zebrafish brain. In these lines, a genetically engineered neurotoxin, botulinum toxin B light chain, was expressed and the fish were assessed in the active avoidance fear conditioning paradigm. The transgenic lines that showed impaired avoidance response were isolated and, in these fish, the Gal4-expressing neurons were analyzed to identify the neuronal circuits involved in avoidance learning.

Key words Active avoidance fear conditioning, Neuronal circuits, Gal4–UAS, Botulinum toxin, Telencephalon

1 Introduction

In teleost, the study of brain function has been done mostly by using surgical ablation, comparative neuroanatomy, histology, and expression data of few genes [1, 2]. Although these studies provide evidence of homology, to understand the functional organization of the zebrafish amygdala, it is crucial to study the structure and function of its distinct populations of neurons in a particular behavioral task. The Gal4–UAS system has been developed to label, visualize, and manipulate specific tissues in the zebrafish embryo, and a collection of Gal4 transgenic lines has been generated using gene trap and enhancer trap approaches [3, 4]. In these transgenic lines, Gal4FF, containing the DNA binding domain from the yeast Gal4 transcriptional activator and two transcription activation

James F. Amatruda et al. (eds.), *Zebrafish: Methods and Protocols*, Methods in Molecular Biology, vol. 2707,
https://doi.org/10.1007/978-1-0716-3401-1_11,

modules from the herpes simplex virus VP16 gene, is expressed in specific cells, tissues, and organs. Gal4FF binds to its specific recognition sequence UAS (upstream activating sequence) and activates transcription of a target gene placed downstream of UAS sequence. The Gal4FF expressing cells can be visualized via GFP linked to UAS sequence, and can be manipulated via an effector gene linked to UAS. Thus, the Gal4–UAS binary gene expression system has been a powerful tool to study neuronal functions.

Here, we describe a method to inhibit a specific neuronal population in the adult zebrafish brain and identify neuronal populations essential for active avoidance fear conditioning [5, 6]. First, we isolated transgenic zebrafish lines expressing Gal4FF transactivator in specific populations of neurons in the adult zebrafish brain. Second, we expressed a genetically engineered neurotoxin, botulinum toxin B light chain, in the Gal4FF expressing cells in these transgenic lines [7]. Third, the double transgenic zebrafish were assessed for active avoidance fear conditioning, and the transgenic fish lines that showed impaired avoidance response were identified. Finally, the Gal4FF expressing neurons in these transgenic fish were analyzed to identify neuronal populations essential for fear learning. This chapter shows an example of structure–function relationship study of a genetically identifiable population of neurons in adult zebrafish brain in fear leaning. Furthermore, a similar approach is applicable in identifying brain regions and neuronal substrates mediating behavioral responses.

2 Materials

2.1 Identification of Brain-Specific Transgenic Zebrafish Lines

1. UAS:GFP and UAS:BoTxBLC:GFP transgenic fish.
2. For Gal4 transgenic fish lines, we used gene and enhancer trap lines created in Kawakami lab [8].
3. Artificial fish water used to maintain the adult fish. Seachem Marin Salt added to reverse osmosis water to achieve the desired conductivity and $NaHCO_3$ is used to adjust pH. In our lab, conductivity and pH are kept at 470 μs/cm (microsiemens per centimeter) and 6.8, respectively.
4. Disposable transfer pipets.
5. 0.025% tricaine in fish water.
6. An epifluorescence microscope (Leica, MZ 16FA).
7. Image capture software. We use DFC300 FX (Leica Microsystems).
8. 4% paraformaldehyde (PFA) (Wako, 163-20145).
9. Silicon plates.
10. Agarose (Iwai Kagaku, 50013R).

11. 1× Phosphate buffered saline.
12. Tissue-tek cryomold.
13. Vibratom 3000 sectioning system.
14. Microblade (Feather Safety Razor Co. Ltd. Micro Feather P-715).
15. Tissue culture plate, 24 (Falcon).
16. Soft paintbrush.
17. Dissection tools—scalpel, forceps, and fine scissors.
18. Micro slide glass (Matsunami, S8111).
19. Cover glass.
20. Aqueous Permafluor mountant (Thermoscientific, FM 90814).
21. Epifluorescence microscope.
22. Image analysis software (ImageJ).
23. Zebrafish brain atlas [9].

2.2 Active Avoidance Fear Conditioning

1. The zebrafish shuttle box apparatus. A white opaque tank (length: 41 cm width: 17 cm × depth: 12 cm) with a trapezoidal wedge (widths: 10 and 20 cm, height: 5 cm) that divides the tank into two equal compartments.
2. Green LED lamps.
3. 12/9 V voltage source.
4. Visible or Infrared (IR) camera. For IR camera, IR light source is needed.
5. Platinum/stainless steel electrode.
6. Sound isolation unit or behavioral chamber (height: 125 cm × width: 100 cm × depth: 70 cm).
7. Background illumination (white LED).
8. Computer with video capture and stimulus application software. We used customized software using LabView (National Instrument).
9. Gal4 and UAS:zBoTxBLCGFP double transgenic fish [4].
10. Artificial fish water used to maintain the adult fish (as above).
11. E3 water (5 mM NaCl, 0.17 mM KCl, 0.33 mM $CaCl_2$, 0.33 mM $MgSO_4$, 0.00001% Methylene blue).
12. Nine-well glass depression plate (Corning).

2.3 Analysis of Neural Circuits Mediating Active Avoidance Fear Conditioning

1. Selected Gal4 gene and enhancer trap transgenic fish.
2. UAS:GFP and UAS:BoTxBLC:GFP transgenic fish.
3. 0.025% tricaine in fish water.
4. 4% paraformaldehyde (PFA) (Wako, 163-20145).

5. 2% agarose (Iwai Kagaku, 50013R) in 1× PBS.
24. 1× phosphate buffered saline.
25. Tissue-tek cryomold.
6. Vibratom 3000 sectioning system.
7. Microblade (Feather Safety Razor Co. Ltd. Micro Feather P-715).
8. 24-well plates and inserts with porous membrane.
9. Rabbit anti-GFP polyclonal (Invitrogen, A6455, 1:500) primary antibody.
10. AlexaFluor 488 conjugated goat anti-rabbit IgG (Invitrogen, A11008, 1:400) secondary antibody.
11. Blocking buffer—0.5% Tween-20 and 3% bovine serum albumin (BSA) in PBS.
12. Micro slide glass (Matsunami, S8111).
13. Aqueous Permafluor mountant (Thermoscientific, FM 90814).
14. Epifluorescence microscope.

3 Methods

3.1 Identification of Brain-Specific Transgenic Zebrafish Lines

1. We used 5 months to 1.5 years old adult zebrafish for the experiments.
2. Anesthetize the gene and enhancer trap Gal4 transgenic fish (gt/et-Gal4FF;UAS:GFP) in tricaine methane sulfonate (0.025% tricaine in fish water).
3. Place the fish in a Petri dish, supported by sponges, carefully to observe dorsal view of the head.
4. Observe the head under a fluorescence microscope (Leica, MZ 16FA) and take images using DFC300 FX (Leica Microsystems) for record.
5. Select the transgenic fish that showed green fluorescence in the head and anesthetized the selected fish in tricaine methane sulfonate (0.025% tricaine in fish water).
6. Decapitate by scalpel and transfer the head into cooled 1× PBS in a silicon plate. Remove jaw, eyes, and soft tissues from the ventral side.
7. Transfer the head in a tube containing 2 mL 4% PFA (Wako, 163-20145). Incubate at 4 °C O/N.
8. Wash the head 6–8 times with 1× PBS (4 °C).
9. Dissect the brain out of the skull using forceps and fine scissors.

10. Place the isolated brain in a tissue-tek cryomold. Apply melted 2% agarose in 1× PBS. The agarose temperature is recommended between 40 and 50 °C. Brain sample is placed in an appropriate orientation for desired slicing plane.
11. Make 100 μm thick coronal sections using vibrating microtome (Vibratom 3000 sectioning system) and with Microblade. Tissue collection tray is filled with 1× PBS.
12. Collect and transfer the slices to a 24-well plates containing 1× PBS using a soft paintbrush.
13. Mount the slices on micro slide glass (Matsunami, S8111) using Aqueous Permafluor mountant (Thermoscientific, FM 90814) for microscopic observation.
14. Image the serial sections of zebrafish brain under an epifluorescence microscope. We used Imager.Z1 (Zeiss) using AxioCam MRc5 (Zeiss) camera. Images are analyzed using Axio Vision Ver4.1 imaging software. Serial coronal sections of SAGFF120A/UAS:GFP are shown in Fig. 1 [5].
15. Compare the expression pattern with the zebrafish reference brain atlas and map the transgene expression in specific regions on the brain (*see* **Note 1**).

3.2 Active Avoidance Fear Conditioning

We designed the zebrafish shuttle box apparatus for active avoidance fear conditioning based on the previously reported similar system for goldfish avoidance response assay [10]. Before the behavioral assay, fish are moved from a fish room to a behavioral assay room and are kept in isolation to reduce interference from background noise and disturbance during the behavioral assay. For this analyses, zebrafish of age about 5 months to 1.5 years with at least 2.5 cm body length are recommended.

3.2.1 Setting Up the Behavior Arena

1. Set up the apparatus using an opaque tank (41 cm × 17 cm) and a trapezoidal wedge (width: 10 and 20 cm, height: 5 cm). The trapezoidal wedge divides the tank into two equal compartments (Fig. 2a).
2. Behavior chamber is placed within the fish facility. The camera (Pike F-032B) is mounted on the ceiling of the behavioral chamber (Fig. 2a).
3. Video recording starts at least 2 min before the behavioral tests. Recording is performed at 7.5 frames per second (fps). At least 5 fps is necessary to observe its body movement and for effective locomotive behavior analyses.
4. Place the green LED lights at each long end of the compartment. Light (green LED) is used as conditioned stimulus (CS).
5. Place the platinum/steel mesh electrodes in the two compartments. Connect each pair of electrodes in each compartment to a 12 V voltage source. Electric stimulus is used as unconditioned stimulus (US).

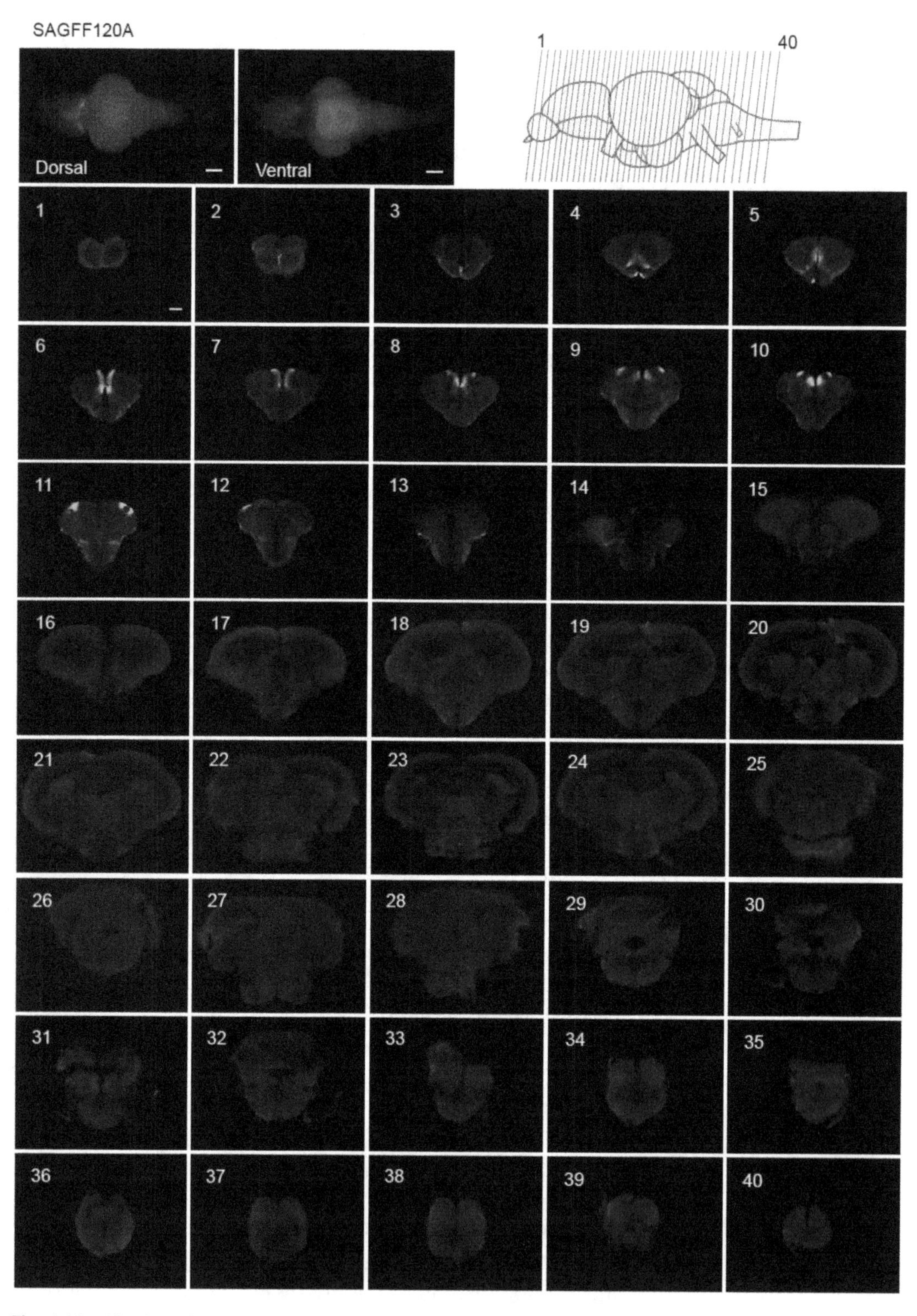

Fig. 1 Identification of a brain-specific transgenic zebrafish line. An example of SAFFF120A transgenic fish line. First, GFP expression is observed by using an epifluorescence stereoscope. Then, serial coronal sections are created and observed under an epifluorescence upright microscope. Scale bars in whole brain images: 500 μm. Scale bars in coronal section images: 200 μm

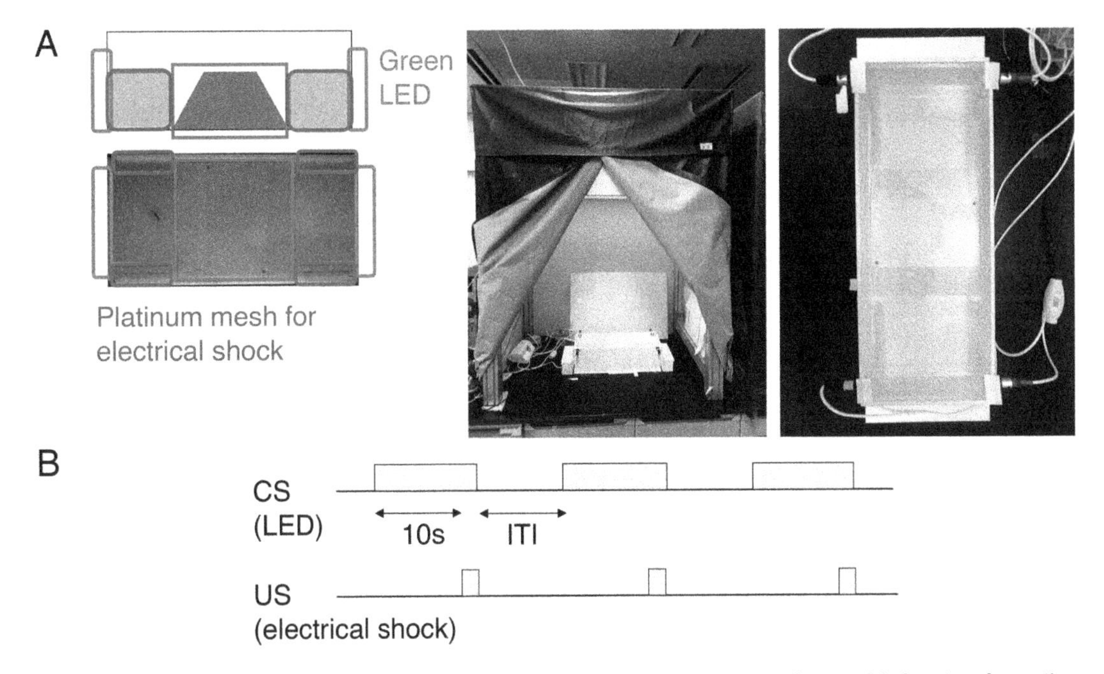

Fig. 2 Setup for active avoidance fear conditioning and an experimental scheme. (**a**) A setup for active avoidance fear conditioning. A white opaque tank (41 cm × 17 cm × 12 cm) is placed in a behavioral chamber (125 cm × 100 cm × 70 cm). (**b**) A paradigm for active avoidance fear conditioning is shown

6. Fill the behavior tank with system water (28 °C). Keep the water level about 1.5 cm above the wedge. This allows free swimming across the two compartments.
7. Use computer-controlled systems for stimulus delivery (light and electric stimuli). Fish position is tracked in real time with an image processing software. We used a custom-made semiautomated LabView 8.6 (National Instruments) based program for tracking the fish and deliver stimuli to the fish.

3.2.2 Habituation

1. Before starting training protocol, keep zebrafish in isolation in a 2 L home tank (with constant water flow) for 7 days (*see* **Note 2**). White paper is used as separator between the tanks for fish not to see the outside.
2. To habituate the fish with the behavioral apparatus, gently transfer them into the apparatus and leave them for 15 min per day for next 2 days (*See* **Note 3**).
3. On the day of training, place the fish on the wedge as initial starting point and let it swim freely for at least 2 min.

3.2.3 Training Protocol

1. Recorded the zebrafish behavior through the camera attached to the ceiling.
2. Turn on the green light (CS) of the compartment where the fish stays.

3. The light is kept on for 10 s and then apply 12-V electric stimulus (US) for a maximum of 5 s (Fig. 2b).
4. Turn off the green light at the end of the electric stimulus.
5. During this process, if the fish swims out of the light-on compartment during the electric stimulus, turn off both electric stimulus and light simultaneously.
6. If the fish swims out of the light-on compartment before the application of the electric stimulus, the electric stimulus is not given and turn off the light 10 s after the light-on.
7. Repeat this process 10 times with inter trial interval (ITI) of 25 ± 5 s (Fig. 2b).
8. After training, fish are kept in the tank for at least 2 min.
9. These constitutes one session of training. Conduct one training session per day for each fish over 5 consecutive days (*see* **Note 4**).
10. The procedure, where US is on as well as CS is on, is called "non-trace active avoidance fear conditioning."
11. As control, perform the above training protocol without electric stimulus.

3.2.4 Data Analysis

1. The escape after light-on (CS) and before the electric shock (US) is called avoidance response.
2. Count the number of the avoidance responses and calculate the ratio (percentage) of each session (10 trials per day) (Fig. 3a).
3. The ratio is improved day by day: namely, on day 1 they show about 14% avoidance performance in average. On day 2, the avoidance performance increases to about 40%. On day 5, wild-type fish usually show more than 70% avoidance performance.
4. The statistical analysis is carried out by using GraphPad Prism8 (GraphPad, MA).
5. Go to "Results" > "new analysis." Select "one-way ANOVA" (Fig. 3b).
6. Go to "Experimental design." Select "Yes" to "Assume Gaussian distribution," and "No" to "Assume equal SDs" (Fig. 3c).
7. Go to "multiple comparisons." Select "Compare the mean of each column with the mean of a control column."
8. The results of one-way ANOVA test and Dunnett's multiple comparison test will be shown.
9. Go to "Graph family" and select "column" (Fig. 3d). Draw a graph.

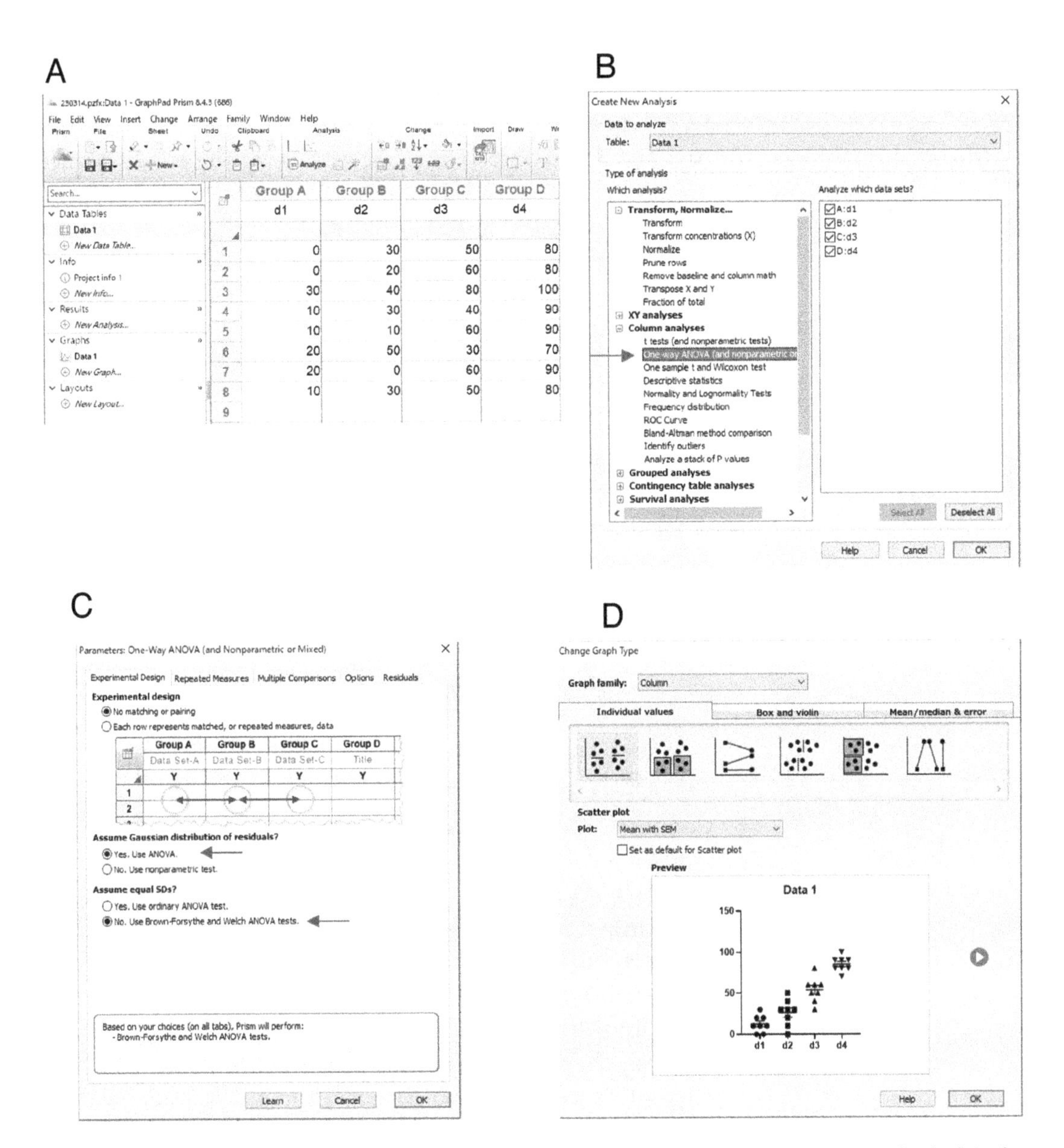

	Group A	Group B	Group C	Group D
	d1	d2	d3	d4
1	0	30	50	80
2	0	20	60	80
3	30	40	80	100
4	10	30	40	90
5	10	10	60	90
6	20	50	30	70
7	20	0	60	90
8	10	30	50	80
9				

Fig. 3 Data analysis using GraphPad Prism8. (**a**) Count the number of avoidance responses and calculate the ratio (percentage) of each session (10 trials per day) (**b**) Go to "Results" > "new analysis." Select "one-way ANOVA." (**c**) Go to "Experimental design." Select "Yes" to "Assume Gaussian distribution," and "No" to "Assume equal SDs." (**d**) Go to "Graph family" and select "column"

3.2.5 Inhibiting Neuronal Activity in Adult Zebrafish Brain

To identify functional neurons involved in the learning process, the Gal4 driver lines are crossed with the UAS:BoTxBLCGFP line, in which engineered botulinum neurotoxin B light chain expressed in the Gal4 positive neuronal cells [7].

1. Set up crosses of a selected Gal4 line (*see* **Note 1**) and the UAS: BoTxBLCGFP line (homozygous).
2. Collect eggs and raise them in E3 water at 28 °C.
3. At 1 and 5 dpf, observe the embryos under an epifluorescence microscope and select GFP positive embryos. We use Leica MZ 16FA epifluorescence microscope.
4. At 5 dpf, observe locomotive behaviors and the touch response by using a pipette tip under a microscope. If the embryos show defect in locomotive behavior (e.g., no movement or paralysis, loss of posture control), the transgenic lines will not survive to adulthood and may not be used for behavioral studies in adult stages.
5. Select the fish lines that do not show gross locomotive defects and raise them to adulthood.
6. When we crossed 38 lines having Gal4 expression in embryonic central nervous system (CNS) with the UAS:BoTxBLCGFP line, we observed motor defects or paralysis in ten double transgenic lines. Nine out of the ten lines could not survive to the adulthood [5].
7. As describe in Subheading 3.2.2, before starting the training protocol, keep zebrafish in isolation in a 2 L tank (with constant water flow) and then habituate the fish with the behavioral apparatus (*see* **Note 2**).
8. Conduct active avoidance conditioning as described in Subheading 3.2.3 (*see* **Note 5**).
9. Calculate the avoidance response performance of the transgenic fish. Compare the increase in avoidance performance from session 1 to each session day. Also compare the increase in avoidance performance of the transgenic fish with that of wild-type fish.
10. We analyzed the avoidance response performance in 30 of the selected 39 transgenic fish lines and found defect in avoidance response in 18 out of 30 analyzed lines [5].
11. Select the transgenic fish lines with impaired active avoidance response performance for anatomical study.

3.3 Analysis of Neural Circuits Mediating Avoidance Fear Learning

1. Transgenic fish lines that show defects in active avoidance fear conditioning will be selected to identify the projections of Gal4 expressing neurons.
2. Fix the brain of the selected transgenic fish as described in Subheading 3.1.
3. Embed the fixed brains in 2% agarose in 1× PBS.

4. Make 100 μm thick serial coronal and sagittal slices using vibrating microtome (Vibratom 3000 sectioning system). The tissue slices are collected in a tube containing 1× PBS.
5. Remove the PBS and add blocking buffer containing 0.3% Triton X-100 and 3% bovine serum albumin (BSA) in 1× PBS. Incubate the sections for 1 h at room temperature.
6. Incubate the slices with primary antibodies diluted in blocking buffer at 4 °C overnight or at room temperature for 4 h. We used rabbit anti-GFP polyclonal (Invitrogen, A6455, 1:500) primary antibody.
7. Wash (4–6 times) the samples with 0.3% Triton X-100 in 1× PBS.
8. Incubate with secondary antibody at 4 °C overnight or at room temperature for 1–2 h. We used rabbit AlexaFluor 488 conjugated goat anti-rabbit IgG (Invitrogen, A11008, 1:400) secondary antibody.
9. Wash (4–6 times) the samples with 0.3% Triton X-100 in 1× PBS.
10. Mount the slices on micro slide glass (Matsunami, S8111) using Aqueous Permafluor mountant (Thermoscientific, FM 90814) for microscopic observation.
11. Perform laser scanning confocal microscopy. We used an Olympus FV-1000-D microscope. Image editing is performed using ImageJ and Adobe Photoshop CS5. Images of SAGFF120A/UAS:GFP [5] are shown in Fig. 4 as examples.

4 Notes

1. Choice of Gal4 line depends on the neurons or the brain regions that you are interested in analyzing.
2. Fish are housed alone during the course of study. This duration is limited to an experimental period under welfare consideration.
3. Minimum habituation time of 2 × 10 min each is necessary for effective avoidance response behavior. Performance decreases in the absence of habituation.
4. Since there are no circulation in the behavior chamber, water should be exchanged for each experiment.
5. Double blind test is recommended for the behavioral test. During the double blind test, the genotype of the fish should not be known.

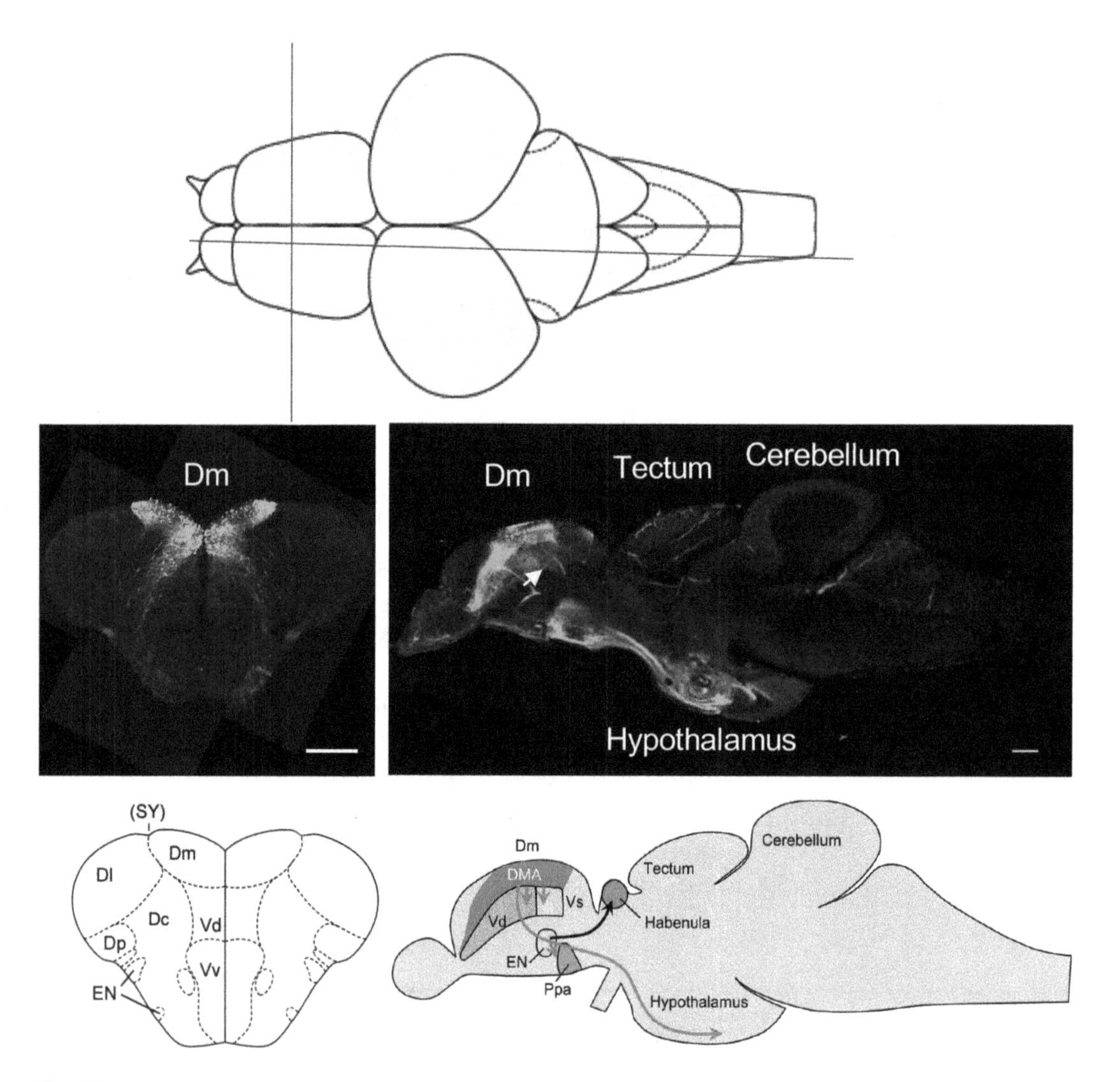

Fig. 4 Transverse and sagittal sections of the zebrafish adult brain. Images of SAGFF120A transgenic zebrafish are shown in which neurons in Dm (medial zone of the dorsal telencephalon) essential to perform active avoidance fear conditioning are labeled with Gal4FF. Scale bar: 200 μm

Acknowledgments

This work was supported by internal support from NORCE Norwegian Research Centre AS to P.L. and JSPS Kakenhi 21H02463 and NBRP from AMED to K.K.

References

1. Wullimann MF, Rupp B, Reichert H (1996) Neuroanatomy of the zebrafish brain. Birkhäuser Verlag, Basel
2. Wullimann MF, Mueller T (2004) Teleostean and mammalian forebrains contrasted: evidence from genes to behavior. J Comp Neurol 475(2):143–162
3. Asakawa K, Suster ML, Mizusawa K, Nagayoshi S, Kotani T, Urasaki A et al (2008) Genetic dissection of neural circuits by Tol2

transposon-mediated Gal4 gene and enhancer trapping in zebrafish. Proc Natl Acad Sci U S A 105(4):1255–1260

4. Kawakami K, Asakawa K, Hibi M, Itoh M, Muto A, Wada H (2016) Gal4 driver transgenic zebrafish: powerful tools to study developmental biology, organogenesis, and neuroscience. Adv Genet 95:65–87
5. Lal P, Tanabe H, Suster ML, Ailani D, Kotani Y, Muto A et al (2018) Identification of a neuronal population in the telencephalon essential for fear conditioning in zebrafish. BMC Biol 16(1):45
6. Lal P, Kawakami K (2022) Integrated behavioral, genetic and brain circuit visualization methods to unravel functional anatomy of zebrafish amygdala. Front Neuroanat 16:837527
7. Sternberg JR, Severi KE, Fidelin K, Gomez J, Ihara H, Alcheikh Y et al (2016) Optimization of a neurotoxin to investigate the contribution of excitatory interneurons to speed modulation in vivo. Curr Biol 26(17):2319–2328
8. Kawakami K, Abe G, Asada T, Asakawa K, Fukuda R, Ito A et al (2010) zTrap: zebrafish gene trap and enhancer trap database. BMC Dev Biol 10:105
9. Wullimann MF, Rupp B, Reichert H (1996) Neuroanatomy of the zebrafish brain. Birkhauser, Basel
10. Portavella M, Torres B, Salas C (2004) Avoidance response in goldfish: emotional and temporal involvement of medial and lateral telencephalic pallium. J Neurosci 24(9): 2335–2342

Part III

Regeneration

Chapter 12

Quantitative Live Imaging of Zebrafish Scale Regeneration: From Adult Fish to Signaling Patterns and Tissue Flows

Alessandro De Simone

Abstract

In regeneration, a damaged body part grows back to its original form. Understanding the mechanisms and physical principles underlying this process has been limited by the difficulties of visualizing cell signals and behaviors in regeneration. Zebrafish scales are emerging as a model system to investigate morphogenesis during vertebrate regeneration using quantitative live imaging. Scales are millimeter-sized dermal bone disks forming a skeletal armor on the body of the fish. The scale bone is deposited by an adjacent monolayer of osteoblasts that, after scale loss, regenerates in about 2 weeks. This intriguing regenerative process is accessible to live confocal microscopy, quantifications, and mathematical modeling. Here, I describe methods to image scale regeneration live, tissue-wide and at sub-cellular resolution. Furthermore, I describe methods to process the resulting images and quantify cell, tissue, and signal dynamics.

Key words Regeneration, Zebrafish, Scale, Bone, Tissue growth, Erk signaling, Tissue mechanics, Morphogenesis, Physics of biology

1 Introduction

In regeneration, choreographed cellular events restore a body part to its original form [1–4]. In this complex process, cells face strong developmental challenges, such as communicating over large distances and long timescales and adapting the regenerative response to variable initial injuries. Addressing how highly regeneration-capable animals overcome these challenges is revealing fundamental principles of organization in multicellular systems.

Quantitative live imaging of signals and cell behaviors can tackle this question [5–8]. This approach has been applied successfully to study ontogenesis in many systems, for example in body plan patterning [9, 10], in the growth of the fly wing disk [11, 12], and in pectoral fin growth [13], somitogenesis [14], and lateral line migration in zebrafish [15, 16]. Although there has been progress in applying live imaging to the study of regeneration, it has been

James F. Amatruda et al. (eds.), *Zebrafish: Methods and Protocols*, Methods in Molecular Biology, vol. 2707,
https://doi.org/10.1007/978-1-0716-3401-1_12,

limited by the size, location, and architecture of adult regenerating tissues [6]. To overcome these issues, zebrafish scales are emerging as a model system that allows quantitative live imaging of vertebrate bone regeneration [17–21].

Zebrafish scales are millimeter-sized dermal bone disks (Fig. 1a–c) [22–24]. They are arranged in an array on the body of the fish to form a skeletal armor (Fig. 1b, c). After loss, scales regenerate completely within 2 weeks (Fig. 1c, d) [19, 20, 25]. The newly formed bone is deposited by a monolayer of adjacent osteoblasts (Fig. 1a) [22, 23, 25] that regenerate first by differentiation of an unknown progenitor, then proliferation, and finally hypertrophic growth (Fig. 1d) [18]. The simple organization, accessibility to live imaging, and intriguing regenerative biology make scale osteoblasts a powerful system to study morphogenesis in vertebrate regeneration. Markers and reporters can be expressed in osteoblasts using random insertion transgenes under control of an *sp7* (*osx*) promoter [17–20, 26–28] or, alternatively, knock-ins. Live imaging has also been applied successfully to study the regeneration of the zebrafish scale epidermis [29], nerves and vasculature [21] and osteoclasts [30]. Using this system and a transgenic Erk Kinase Translocation Reporter (KTR), we have recently discovered Erk activity waves that travel across the scale osteoblast tissue, from the scale core to its edge, and coordinate osteoblast tissue growth (Fig. 1d, e) [19].

Here, I describe methods to image and quantify the regeneration of the scale osteoblast tissue [18, 19]. First, I describe how to pluck scales to induce regeneration (Subheading 3.1). Second, I describe a protocol to perform time-lapse imaging of the regeneration of the osteoblast mono-layer in anesthetized adult zebrafish (Subheading 3.2). A similar method can be applied to photoconvert and track nuclei tagged with photoconvertible fluorescent proteins (Subheading 3.3). Then, I describe the image processing steps needed to obtain a single and manageable *z*-stack from raw images (Subheading 3.4). Finally, I describe how to quantify the spatial-temporal pattern of signals (Subheading 3.5) and tissue movements (tissue flows—Subheading 3.6).

2 Materials

2.1 Scale Injury

1. Adult transgenic fish (3–18 months old) expressing a fluorescent marker or reporter in the osteoblast tissue (*see* **Note 1**).
2. Fish system water.
3. Fish net.
4. Finger bowl.
5. Plastic spoon.

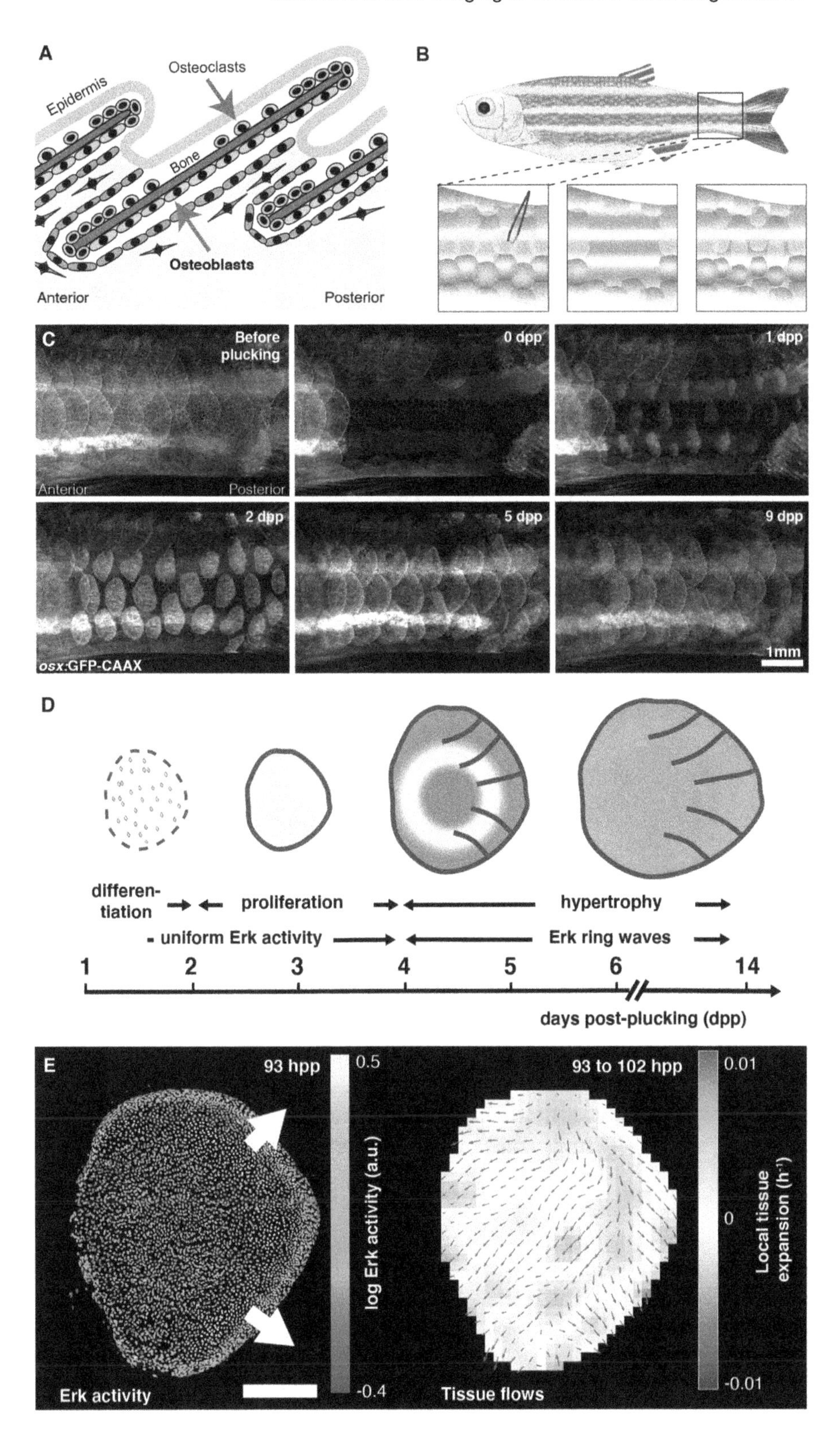

Fig. 1 Quantitative live imaging of scale regeneration in zebrafish. (**a**) Zebrafish scale morphology. The scale bone is deposited by an adjacent osteoblast monolayer. Each scale sits in a dermal pocket and is covered by

6. 90 mm Petri dish lid.
7. Anesthetic solution: 0.075% 2-phenoxyethanol (*see* **Note 2**). Dilute 750 μL of 2-phenoxyethanol in 1 L of fish water.
8. Plastic transfer pipette.
9. Fine-tip straight forceps.
10. Fluorescence dissecting scope.

2.2 Scale Regeneration Imaging

1. Adult transgenic fish (3–18 months old) expressing (*see* **Note 1**):
   (a) The osteoblast nuclear marker *osx:*H2A-mCherry [18].
   (b) A transgenic reporter. Here, we will describe an example using an Erk activity sensor Erk KTR (Kinase Translocation Reporter; *osx:*ErkKTR-mCerulean [19]) and a reporter localized in the nucleus, i.e., the Cdh1 inactivation reporter *osx:*Venus-hGeminin [18].
2. Confocal microscope allowing imaging multiple overlapping *z*-stacks.
3. 25× water immersion objective with 2.5 mm working distance.
4. Fish system water.
5. Fish net.
6. Finger bowl.
7. Plastic spoon.
8. 600 mL beaker.
9. 250 mL flask.
10. Plastic transfer pipettes.
11. Nontoxic modeling clay.
12. Heated water bath set at 40 °C.
13. Anesthetic solution and imaging medium: 0.01% Tricaine (MS-222) solution in distilled water. Prepare a 0.4% stock solution dissolving 400 mg Tricaine powder in 97.9 mL double-distilled water and add 2.1 mL 1 M Tris (pH 9). Adjust pH to ~7. Store the stock solution in the freezer. At the beginning of the experimental session, dilute 8 mL of Tricaine stock solution in 300 mL of fish water in the 600 mL beaker.

Fig. 1 (continued) epidermis. (**b**) Zebrafish scales are disk-shaped dermal bones covering the body of the fish. An array of scales is removed using forceps and regenerates in about 2 weeks. (**c**) Regenerating scales on the trunk of a fish imaged over time. (**d**) Phases of osteoblast tissue regeneration (yellow: high Erk activity). (**e**) Erk activity waves pattern tissue growth. Quantification of Erk activity as reported by Erk KTR (left). Tissue velocity field (tissue flow, blue arrows) and its divergence (heat-map), indicating tissue expansion and contraction (right). Dpp (hpp): days (hours) post plucking. Scale bar: 250 μm. Panel A is adapted from [18] with permission from Elsevier. Panels B, C, E are adapted from [19]; Copyright © 2021, The Author(s), under exclusive license to Springer Nature Limited

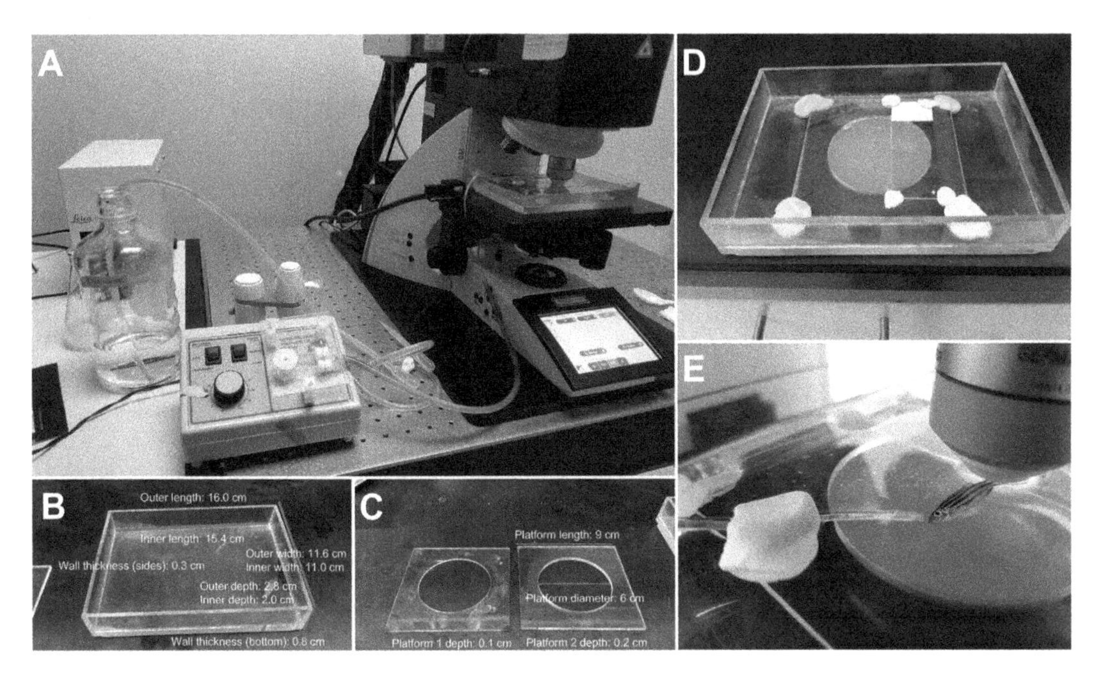

Fig. 2 Imaging platform for adult zebrafish. (**a**) An anesthetized fish is mounted on a plexiglass tray and imaged using a confocal microscope. A peristaltic pump loaded with system water is used to stimulate fish if operculum movements slow or stop. (**b**) Plexiglass dish. (**c**) Plexiglass plates, used to mold an agarose pad. (**d**) Prepared imaging dish, with plexiglass plates, agarose pad, and glass slide. (**e**) Anesthetized fish lying on a 1% agarose bed with caudal fin raised on glass slide to bring the imaged region as parallel as possible to the stage. Cooling 1% agarose is applied immediately prior to solidification on the caudal fin, the trunk of the fish anterior and posterior to the scale of interest and the parts of the agarose pad dorsal and ventral of the trunk. The outflow tubing of the peristaltic pump is placed near the mouth of the fish and held in place using nontoxic modeling clay. Adapted from [18] with permission from Elsevier

14. Melted agarose: 1% agarose solution in fish water. Weigh 0.75 g of agarose and transfer it to 75 mL of fish water in the 250 mL flask. Dissolve agarose by microwaving the solution for 30 s, stir, microwave for 15 s, and again stir and repeat microwaving for 15 s until the agarose is completely dissolved. Put the flask containing the melted agarose in a heated water bath at 40 °C and let it cool down. You will use this agarose solution to prepare an agarose pad in the plexiglass tray (see below) and to mount fish (*see* Subheading 3.2).
15. Fish mount (Fig. 2a) composed of a plexiglass tray (Fig. 2b), plexiglass plate(s) with a hole 6 cm in diameter (Fig. 2c), and a microscope slide (Fig. 2d). Place the plexiglass plates in the center of the plexiglass tray and fix it with nontoxic modeling clay (Fig. 2d). Place a glass microscope slide on one end of the hole (this will allow elevation of the caudal fin) and attach it with modeling clay (Fig. 2d). Fill the hole of the plexiglass plate with 1% melted agarose in system water. Let the agarose pad cool down.

2.3 Laser-Mediated Photoconversion

1. Adult transgenic fish (3–18 months old) expressing the osteoblast nuclear marker *osx:*H2A-mEos2 [19].
2. Confocal microscope equipped with a photoconverting laser (typically a UV 405 nm laser) that can be restricted to a select region of interest (ROI). In our microscope, this is possible using a FRAP module.
3. The same imaging setup described in Subheading 2.2.

2.4 Scale Image Processing

1. Images from adult transgenic fish (3–18 months old) expressing the osteoblast nuclear marker *osx:*H2A-mCherry [18] (*see* **Note 3**).
2. MATLAB (Mathworks; 2016b) or alternative image processing software. The reader can write custom-image processing code following the steps described in the Methods. Alternatively, the reader can use our code [19] that can be downloaded at: https://github.com/desimonea/DeSimoneErkwaves2020.
3. Fiji [31] or equivalent image visualization and processing software. Fiji can be downloaded at: https://imagej.net/software/fiji/.

2.5 Scale Image Quantification: Nuclear Segmentation and Reporter Signal Quantification

1. Images from adult transgenic fish (3–18 months old) expressing:
   (a) The osteoblast nuclear marker *osx:*H2A-mCherry [18] (*see* **Note 3**).
   (b) A transgenic reporter. Here, we will describe an example using an Erk activity sensor Erk KTR (Kinase Translocation Reporter; *osx:*ErkKTR-mCerulean [19]) and a reporter localized in the nucleus, i.e., the Cdh1 inactivation reporter *osx:*Venus-hGeminin [18].
2. MATLAB (Mathworks; 2016b) or alternative image processing software. The reader can write custom-image processing code following the steps described in the Methods. Alternatively, the reader can use our code [19] that can be downloaded at: https://github.com/desimonea/DeSimoneErkwaves2020.
3. Nuclear segmentation code. We use TGMM 1.0 software [32] that can be found at: https://sourceforge.net/projects/tgmm.

2.6 Scale Image Quantification: Tissue Flows

1. Images from adult transgenic fish (3–18 months old) expressing the osteoblast nuclear marker *osx:*H2A-mCherry [18] (*see* **Note 4**).
2. MATLAB (Mathworks; 2016b or later) or alternative image processing software. The reader can write custom-image processing code following the steps described in the Methods. Alternatively, the reader can use our code [19] that can be

downloaded at: https://github.com/desimonea/DeSimoneErkwaves2020.

3. Nuclear segmentation and tracking code. We use Ilastik software 1.3.3 [33] that can be downloaded at: https://www.ilastik.org/.

3 Methods

3.1 Scale Injury

1. Fill half of a finger bowl with 0.075% 2-phenoxyethanol anesthetic solution (*see* **Note 2**).
2. Transfer an adult fish to the finger bowl using a fish net. Wait until swimming ceases and operculum movements slow down, but do not halt. Check reaction to tail pinch to confirm appropriate depth of anesthesia.
3. Transfer the anesthetized fish to a Petri dish lid using a plastic spoon. Using a transfer pipette, put some drops of 0.075% 2-phenoxyethanol anesthetic solution around the head of the fish. Move the lid with the anesthetized fish on the stage of a fluorescence dissecting scope.
4. Turn fluorescent light on and use the eye piece of the dissecting scope to locate the scale you want to pluck. Typically, we pluck scales close to caudal peduncle, which are easier to image, as they are smaller in size and at low inclination with respect to the stage.
5. Pick the caudal end of the scale using forceps and pull gently in the direction of the caudal fin (Fig. 1b). The scale will easily come off. We usually pluck 5–15 scales from three adjacent rows (Fig. 1c).
6. Return the fish to a tank filled with system water. Gently flush system water using a transfer pipette to recover the fish from anesthesia. Monitor full fish recovery.

3.2 Scale Regeneration Imaging

1. Pluck scales as described in Subheading 3.1.
2. Pour 100 mL of 0.01% Tricaine anesthetic solution in a finger bowl.
3. Transfer an adult fish to the anesthetic solution in the finger bowl for approximately 15–30 min or until swimming ceases and operculum movement slow down, but do not halt. Check reaction to tail pinch to confirm appropriate depth of anesthesia.
4. Pick the fish using a plastic spoon and place it on the agarose bed of the fish mount (Fig. 2e). Position the caudal fin on a glass slide to bring the caudal peduncle as parallel as possible to the platform.

5. Place several drops of diluted Tricaine near the head of the fish. Use light-duty wipes to dry up the agarose bed around the fish.
6. Using a transfer pipette, take 1% melted agarose and let it cool down briefly. Shortly before agarose solidifies, apply the cooling agarose on the caudal fin, the trunk anterior and posterior of the scale of interest, and the areas dorsal and ventral to the fish, avoiding placement of agarose on the scale of interest. Let agarose solidify completely (*see* **Note 5**).
7. Load a peristaltic pump with fish water and attach the outflow tubing to fit near the mouth of the fish. You can hold the tubing using nontoxic modeling clay or a small clamp.
8. Pour 120 mL of 0.01% Tricaine imaging medium in the fish tray, or just enough to cover the fish and immerse the objective lens.
9. From now on, monitor the movements of the visible operculum of the fish. When movements slow, apply system water (typically 3.5 mL/min) until regular rhythm is restored.
10. Using the eye piece and fluorescent light, move the objective to the posterior side of the scale of interest.
11. Image the scale. We typically acquire 1024 × 1024 pixels images with 0.606 μm pixel size and a *z*-step of 0.606 μm (100–300 planes per stack) (*see* **Note 6**). We find that 8-bit images are sufficient for quantifications. As scales are typically larger than the field-of-view of confocal microscopes, multiple overlapping stacks (hereafter, referred to as "Positions") are needed. Start setting up a stack at the posterior side of the scale, then proceed adding overlapping stacks until you have covered the entire volume. Stacks will have variable *z*-position and depth owing to the inclination and curvature of scales. One to nine overlapping stacks are usually needed to image the entire osteoblast tissue.
12. Adjust laser power to reach the maximum of the dynamic range in some sparse points in the posterior of the scale. The anterior of the scale will be dimmer, but it can still be quantified and/or equalized for visualization.
13. Image the scale (*see* **Note 7**). While imaging, monitor the fish and, if needed, provide fresh fish water as described in **step 9** (you may need to pause imaging while doing so).
14. Export stacks as compressed TIFF files.
15. Repeat the procedure for each scale of interest in the anesthetized fish.
16. Unmount the fish. First, activate the peristaltic pump to apply system water until operculum movements quicken. Then, release the fish from agarose and return it to system water.

Gently flush system water using a transfer pipette until the fish recovers from anesthesia and restarts swimming. Monitor full fish recovery during the day and every 24 h from then.

17. Repeat the procedure for the other fish you are imaging.
18. Repeat the procedure at each time-point. We typically perform 1 time-point every 0.5, 3, 6, 12, or 24 h (*see* **Note 8**). If you are performing high time-resolution imaging (>1 time-point every 30 min), maintain the fish mounted in between time-points and use the time-lapse function of the microscope. Check routinely that the scale has not moved out of focus. Check that the movements of the visible operculum are not slowed down or halted (**step 9**). If the scale moves out of focus, reset stacks *z*-position. Replace the imaging medium with freshly prepared 0.01% Tricaine imaging medium every 4 h using a serological pipette.

3.3 Laser-Mediated Photoconversion

1. Pluck scales as described in Subheading 3.1. Anesthetize and mount a fish as described in Subheading 3.2. Locate the scale of interest.
2. Open the photoconversion tool of the microscope. Zoom to region of 100–200 μm size. Locate a nucleus of interest. Draw a region of interest (ROI) on the nucleus of interest. If possible, select "Zoom in" to concentrate laser power in the ROI.
3. Set the photoconversion laser to a high power, potentially 100%. Scan multiple times the ROI (e.g., 3 rounds of 10 iterations).
4. Check whether the desired level of photoconversion has been achieved (*see* **Note 9**). If not, repeat the photoconversion procedure.
5. Repeat the photoconversion procedure on all nuclei of interest. We typically convert ten nuclei (Fig. 3).
6. Image the photoconverted scale and recover fish from anesthesia as described in Subheading 3.2. Photoconverted nuclear fluorescent proteins can be detected many days after the initial photoconversion (Fig. 3) [19]. If photoconverted signal becomes dimmer over time, the photoconversion procedure can be repeated on the same nucleus.

3.4 Scale Image Processing

After image acquisition, raw images comprise several overlapping *z*-stacks ("Positions"; Fig. 4a). Scales are mildly curved and positioned with an inclination with respect to the *xy* plane (Fig. 4a). Several scales overlap in the same stack. Each scale includes a monolayer of hyposquamal osteoblasts on the dermal side, here the tissue of interest, and a sparse population of episquamal osteoblasts on the epidermal side. Therefore, several image processing steps are required to isolate the hyposquamal layer of a single scale

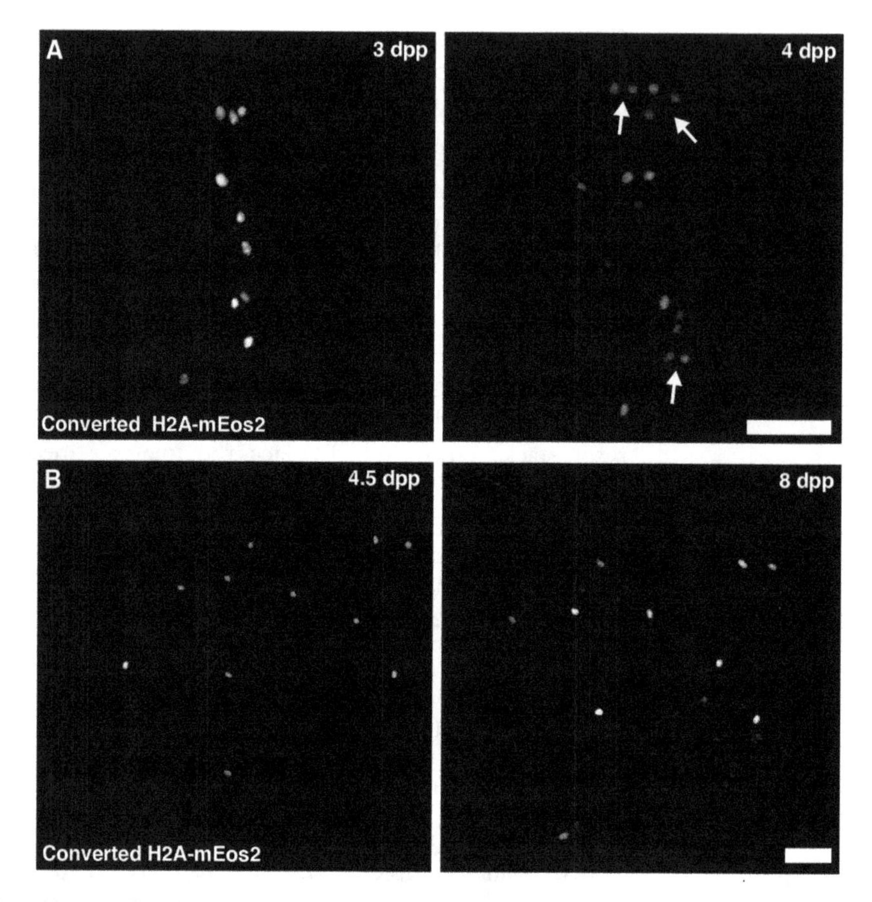

Fig. 3 Osteoblast nuclei photoconversion. (**a**) Osteoblast nuclei tagged with the photoconvertible protein mEos2 are photoconverted during the proliferative phase (3 dpp). Nuclei are imaged after 1 day and tracked. Cell division events can be detected (white arrows). (**b**) Osteoblast nuclei tagged with the photoconvertible protein mEos2 are photoconverted during the hypertrophic phase (4.5 dpp). Nuclei are imaged and tracked thereafter; nuclei can be detected 4 days later. Scale bars: 50 μm. Dpp: days post plucking. Adapted from [19]; Copyright © 2021, The Author(s), under exclusive license to Springer Nature Limited

computationally and obtain a single stack that is manageable and accessible to quantifications [19]. Here, I will describe the steps that are needed for this image processing task, which the reader can use as guideline to write an image processing code. Alternatively, the reader can use our MATLAB code (*see* **Materials**). Please, refer to the documentation of the code for installation and usage.

At each step, the nuclear histone marker *osx*:H2A-mCherry is used as a reference channel (*see* **Note 3**) to compute the required image transformations. Then, we apply those transformation to the other channels. At each intermediate step, we create a new folder in which we save the processed stacks (LZW compressed TIFF files).

1. Import the metadata of the images (*see* **Note 10**).
2. Stitch individual overlapping stacks ("Positions") into a single stack (Fig. 4b). The relative 3D coordinates of each Position

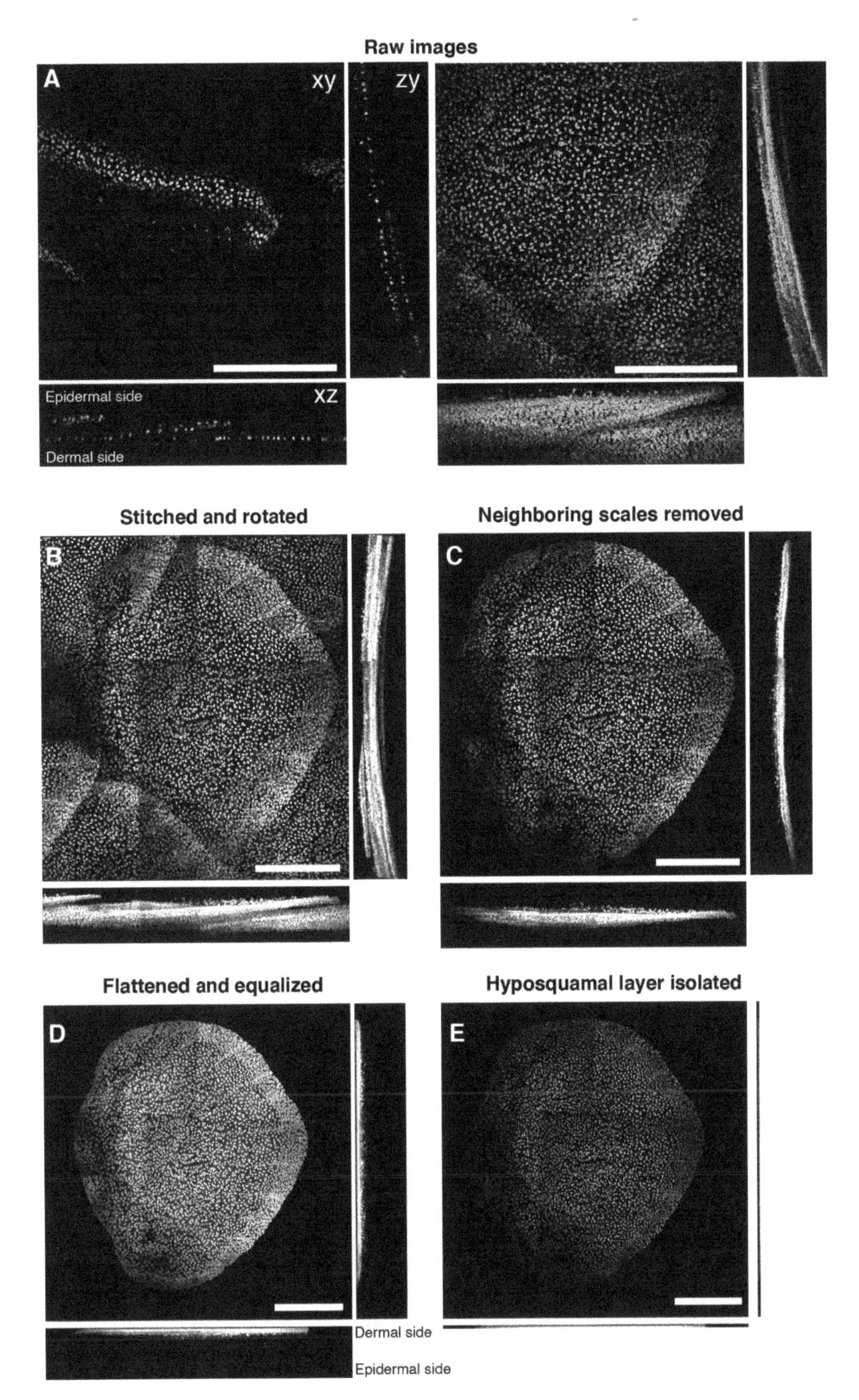

Fig. 4 Image processing of regenerating scales. (**a**) Example of an individual slice (left) and maximum projection (right) of a single raw stack ("Position") from a regenerating scale at 4 days post plucking. Here

can be calculated using cross-correlation between those Positions in the reference channel (*see* **Note 11**). Stitch the other channels using the same parameters calculated for the reference channel. This step can be performed using the function stitchScale in our code.

3. In **steps 4–9**, you will rotate the stack to position the osteoblast tissue as parallel as possible to the *xy* axes; then, you will crop unnecessary portions of the stack (Fig. 4b). You will calculate rotation angles and cropping rectangles using the reference stack and then apply those transformations to all channels. These steps can be performed using the function rotateScale in our code.
4. Crop the stack to remove neighboring scales or empty regions. To do this, perform max *z*-projection of the stack and select a rectangle ROI manually (*see* **Note 12**); crop the stack using this rectangle ROI.
5. Permute the axes to turn the *z*-stack into an *x*-stack composed of *yz* planes and perform a max *x*-projection. Fit the max *x*-projection with a line (use pixel intensities as fit weights). Calculate the inclination angle θ_{yz} of this line with respect to the *y* axis. Rotate *yz x*-stack of $-\theta_{yz}$ around the *x* axis. Adjust the size of the stack to accommodate for the rotated volume.
6. Crop the *x*-stack to remove neighboring scales or empty regions. To do this, perform a max *x*-projection of the *x*-stack and select a rectangle ROI manually (*see* **Note 12**); crop the *x*-stack using this rectangle ROI.
7. Permute the axes to turn the *x*-stack into a *y*-stack composed of *xz* planes and perform a max *y*-projection. Fit the max *y*-projection with a line (use pixel intensities as fit weights). Calculate the inclination angle θ_{xz} of this line with respect to the *x* axis. Rotate the *y*-stack of $-\theta_{xz}$ around the *y* axis. Adjust the size of the stack to accommodate for the rotated volume.
8. Crop the *y*-stack to remove neighboring scales or empty regions. To do this, perform a max *y*-projection of the *y*-stack

Fig. 4 (continued) and hereafter, *xy*, *xz*, and *zy* views are shown. (**b**) Individual positions are stitched and the stack is rotated in 3D to position the osteoblast tissues as parallel as possible to the *xy* axes. Max projections are shown. (**c**) The osteoblast tissue of an individual scale is isolated by removing neighboring scales from the stack computationally. Max projections are shown. (**d**) After a second rotation, the stack is deformed in *z* to bring the dermal end of the osteoblast tissue on the same plane. The *z*-axis is inverted in this operation, so that the dermal side of the scale is on the top part of the *xz* projection. Max projections are shown (equalization was applied—*see* Subheading 3.5, **step 2**). (**e**) A slice of the stack is selected to isolate the hyposquamal layer of an individual scale. Max projections are shown (equalization was applied—*see* Subheading 3.5, **step 2**). Scale bars: 250 μm. Data in this image were published in [19]

and select a rectangle ROI manually (*see* **Note 12**); crop the *y*-stack using this rectangle ROI.

9. Permute the axes to turn the *y*-stack into a *z*-stack composed of *xy* planes. Perform a max *z*-projection of the *z*-stack and select a rectangle ROI manually (*see* **Note 12**); crop the *z*-stack using this rectangle ROI.
10. In **steps 4–9**, rotation and cropping result in the removal of portions of neighboring scales from the stack. In this step, you will continue removing neighboring scales computationally. You will calculate regions using the reference stack and then apply those transformations to all channels. Perform a max *z*-projection of the *z*-stack. Select a polygonal ROI around the scale of interest in the *z*-projection (*see* **Note 12**). Set voxel intensity to zero in the regions of the *z*-stack having *xy* coordinates outside the ROI. Permute the axes and repeat the same procedure to select a *xz* ROI using max *y*-projections. Set voxel intensity to zero in the regions of the *y*-stack having *xz* coordinates outside the *xz* ROI. Permute the axes and repeat the same procedure to select a *yz* ROI using max *x*-projections. Set voxel intensity to zero in the regions of the *x*-stack having *yz* coordinates outside the *yz* ROI. Create a 3D mask "ROI1" corresponding to the intersection of the *xy*, *xz*, and *yz* ROIs. This step can be performed using the function cleanupScale in our code.
11. After the previous step, portions of neighboring scales will likely remain in the stack; in this step, you will remove them proceeding plane-by-plane (Fig. 4c). Add one intensity value to each voxel of the partially cleaned reference stack and save it as ROI2. Open the stack ROI2 in Fiji. Proceed plane-by-plane to manually select polygonal regions corresponding to neighboring scales and set them to zero intensity. This is a tedious operation and a mouse with a good scroll-wheel will help. Good music too. When you have eliminated all the neighboring scales, save ROI2.
12. Create a 3D mask ROI3 that is the intersection of ROI1 and ROI2 $>$ 0. Set intensity values to zero in all regions outside ROI3 in all channels. These steps can be performed using the function cleanupScale in our code.
13. Repeat the rotation **steps 3–9** to further bring the select scale tissue parallel to the *xy* plane.
14. If you have imaged a time-series, register the stacks in 3D using a cross-correlation approach (*see* **Note 13**). This is needed for nuclei tracking and tissue flow measurements (*see* Subheading 3.6). These steps can be performed using the function imregisterStackSeriesXY3D in our code.
15. Segment the osteoblast tissue in the reference channel using intensity-based thresholding and morphological operations.

This step can be performed using the function segmentMyScale in our code.

16. In this step, you will deform the stack in the *z* direction to bring the osteoblast tissue parallel to the *xy* plane ("flattening"; Fig. 4d). This deformation will not change dramatically the geometry of the scale as they typically extend 10 times more in *xy* than in *z*. First, use the scale segmentation mask to calculate the *z*-position of the dermal end of the osteoblast tissue for each *xy* position. Then, for each *xy* coordinate, shift the *z*-coordinate to bring the dermal end on a select plane (this plane will be the same for each *xy* position). Apply this operation to all channels. These steps can be performed using the function flattenLayersScaleSegm in our code.
17. In this step, you will select a *z*-slice of the flattened stack to isolate the hyposquamal osteoblast tissue (Fig. 4e). To compute the thickness of the slice, calculate the total intensity of each *z*-plane of the flattened reference stack; since the hyposquamal tissue contains the majority of osteoblasts in 4 dpp to mature scales, the *z* coordinate of its centroid corresponds roughly to the peak of the *z*-intensity profile. Thus, select a slice starting at the dermal end of the flattened scale until the peak of the total intensity *z*-profile (a thicker slice would include cells from the episquamal layer). Apply this procedure to all channels. These steps can be performed using the function divideLayersScaleSegm in our code.

At the end of this procedure, you will obtain a relatively small *z*-stack (<20 μm; thereafter, the "hyposquamal stack"), including the entire hyposquamal osteoblast tissue of a single regenerating scale (Fig. 4e). The stack is now ready for quantification.

3.5 Image Quantification: Nuclei Segmentation and Reporter Signal Quantification

Here, I describe a method to quantify Erk activity and Cdh1 inactivation using the *osx*:ErkKTR-mCerulean KTR sensor (Figs. 1e and 5a) and the *osx*:Venus-hGeminin nuclear reporter [19]. Use the *osx*:H2A-mCherry nuclear marker as reference stack for nuclear segmentation (*see* **Note 4**). The same approach can be applied to any KTR or nuclear reporter.

1. Process images as described in Subheading 3.4.
2. Apply contrast-limited adaptive histogram equalization to the hyposquamal nuclear marker stack obtained in Subheading 3.4. This step can be performed using the function equalizeScale in our code.
3. Segment nuclei in 3D using TGMM software on the equalized hyposquamal nuclear marker stack (*see* **Note 14**). Please refer to TGMM documentation for its usage. A sample

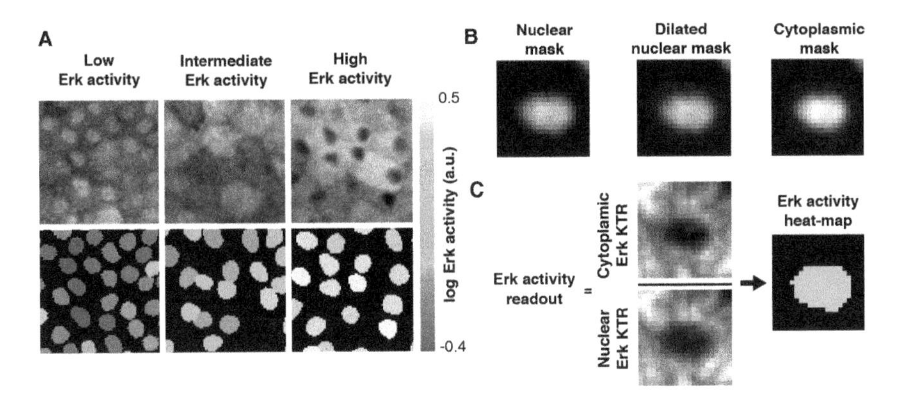

Fig. 5 Quantification of the Erk Kinase Translocation Reporter (KTR). (**a**) Quantified Erk activity in scale osteoblasts. (**b**) Nuclear and cytoplasmic masks for Erk KTR signal calculation (red-shaded areas). The cytoplasmic mask is calculated as the difference between a dilated nuclear mask and the nuclear mask itself. (**c**) The KTR readout (Erk activity) is calculated as the ratio of the average Erk KTR signal in the cytoplasmic and nuclear mask regions. Erk activity maps (Figs. 1e left and 5a) are generated by representing each nucleus with a Look Up Table (LUT) that depends on its measured Erk activity. Panel A is adapted from [19]; Copyright © 2021, The Author(s), under exclusive license to Springer Nature Limited, where data in this image were published

configuration file is provided in our GitHub repository https://github.com/desimonea/DeSimoneErkwaves2020.

4. Import the nuclear segmentation in MATLAB using the TGMM-provided MATLAB functions readXMLmixtureGaussians and readListSupervoxelsFromBinaryFile. In **steps 5–10**, you will measure KTR and nuclear reporter signals. These steps can be performed using the function assignKTRReduced in our code.
5. For a given segmented nucleus, create a sub-stack that roughly includes its entire cell and portions of the neighbors. Use the nuclear segmentation to create a 3D nuclear mask (only one nucleus) corresponding to the sub-stack.
6. Calculate the *xy*-total intensity *z*-profile of the 3D mask. Select the plane with highest *z*-intensity profile; you will use this plane for quantifications. Hereafter, I will refer to the select plane of the 3D nuclear mask as "2D nuclear mask" (Fig. 5b "Nuclear mask"; *see* **Note 15**).
7. Dilate the 2D nuclear mask of a few pixels (e.g., using a 3 μm × 3 μm flat structuring element; Fig. 5b "Dilated nuclear mask"; *see* **Note 16**). Subtract the 2D nuclear mask from the dilated 2D nuclear mask to obtain a 2D cytoplasmic mask (Fig. 5b "Cytoplasmic mask"); this mask will not include the entire cytoplasm, but just a portion of it.
8. For Kinase Translocation Reporter (KTR) quantification, average the KTR channel in the regions corresponding to the

cytoplasmic and nuclear masks; the KTR read-out will be the ratio of the signals calculated in the cytoplasmic and nuclear regions (Fig. 5c "Erk activity readout").

9. For nuclear reporter quantification, average the reporter signal in the region corresponding to the nuclear mask. As scale stacks have illumination gradients due to variable tissue depth and overlap with other scales, we usually normalize this nuclear signal by the cytoplasmic signal (hence the inverse of what we do for KTRs).
10. Repeat the procedure for all nuclei.
11. You can generate heat-maps in which each nucleus is colored using a Look Up Table (LUT) that depends on its KTR or nuclear reporter intensity (Fig. 5c "Erk activity heat-map"). This step can be performed using the function visualizeERKKTR in our code.

3.6 Image Quantification: Tissue Flows

Here, I describe a method to measure tissue movements (flows) using the nuclear marker *osx:*H2A-mCherry [19] (Fig. 1e "Tissue flows"; *see* **Note 4**). For these quantification, time-courses spanning 9–15 h are needed. For a 4–5 dpp scale, a time resolution of 1 time-point every 3 h allows one to track nuclei (*see* **Note 17**).

1. Process images as described in Subheading 3.4, including stack registration for time-points of the same scale.
2. Apply contrast-limited adaptive histogram equalization to the hyposquamal nuclear marker stack obtained in Subheading 3.4. This step can be performed using the function equalizeScale in our code.
3. Segment and track osteoblast nuclei using Ilastik software and the equalized hyposquamal stack (*see* **Note 18**). Please refer to the Ilastik documentation for its installation and usage.
4. Import tracking results in MATLAB. Calculate the *xy* velocities of each nucleus using its *xy* displacement from the first to the last time-point of interest (often, the whole time-series).
5. Bin nuclei in a 30 μm-sized grid and average nuclear velocities within the same bin to obtain a tissue velocity map.
6. Smoothen the velocity map using an averaging filter (e.g., size: 60 μm; *see* **Notes 19** and **20**).
7. Calculate the divergence and curl (i.e., vorticity) of the velocity map and smoothen it using a Gaussian filter (e.g., SD: 91 μm; size: 152 μm; *see* **Notes 19** and **20**). The divergence map reveals regions of tissue expansion (positive divergence) and contractions (negative divergence). The curl map reveals regions in which the flow is rotating locally.
8. Visualize velocity fields using quiver plots and the divergence/curl using heat-maps.

4 Notes

1. Many fluorescent markers and reporters expressed in scale osteoblasts are available, under control of the zebrafish or medaka *sp7* (*osx*) promoters [17–20, 26–28]. These include the cytoplasmic markers *osx:*EGFP [26] and *osx:*mCherry [28], the nuclear markers *osx:*NLS-GFP [28] and *osx:*H2A-mCherry [18], the membrane markers *osx:*EGFP-CAAX [34] and *osx:*GFP-CAAX [20], the cell-cycle reporter FUCCI (*osx:* mCherry-zCdt1 *osx:*Venus-hGeminin) [18], the photoconvertible cytoplasmic marker *osx:*kaede [18], the photoconvertible nuclear marker *osx:*H2A-mEos2 [19], the conditional osteoblast ablation tool *osx:*mCherry-NTR [35], and the Erk activity sensor *osx:*ErkKTR-mCerulean [19].
2. We successfully anesthetized fish in a 0.07–0.09% 2-phenoxyethanol range.
3. An alternative nuclear or cytoplasmic marker can be used as reference channel to calculate image processing transformations. For FUCCI fish (*osx:*Venus-hGeminin *osx:*mCherry-zCdt1), we use the combined Venus-hGeminin and mCherry-zCdt1 signals as nuclear osteoblast markers.
4. An alternative nuclear marker can be used to segment or track nuclei. For FUCCI fish (*osx:Venus-hGeminin osx:mCherry-zCdt1* fish), we use the combined Venus-hGeminin and mCherry-zCdt1 signals as nuclear osteoblast markers.
5. It is important that the fish does not move at this stage; if it does, this will leave space in the agarose, allowing the fish to perform small imaging-destructive movements. If the fish moves, unmount it as described in Subheading 3.2 **Step 16** and mount it again.
6. Imaging scales with a 0.606 μm voxel size offers a good compromise between spatial resolution, acquisition time, and image size. For faster imaging, 1.2 μm voxel *xy*-size and 1.2–1.7 μm voxel *z*-size can be used; this lower resolution allows quantifications in scales past 4 dpp, but the quality of results may depend on cell density. Dim signals may require acquiring images at 12 or 16 bits. For faster imaging, we often image multiple fluorescence channels at the same time, if they are spectrally compatible on our microscope.
7. Imaging one scale can take 5–45 min depending on the size of the stack, the number of positions, and the number of channels imaged sequentially.
8. Based on comparison of signaling and cell dynamics from time-lapse at different frame rates, it is possible that regeneration kinetics are slowed by long-term anesthesia [18].

9. It is important that fish do not move during photoconversion, otherwise photoconversion will be less efficient and will affect multiple neighboring nuclei, which may confound tracking.
10. Our code can import Metadata from a Leica SP8 microscope (stitching step; see Materials). If you are using a different microscope, please write a Metadata import code.
11. In principle, stage coordinates reported in the Metadata provide information about the overlap between Positions. However, we find that using those coordinates does not provide a good stitching, likely owing to slight movements of the stage, the agarose, or the fish. Nevertheless, one can reduce computational time by restricting the calculation of cross-correlation between two Positions to their overlapping portion, as calculated from stage coordinates. We stitch one Position at the time. As 3D cross-correlation can be computationally expensive, we use a 2D *xy* correlation-based approach. In brief, we calculate the 2D *xy* correlation of a select plane (e.g., the one with max *z*-intensity) of the first stack with all the planes of the second stack, then pick the *z*-shift with highest *xy* correlation; we proceed iteratively to add all Positions to the same stack.
12. It may be difficult to locate the boundary of the scale in some part of the max projections. In this case, give enough room to ensure to not remove portions of the scale of interest.
13. We use 4–5 iterations of the MATLAB registration function imregtform using different metrics and/or reference channels at each iteration.
14. You can use any nuclear segmentation software that works well with your images.
15. Optionally, you can refine the 2D nuclear mask using a closing operation with a small filter (e.g., using a 3 μm × 3 μm flat structuring element).
16. When calculating the 2D cytoplasmic mask, you can apply or not apply the closing step to the subtracted 2D nuclear mask.
17. Typically, we image 3–5 fish during each 3 h time-point.
18. You can use any nuclear segmentation and tracking software that works well with your images, including TGMM.
19. When applying an averaging or gaussian filter for smoothening, the boundary of the scale requires special treatment. We usually set the velocity in the region outside the scale to NaN and use the function nanconv (without edge correction) for filtering.
20. The averaging and gaussian smoothening filter will introduce spatial autocorrelations in the velocity field. Thus, the size of the filter must be chosen to be smaller than the typical size of the pattern of interest. For calculations of spatial autocorrelations, the unsmoothed velocity field should be used.

Acknowledgments

I thank J. Burris, S. Miller, K. Oliveri, C. Dolan, L. Frauen, and D. Stutts for zebrafish care; B. Cox for sharing transgenic animals and protocols; S. Di Talia and K. Poss for their supervision during the development of the methods presented here; P. Adhyapok, S. Di Talia, M. Evanitsky, Z. Lu, and A. Rich for critical reading of the manuscript. A.D. was supported by Early (P2ELP3_172293) and Advanced (P300PA_177838) Postdoc.Mobility fellowships from the Swiss National Science Foundation, as well as an Innovation in Stem Cell Science Award from the Shipley Foundation, Inc. to Stefano Di Talia.

References

1. Yoshinari N, Kawakami A (2011) Mature and juvenile tissue models of regeneration in small fish species. Biol Bull 221:62–78
2. Poss KD (2010) Advances in understanding tissue regenerative capacity and mechanisms in animals. Nat Rev Genet 11:710–722
3. Tanaka EM, Reddien PW (2011) The cellular basis for animal regeneration. Dev Cell 21: 172–185
4. Tornini VA, Poss KD (2014) Keeping at arm's length during regeneration. Dev Cell 29:139–145
5. Di Talia S, Poss KD (2016) Monitoring tissue regeneration at single-cell resolution. Cell Stem Cell 19:428–431
6. Grillo M, Konstantinides N, Averof M (2016) Old questions, new models: unraveling complex organ regeneration with new experimental approaches. Curr Opin Genet Dev 40:23–31
7. Masselink W, Tanaka EM (2021) Toward whole tissue imaging of axolotl regeneration. Dev Dyn 250:800–806
8. Oates AC, Gorfinkiel N, Gonzalez-Gaitan M, Heisenberg CP (2009) Quantitative approaches in developmental biology. Nat Rev Genet 10:517–530
9. Gregor T, Bialek W, van Steveninck RRDR, Tank DW, Wieschaus EF (2005) Diffusion and scaling during early embryonic pattern formation. Proc Natl Acad Sci USA 102:18403–18407
10. Muller P et al (2012) Differential diffusivity of nodal and lefty underlies a reaction-diffusion patterning system. Science 336:721–724
11. Wartlick O et al (2011) Dynamics of Dpp signaling and proliferation control. Science 331: 1154–1159
12. Kicheva A et al (2007) Kinetics of morphogen gradient formation. Science 315:521–525
13. Mateus R et al (2020) BMP Signaling gradient scaling in the zebrafish pectoral fin. Cell Rep 30:4292
14. Soroldoni D et al (2014) Genetic oscillations. A Doppler effect in embryonic pattern formation. Science 345:222–225
15. Dona E et al (2013) Directional tissue migration through a self-generated chemokine gradient. Nature 503:285
16. Durdu S et al (2014) Luminal signalling links cell communication to tissue architecture during organogenesis. Nature 515:120
17. Aman AJ, Fulbright AN, Parichy DM (2018) Wnt/beta-catenin regulates an ancient signaling network during zebrafish scale development. elife 7
18. Cox BD et al (2018) In Toto imaging of dynamic osteoblast behaviors in regenerating skeletal bone. Curr Biol 28:3937–3947. e3934
19. De Simone A et al (2021) Control of osteoblast regeneration by a train of Erk activity waves. Nature 590:129–133
20. Iwasaki M, Kuroda J, Kawakami K, Wada H (2018) Epidermal regulation of bone morphogenesis through the development and regeneration of osteoblasts in the zebrafish scale. Dev Biol 437:105–119
21. Rasmussen JP, Vo NT, Sagasti A (2018) Fish scales dictate the pattern of adult skin innervation and vascularization. Dev Cell 46:344–359. e344
22. Metz JR, de Vrieze E, Lock E-J, Schulten IE, Flik G (2012) Elasmoid scales of fishes as model in biomedical bone research. J Appl Ichthyol 28:382–387

23. Pasqualetti S, Banfi G, Mariotti M (2012) The zebrafish scale as model to study the bone mineralization process. J Mol Histol 43:589–595
24. Sire JY, Akimenko MA (2004) Scale development in fish: a review, with description of sonic hedgehog (shh) expression in the zebrafish (Danio rerio). Int J Dev Biol 48:233–247
25. Bereiter-Hahn LZJ (1993) Regeneration of teleost fish scale. Comp Biochem Physiol A Physiol 105:625–641
26. DeLaurier A et al (2010) Zebrafish sp7:EGFP: a transgenic for studying otic vesicle formation, skeletogenesis, and bone regeneration. Genesis 48:505–511
27. Renn J, Winkler C (2009) Osterix-mCherry transgenic medaka for in vivo imaging of bone formation. Dev Dyn 238:241–248
28. Spoorendonk KM et al (2008) Retinoic acid and Cyp26b1 are critical regulators of osteogenesis in the axial skeleton. Development 135: 3765–3774
29. Richardson R et al (2016) Re-epithelialization of cutaneous wounds in adult zebrafish combines mechanisms of wound closure in embryonic and adult mammals. Development 143: 2077–2088
30. Kobayashi-Sun J et al (2020) Uptake of osteoblast-derived extracellular vesicles promotes the differentiation of osteoclasts in the zebrafish scale. Commun Biol 3:190
31. Schindelin J et al (2012) Fiji: an open-source platform for biological-image analysis. Nat Methods 9:676–682
32. Amat F et al (2014) Fast, accurate reconstruction of cell lineages from large-scale fluorescence microscopy data. Nat Methods 11:951–958
33. Sommer C, Straehle C, Kothe U, Hamprecht FA (2011) Ilastik: interactive learning and segmentation toolkit. IS Biomed Imaging 230–233
34. Nachtrab G, Kikuchi K, Tornini VA, Poss KD (2013) Transcriptional components of anteroposterior positional information during zebrafish fin regeneration. Development 140:3754–3764
35. Singh SP, Holdway JE, Poss KD (2012) Regeneration of amputated zebrafish fin rays from de novo osteoblasts. Dev Cell 22:879–886

Chapter 13

Generation of Conditional Knockout Zebrafish Using an Invertible Gene-Trap Cassette

Masahito Ogawa and Kazu Kikuchi

Abstract

Conditional knockout (cKO) is a genetic technique to inactivate gene expression in specific tissues or cell types in a temporally regulated manner. cKO analysis is essential to investigate gene function while avoiding the confounding effects of global gene deletion. Genetic techniques enabling cKO analysis were developed in mice based on culturable embryonic stem cells that were not generally available in zebrafish, which hampered precise analysis of genetic mechanisms of organ development and regeneration. However, recent advances in genome editing technologies have resolved this limitation, providing a platform for the generation of cKO models in any organism. Here we describe a detailed protocol for the generation of cKO zebrafish using a Cre-dependent genetic switch.

Key words Zebrafish, Conditional knockout, Gene trap, Cardiomyocytes, Heart regeneration

1 Introduction

Conditional knockout (cKO) is a genetic technique to inactivate gene expression in specific tissues or cell types in a temporally regulated manner [1–3]. cKO analysis is essential for investigating gene function if global deletion of the gene causes embryonic lethality or multiple tissue defects. Genetic techniques of the generation of cKO models were developed mainly in mice based on embryonic stem (ES) cells that possess an efficient capacity for homologous recombination [4, 5], thereby providing a platform for genetic engineering of cKO models. The most common approach to performing cKO analysis is to generate a mouse line with loxP-flanking (flox) genomic segments using ES cells and cross this line with a tissue-specific inducible Cre driver line to delete the floxed gene segments in specific tissues by tamoxifen administration. Alternatively, cKO models are generated by inserting a Cre-dependent gene-trap cassette, referred to as FLEX [6], at a target gene locus. The FLEX cassette consists of a splice-acceptor

James F. Amatruda et al. (eds.), *Zebrafish: Methods and Protocols*, Methods in Molecular Biology, vol. 2707,
https://doi.org/10.1007/978-1-0716-3401-1_13,

site and transcriptional stop sequence flanked with tandem loxP and loxP5171 sites. These sites are permanently inverted by Cre-mediated recombination, thus terminating the transcription of a target gene.

Traditionally, genetic engineering of floxed or FLEX alleles requires ES cells, but recent advances in genome editing technologies have opened a new avenue of cKO analysis of any model organism without the need for culturable ES cells. In zebrafish, recent studies successfully achieved the generation of loxP-inserted alleles using TAL effector nucleases (TALENs) or the CRISPR/Cas9 system with donor DNA templates for homology-directed repair [7–11]. Utilizing a FLEX approach, we generated a donor vector for the generation of zebrafish with an invertible gene-trap cassette via homologous recombination (Zwitch) [12]. Using this vector, we generated several cKO zebrafish models and performed cKO analysis in studies of heart regeneration [12, 13]. In this chapter, we describe a detailed protocol for the establishment of a cKO zebrafish line for application in heart regeneration research by referencing our recently published case utilizing the Krüppel-like factor 1 (*klf1*) gene [13].

2 Materials

2.1 Solutions

1. E3 embryo medium: 5 mM NaCl, 0.17 mM KCl, 0.33 mM $CaCl_2$, 0.33 mM $MgSO_4$. Add 100 mL of 1% methylene blue to 1 L E3 embryo medium before use.
2. 0.5% phenol red solution: Dissolve 5.0 g/L of phenol red sodium salt in Dulbecco's phosphate buffered saline and sterilize through a 0.22 μM pore size syringe filter.
3. Injection solution: 0.1 M KCl. Dilute 100 mL of 1 M KCl to 900 mL water. Sterilize through a 0.22 μM pore size syringe filter and store at 4 °C until use. Add 10% volume of 0.5% phenol red solution before injection.
4. Proteinase K stock: Dissolve the proteinase K powder at 40 mg/mL in 100 mM Tris–HCl (pH 8.0), 6 mM $CaCl_2$, and add an equal volume of 100% glycerol to make the final concentration of proteinase K to 20 mg/mL. Store at −20 °C.
5. DNA extraction buffer: 10 mM Tris–HCl (pH 8.0), 10 mM EDTA, 200 mM NaCl, 0.5% (w/v) SDS. Add 20 mg/mL proteinase K stock at final concentration 200 mg/mL before use.
6. Fin digestion buffer: 10 mM Tris–HCl (pH 8.0), 2 mM EDTA, 0.2% NP40. Add 20 mg/mL proteinase K stock at final concentration 200 μg/mL before use.

2.2 Reagents

The reagents used for this study were as follows:

1. Platinum Gate TALEN Kit (Addgene kit # 1000000043).
2. pZwitch2 vector [13].
3. PrimeSTAR GXL DNA Polymerase (Clontech Laboratories, Inc.).
4. mMESSAGE mMACHINE T7 Transcription Kit (Thermo Fisher Scientific).
5. DIG DNA Labeling and Detection Kit (Roche).
6. Prime-a-Gene Labeling System (Promega).

2.3 Equipment

The following equipment was used for this study:

1. Injection plates: Dissolve 3 g agarose in 100 mL E3 embryo medium, boil with a microwave, and pour 10–20 mL melted agarose into a 10 cm Petri dish. Remove bubbles and place a zebrafish injection mold onto the surface of agarose. When the agarose hardens, carefully detach the mold, pour the E3 embryo medium, and store at 4 °C until use.
2. Injection needles: Pull 1.0 OD mm glass capillaries (Harvard Apparatus) using a needle puller.
3. Veriti thermal cycler (Thermo Fisher Scientific).
4. Stereomicroscope SMZ745 (Nikon).
5. Fluorescent microscope SMZ18 (Nikon).
6. Needle puller PC-10 (Narishige).
7. Microinjection apparatus FemtoJet 4i (Eppendorf).
8. MultiNA microchip electrophoresis system (Shimadzu).

3 Methods

3.1 Identification of Insertion Sites

1. The insertion sites for a Zwitch cassette are the intronic regions where TALENs induce DNA double-strand breaks at high efficiency (*see* **Note 1**). Using the Ensembl Genome Browser (Zebrafish Genome), search for a gene of interest and obtain the entire sequence of the intron located after the exon containing the ATG initiation codon (*see* **Note 2**).
2. Use CHOPCHOP (https://chopchop.cbu.uib.no) or TAL Effector-Nucleotide Targeter 2.0 (https://tale-nt.cac.cornell.edu/node/add/talen) software to search for TALEN target sites in the intron sequence.
3. Select several TALENs that have no off-target sites, or the least number of off-target sites. Choose TALENs containing a restriction enzyme site in the spacer sequence. The enzyme site will be utilized later to check the TALEN efficiency via

restriction fragment length polymorphism (RFLP) assay [9, 14] (*see* **Note 3**). To reduce the risk of interfering with splicing machinery, avoid using TALENs that bind to the sequence within 100 bp from the flanking exons.

3.2 Generation and Efficiency Test of TALENs

1. Assemble TALENs using the Platinum Gate TALEN Kit and clone them into ptCMV-153/47-VR and ptCMV-136/63-VR vectors.
2. Synthesize TALEN mRNAs using the mMESSAGE mMACHINE T7 Transcription Kit following the manufacturer's protocol.
3. In the late afternoon or early evening, set up zebrafish breeding pairs in mating tanks with dividers, which separates male and female fish. Leave the tanks undisturbed with the lights off.
4. When the lights turn on in the next morning, remove the dividers and allow female and male zebrafish to mate. While waiting for the fish to spawn eggs, apply TALEN mRNAs into an injection needle and set the needle to the FemtoJet 4i.
5. Collect spawned eggs with a plastic transfer pipette and place them into the trenches of an injection plate containing E3 medium.
6. Inject eggs with approximately 50 pg TALEN mRNAs into one-cell-stage embryos.
7. Collect the injected embryos into a 1.5 mL tube at 1 day postfertilization (dpf). Add 500 mL DNA extraction buffer and incubate at 55 °C for 30 min. Purify DNA by phenol–chloroform extraction followed by ethanol precipitation. Dissolve DNA in 50 mL nuclease-free water. Similarly, purify the DNA of uninjected embryos as a negative control.
8. Amplify the DNA segment containing the TALEN target site (Fig. 1a) and perform an RFLP assay to estimate the TALEN efficiency (Fig. 1b). If no restriction enzyme sites are found in the spacer sequence, estimate the efficiency of TALENs by a heteroduplex mobility shift assay [15–17] (*see* **Note 4**). This assay separates polymerase chain reaction (PCR) products containing mismatches on the basis of the DNA property by which heteroduplexes migrate more slowly than do homoduplexes during electrophoresis (Fig. 1c, d).
9. For subsequent steps, select TALEN pairs with the highest efficiency in inducing DNA double-strand breaks (*see* **Note 5**).

3.3 In Vivo Homologous Recombination

1. To construct a pZwitch2 targeting vector (Fig. 2a), amplify homology arms from the zebrafish genome with PrimeSTAR GXL DNA polymerase. For the left homology arm (LA), amplify the 500–1000 bp DNA segment of the 5′ upstream region to the TALEN target site (Fig. 2b). For the right

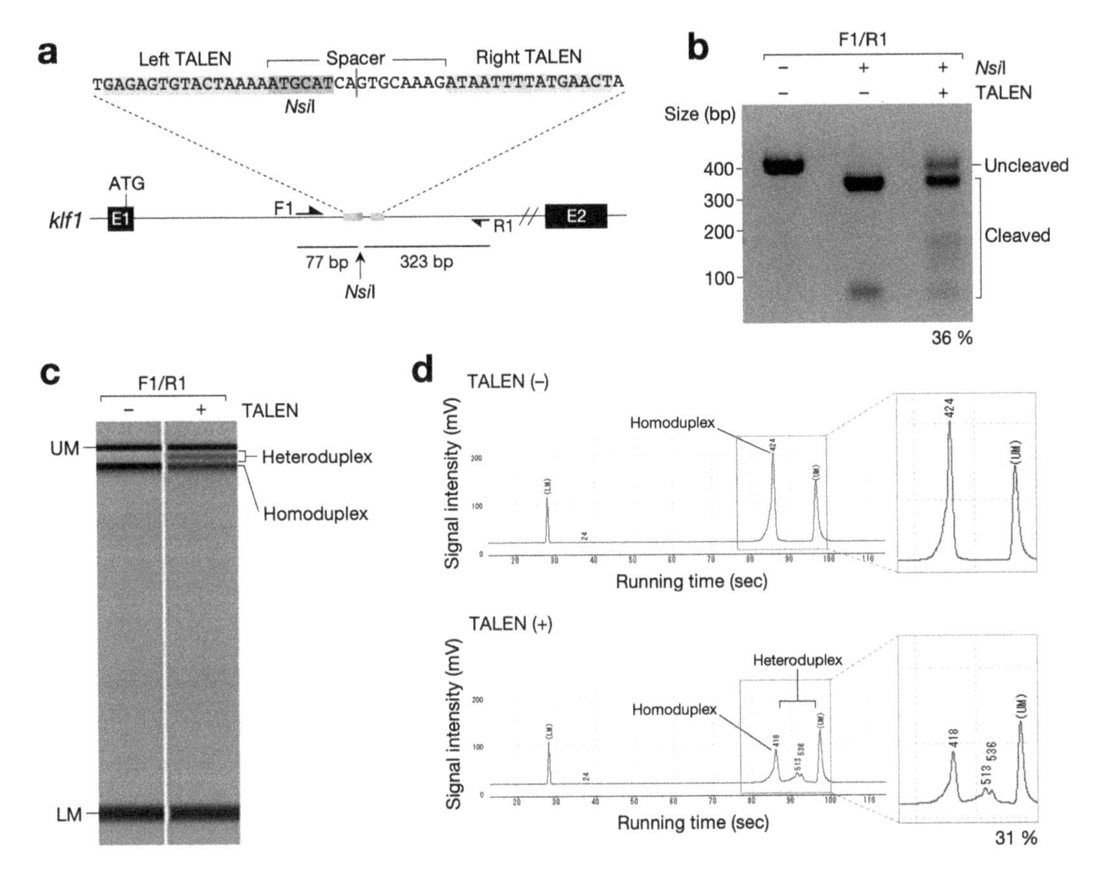

Fig. 1 Efficiency check of TALENs used to generate the *klf1* conditional allele. (**a**) Schematic of the zebrafish *klf1* locus and TALENs used. Binding sites for the TALEN pair (blue) and the *NsiI* site in the spacer region are indicated (gray). E, Exons. (**b**) Efficiency of TALENs in introducing DNA double-strand breaks. The DNA segment was amplified with the F1/R1 primer pair from the genomic DNA of embryos injected with TALEN mRNAs. The polymerase chain reaction (PCR) products were digested with *NsiI*, and the intensities of the uncleaved and cleaved bands were quantified from the gel image by using ImageJ software. The efficiency of the TALEN pair (36%) was calculated as the ratio of the uncleaved band intensity to the total band intensity (uncleaved and cleaved). (**c**) Virtual gel image obtained by MultiNA capillary gel electrophoresis of the PCR products amplified with the F1/R1 pair. UM and LM indicate upper and lower internal standard markers, respectively. (**d**) Electropherograms of (**c**). The approximate areas of the homoduplex and heteroduplex peaks were quantified from the electropherograms using ImageJ software. The efficiency of the TALEN pair (31%) was calculated as the ratio of the heteroduplex area to the total area intensity (homoduplex and heteroduplex). Note that the obtained efficiency (31%) is similar to the efficiency (36%) obtained from the restriction fragment length polymorphism assay in (**b**)

homology arm (RA), amplify the 500–1000 bp DNA segment of the 3′ downstream region (*see* **Note 6**). To amplify the LA, use reverse primers containing the 5′ half of the spacer (left half from the red line; Fig. 1a). Likewise, to amplify the RA, use forward primers containing the 3′ half of the spacer (right half from the red line; Fig. 1a).

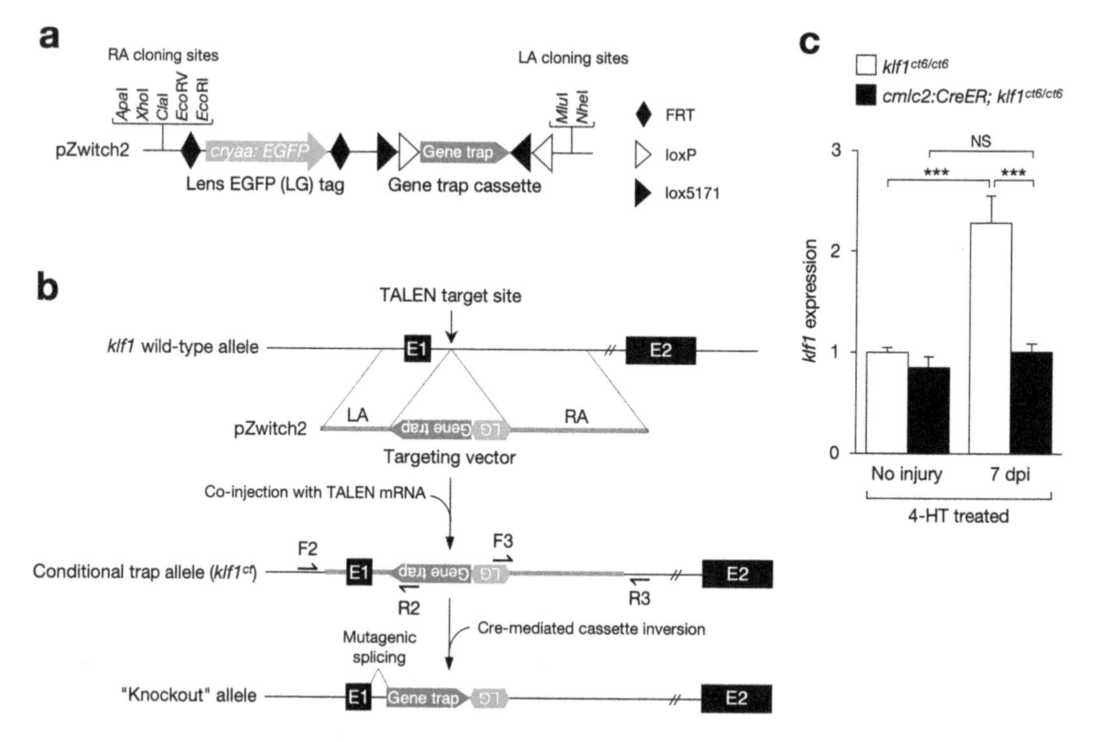

Fig. 2 Targeting vector and genomic engineering strategy. (**a**) Schematic of pZwitch2 vector. The gene-trap cassette consists of a splicing acceptor and three repeats of the bovine growth hormone polyadenylation signal. The *cryaa* indicates the lens-specific *crystallin* αA promoter. (**b**) TALEN-mediated homologous recombination with the pZwitch2 targeting vector and Cre-mediated inhibition of *klf1* gene expression. (**c**) RT-qPCR analysis of *klf1* expression in the uninjured (No injury) and injured (7 dpi) ventricles from Cre-negative control (open column) and cardiac-specific cKO fish (filled column) (mean ± SEM, $n = 3$–4). The data from a second line (ct6) is shown. *Dpi* days post-injury, *NS* not significant. $^{***}p < 0.005$, two-way ANOVA

2. Insert the amplified LA and RA into the corresponding cloning sites of pZwitch2 (*see* **Note 7**) via restriction enzyme cloning (Fig. 2a, b). Verify the resulting vector by DNA sequencing.
3. Inject approximately 50 pg each of the pZwtich2 targeting vector and TALEN mRNA into one-cell-stage embryos as described in Subheading 3.2.

3.4 Founder Fish Identification

1. Screen for lens GFP (LG) expression in injected embryos between 4 and 7 dpf and transfer LG^+ embryos to aquarium tanks.
2. Rescreen LG expression between 30 and 45 dpf and select fish maintaining LG expression uniformly in the eye (*see* **Note 8**).
3. Outcross adult LG^+ fish with the wild-type fish and assess LG expression in the offspring (*see* **Note 9**). Maintain LG^+ fish from different founder fish as a different line.

4. Perform PCR to verify insertion of the targeting vector. Digest LG$^+$ F1 fish with the DNA extraction buffer, followed by phenol–chloroform extraction and ethanol precipitation. Verify the 5′ end of the insertion by PCR with a forward primer binding to the outside of the LA (F2; Fig. 2b) and a reverse primer binding to the inside of the targeting vector (R2; Fig. 2b). Conversely, verify the 3′ end of the insertion by PCR with a forward primer binding to the inside of the targeting vector (F3; Fig. 2b) and a reverse primer binding to the outside of the RA (R3; Fig. 2b).
5. Perform southern blot analysis to verify the single-copy insertion of the targeting vector. Use LA or RA DNA fragments to generate probes using the DIG DNA Labeling and Detection Kit (Roche) or Prime-a-Gene Labeling System following the manufacturer's protocol (*see* **Note 10**). Purify genomic DNA from LG$^+$ fish and digest the DNA with a restriction enzyme for which the recognition site is located at the outside of the homology arms. Electrophoresis, blotting, and detection are performed following a common protocol for southern blot analysis.
6. Verify the entire sequence of the inserted cassette by DNA sequencing.

3.5 Induction of cKO and Assessment of Inactivation Efficiency

1. Propagate LG$^+$ fish and cross with *Tg(cmlc2:CreER)* [18] in which the cardiac myosin light chain 2 (*cmlc2*) promoter drives the expression of a tamoxifen-inducible Cre (CreER). This transgene is linked to a lens DsRedEx (LR) expression cassette to assist in genotyping. Select the LG$^+$LR$^+$ fish, cross them to the LG$^+$ fish, and screen the LG$^+$ fish to obtain the homozygous gene-trap fish.
2. Raise the LG$^+$ fish to adulthood and confirm the genotypes by PCR. Perform fin clipping of LG$^+$ fish and collect a piece of fin tissue into a 96-well PCR plate containing 10 mL of fin DNA extraction buffer in each well. During the collection of fin tissues, keep the plate on ice.
3. Incubate the plate of collected fin tissues with a thermal cycler at 55 °C for 1 h, followed by incubation at 95 °C for 15 min. Take 1 mL of fin tissue extracts and perform PCR to confirm the genotype of the gene-trap allele and the CreER transgene (*see* **Note 11**). Select zebrafish carrying homozygous gene-trap alleles ($klf1^{ct/ct}$) with or without *cmlc2:CreER* transgene. Use *cmlc2:CreER*; $klf1^{ct/ct}$ fish for *klf1* cKO and $klf1^{ct/ct}$ fish as Cre-negative controls.
4. Place zebrafish in a small beaker of aquarium water supplemented with 5 μM 4-hydroxytamoxifen (4-HT) (*see* **Note 12**) for 10–12 h overnight [19]. Place a lid to prevent fish jumping out and leave the beaker undisturbed in dark.

5. In the next morning, rinse zebrafish with fresh aquarium water and return them to the recirculating water system for feeding and resting. Maintain fish in the system until the fish are fed with the last feeding of the day.
6. Repeat **steps 4–5** three times.
7. Two days after the last 4-HT treatment, perform ventricular resection surgery [20] and collect ventricles at 7 days post-injury (dpi). Perform reverse transcription-quantitative polymerase chain reaction (RT-qPCR) with primers for *klf1* cDNA. Compare the *klf1* expression of 7 dpi cKO ventricles to that of uninjured ventricles to determine the efficiency of myocardial inactivation of *klf1* expression (Fig. 2c).

4 Notes

1. We preferred using TALEN over the CRISPR/Cas9 system to induce DNA double-strand breaks, because the latter was not efficient for targeted insertion of a Zwitch cassette in our initial experiments.
2. A Zwitch cassette can be targeted to any introns located downstream of the ATG exon or the exon containing a critical functional domain of a target protein. The ATG exon needs to be included in all transcript variants that are expressed in the cell type of interest. Exon usage may need to be validated experimentally in target cell types or tissues.
3. It is not necessary to include a restriction enzyme site in spacer sequences if the heteroduplex mobility assay is used to determine the TALEN efficiency.
4. This assay may not detect mutations if TALENs induce a difference of several base pairs [21].
5. It is critical to choose TALEN pairs with more than 30% efficiency in the RFLP assay for the success of Zwitch insertion.
6. Template DNA should be obtained from the same strain of zebrafish that are used to produce embryos for Zwitch insertion.
7. The plasmid is available from the corresponding author.
8. Do not select fish exhibiting punctate LG expression as these fish often do not carry the integrated gene-trap cassette.
9. The level of LG expression is usually not very strong, because the gene-trap cassette is inserted at a single copy.
10. We recommend ^{32}P-labeling over DIG-labeling to generate probes due to its higher sensitivity.

11. Although the genotypes can be estimated by the lens fluorescence color and intensity, we recommend performing genomic PCR to confirm the genotypes before performing regeneration analysis.
12. We do not use standard tamoxifen dissolved in corn or peanut oil.

Acknowledgments

We thank T. Sato and C. Yamada for zebrafish care. This work was supported by NHMRC APP1160466 and JSPS KAKENHI JP20H03683 to K.K.

References

1. Lobe CG, Nagy A (1998) Conditional genome alteration in mice. BioEssays 20(3):200–208
2. Rajewsky K et al (1996) Conditional gene targeting. J Clin Invest 98(3):600–603
3. Rossant J, McMahon A (1999) "Cre"-ating mouse mutants-a meeting review on conditional mouse genetics. Genes Dev 13(2): 142–145
4. Koller BH, Smithies O (1992) Altering genes in animals by gene targeting. Annu Rev Immunol 10:705–730
5. Thomas KR, Capecchi MR (1987) Site-directed mutagenesis by gene targeting in mouse embryo-derived stem cells. Cell 51(3): 503–512
6. Schnütgen F et al (2003) A directional strategy for monitoring Cre-mediated recombination at the cellular level in the mouse. Nat Biotechnol 21(5):562–565
7. Hoshijima K, Jurynec MJ, Grunwald DJ (2016) Precise editing of the zebrafish genome made simple and efficient. Dev Cell 36(6): 654–667
8. Bedell VM et al (2012) In vivo genome editing using a high-efficiency TALEN system. Nature 491(7422):114–118
9. Zu Y et al (2013) TALEN-mediated precise genome modification by homologous recombination in zebrafish. Nat Methods 10(4): 329–331
10. Li W et al (2019) One-step efficient generation of dual-function conditional knockout and geno-tagging alleles in zebrafish. elife 8: e48081
11. Burg L et al (2018) Conditional mutagenesis by oligonucleotide-mediated integration of loxP sites in zebrafish. PLoS Genet 14(11): e1007754
12. Sugimoto K et al (2017) Dissection of zebrafish shha function using site-specific targeting with a Cre-dependent genetic switch. elife 6
13. Ogawa M et al (2021) Krüppel-like factor 1 is a core cardiomyogenic trigger in zebrafish. Science 372(6538):201–205
14. Ma AC et al (2013) High efficiency in vivo genome engineering with a simplified 15-RVD GoldyTALEN design. PLoS One 8(5):e65259
15. Ota S et al (2013) Efficient identification of TALEN-mediated genome modifications using heteroduplex mobility assays. Genes Cells 18(6):450–458
16. Ansai S et al (2014) Design, evaluation, and screening methods for efficient targeted mutagenesis with transcription activator-like effector nucleases in medaka. Develop Growth Differ 56(1):98–107
17. Ansai S, Kinoshita M (2014) Targeted mutagenesis using CRISPR/Cas system in medaka. Biol Open 3(5):362–371
18. Kikuchi K et al (2010) Primary contribution to zebrafish heart regeneration by gata4(+) cardiomyocytes. Nature 464(7288):601–605
19. Kikuchi K et al (2011) tcf21+ epicardial cells adopt non-myocardial fates during zebrafish heart development and regeneration. Development 138(14):2895–2902

20. Poss KD, Wilson LG, Keating MT (2002) Heart regeneration in zebrafish. Science 298(5601):2188–2190
21. Sogabe Y (2020) Detection of multi-base mutation by genome editing using MultiNA (Application News No. B110). Retrieved from Shimadzu website: https://www.shimadzu.com/an/sites/shimadzu.com.an/files/pim/pim_document_file/applications/application_note/13139/jpu220002.pdf. Shimadzu

Chapter 14

Spinal Cord Injury and Assays for Regeneration

Brooke Burris and Mayssa H. Mokalled

Abstract

Due to their renowned regenerative capacity, adult zebrafish are a premier vertebrate model to interrogate mechanisms of innate spinal cord regeneration. Following complete transection to their spinal cord, zebrafish extend glial and axonal bridges across severed tissue, regenerate neurons proximal to the lesion, and regain swim capacity within 8 weeks of injury. Here, we describe methods to perform complete spinal cord transections and to assess functional and cellular recovery during regeneration. For spinal cord injury, a complete transection is performed 4 mm caudal to the brainstem. Swim endurance is quantified as a central readout of functional spinal cord repair. For swim endurance, zebrafish are subjected to a constantly increasing water current velocity until exhaustion, and time at exhaustion is reported. To assess cellular regeneration, histological examination is performed to analyze the extents of glial and axonal bridging across the lesion.

Key words Spinal cord injury, Swim endurance assay, Spinal cord histology, Glial bridging quantification, Axon tracing, Axonal bridging quantification

1 Introduction

Adult zebrafish reverse paralysis within 8 weeks of complete spinal cord transection [1, 2]. Unlike poorly regenerative mammals, zebrafish display pro-regenerative immune, neuronal, and glial injury responses that are required for functional spinal cord repair [3–7]. Subheading 1 of this protocol describes a surgical procedure to perform complete spinal cord transections in zebrafish [5].

Complete spinal cord transections result in complete paralysis caudal to the injury site. After injury, paralyzed animals display compensatory, short, and frequent swim bursts by overusing their pectoral fins, which lie rostral to the lesion. This compensatory swim strategy results in rapid exhaustion and lower swim capacity. As the spinal cord regenerates, animals regain a smooth oscillatory swim function caudal to the lesion, which increases their swim endurance. Subheadings 2 and 3 of this protocol describe methods to quantify zebrafish swim endurance at increasing water current

James F. Amatruda et al. (eds.), *Zebrafish: Methods and Protocols*, Methods in Molecular Biology, vol. 2707, https://doi.org/10.1007/978-1-0716-3401-1_14,

velocities inside an enclosed swim tunnel. These methods provide a reliable readout of functional spinal cord repair [8].

Zebrafish extend regenerative glial and axonal bridges across the lesion. Subheadings 4, 5, and 6 of this protocol describe methods to assess glial bridging and to trace axonal growth across the lesion. These methods provide cellular readouts of spinal cord regeneration [7].

2 Materials

2.1 Spinal Cord Injury

1. Surgical scissors.
2. Tissues.
3. Surgery Sponge; modified into a 5.5 cm × 4 cm rectangle with a crevice carved to about the length of the animal.
4. Forceps.
5. Glass Dish.
6. Fish Tanks.
7. Tricaine solution; 0.1% MS-222.
8. Plastic Spoon.
9. Transfer Pipet.
10. Zebrafish system water; alkalinity, 50–150 mg/L $CaCO_3$; pH, 6.8–7.5; temperature, 26–28.5 °C; nitrate <50 mg/L; nitrite <0.1 mg/L; salinity <0.5–1 g/L.

2.2 Swim Endurance Assay

1. Swim Tunnel (Loligo Systems).
2. Flow-Through Tank.
3. Flush Pump.
4. Flow-Through Pump.
5. PVC tubes.
6. Autoswim software. The use of a flow velocity control program in this section is optional. The alternative is to manually control the water current motor.
7. Calibration Lid: Modified lid with reinforced hole in the center just large enough to allow the calibration probe to fit snuggly.
8. Customized Endurance Lid: Modified lid with a window at the back of the tunnel away from the source of flow to allow for easy access and removal of fish from the assay without disrupting the rest of the animals still being measured.
9. Flow Therm Probe.
10. Collection Tank.
11. Binder Clip.

2.3 Histology

1. Gelfoam sponge.
2. Biocytin; 40 μg/μL in 1× PBS.
3. Vetbond.
4. Paraformaldehyde; 4%.
5. Sucrose; 30% in 1× PBS.
6. Optimal Cutting Temperature (OCT) compound.
7. Cryostat.
8. PBS-Tween; 0.1% Tween-20 in 1× PBS PB.
9. Block solution; 5% goat serum in PBS-Tween.
10. Mouse Anti-GFAP; 1:500 dilution in Block solution.
11. Alexa Fluro 488 Goat Anti-Mouse; 1:250 dilution in Block solution.
12. Streptavidin; Fluofore conjugated; 1:100 in 1% Triton-X in PBS.
13. Hoechst; 1:1000 in 1× PBS.
14. Fluoromount G.

3 Methods

3.1 Spinal Cord Injury

1. Ensure the surgeon is blinded to the identity of the experimental cohorts prior to the surgery and regeneration assays.
2. Prepare surgical scissors, forceps, microscope, tissue wipers, and paper towels. Prepare clean recovery tanks for postoperative care of injured fish.
3. Soak the surgery sponge with zebrafish system water and place it under the dissecting microscope.
4. Place one fish at a time in the Tricaine solution until deeply anaesthetized. Watch as the gill movement slows down.
5. Transfer the fish with a plastic spoon to the surgery sponge. Place the fish upside down in the surgery sponge.
6. Locate the lesion site 4 mm caudal to the brainstem region. Use forceps to descale the located lesion site.
7. Perform a longitudinal incision perpendicular to the midline to transect the dorsal muscle overlaying the spinal cord. This step will expose the underlying spinal cord tissue. Perform a full transection of the spinal cord.
8. Use the forceps to encourage the skin and tissue around the lesion site to come back together.
9. Place the fish in a recovery tank. Use a transfer pipet to gently squirt zebrafish system water over the gills until regular gill movement is recovered.

10. Repeat **steps 3.1.4–3.1.9** for the remaining fish.
11. To allow for adequate external wound healing, fish are maintained under a slow flow of circulating system water with no food for 2 days after injury. Water flow is gradually increased and feeding is resumed 3 days after injury.

3.2 Swim Tunnel Preparation and Calibration

The swim tunnel consists of a swim chamber that houses experimental fish during the swim assay, and a surrounding buffer tank that circulates fresh system water into the enclosed swim chamber. This protocol employs an optional "flow velocity control software" for automated manipulation of flow velocity. The alternative is to manually control flow velocity.

1. Fill the swim tunnel, the surrounding buffer tank, and an additional flow-through tank with zebrafish system water.
2. Employ water pumps and PVC tubing to ensure system water is circulating inside the swim tunnel and surrounding buffer tank. A dual pump system ensures continuous water flow into the swim chamber (from the flush pump) and into the buffer tank (from the flow-through pump).
3. Turn on the flush and flow-through pumps to begin water circulation.
4. Clear any air bubbles trapped inside the swim chamber by gradually increasing the water current velocity from 10 to 100 cm/s in intervals of 10 cm/s. Then decrease the flow to 0 cm/s in intervals of 10 cm/s.
5. Use the calibration lid to close the swim tunnel. The calibration lid is customized with a reinforced central opening that fits the flow meter probe used for calibration.
6. Place the flow meter probe inside the swim tunnel via the calibration lid. Position the blades of the flow meter probe perpendicular to the direction of flow.
7. Calibrate the output of the swim tunnel motor with the flow velocity software using the digital flow meter. To calibrate, open the flow velocity control software and click "Calibration." Increase the flow velocity in 5 cm/s increments, starting from 0 cm/s and ending at 100 cm/s. At each step, click the "+" button and record the current velocity indicated by the digital flow meter. The resulting linear relationship should have an R^2 value close to 1.
8. To confirm calibration, increase water current velocities from 0 to 10, 25, 50, 75, and 100 cm/s, then decrease velocities to 75, 50, 25, and 10 cm/s. At each velocity (from the swim tunnel software), measure and record the corresponding velocity indicated by the digital flow meter.

9. Consider the tunnel calibrated and accurate if the measured water current velocities are within a deviation of ±2 cm/s. If the deviation is beyond ±2 cm/s, repeat **steps 2.3.3** and **2.3.4** to ensure proper calibration.

3.3 Assessment of Swim Endurance

For swim endurance, experimental cohorts are split into groups of ten animals or less.

The use of a flow velocity control program in this section is optional. For manual water current control, proceed to **step 3.4** and manually increase water current velocity by 2 cm/s every minute in **step 3.9**.

1. Set up the flow velocity control software. Click “Experiment” and “Uwater [cm/s]” to adjust water current velocities.
2. To begin an automated protocol, click “Start logging” and choose “automated.” To choose a previously saved protocol file, click “Protocol File” to open the desired protocol.
3. Set up a “fish collection tank” to house exhausted fish after their removal from the swim tunnel. Fill the collection tank with zebrafish system water. Fill a long PVC tube with zebrafish system water. Place one end (end 1) of the prefilled PVC tube in the collection tank and the other end (end 2) in the buffer tank. Make sure water can freely flow from the buffer tank into the fish collection tank. Clamp the upper end of the PVC tube (end 2) with a binder clip to prevent water flow. Use the binder clip to control the outflow of water as needed.
4. Close the swim tunnel using the swim endurance lid.
5. Place one group of fish inside the swim tunnel. Start a split lap timer while adjusting current velocity to 0 cm/s for 5 min, 9 cm/s for 5 min, and 10 cm/s for 5 min by typing these values in the “Uwater [cm/s]” section of the flow velocity control software. This step will acclimate animals to the swim tunnel, flow direction, and flow velocity changes.
6. Following acclimation, start the automated flow velocity control program to increase water current velocity by 2 cm/s every min.
7. When a fish is exhausted, unclamp the fish collection tube and collect the fish into the fish collection tank. Record the time at exhaustion.
8. Repeat **steps 3.7** and **3.8** until all the fish in the cohort are exhausted and collected in the collection tank.

3.4 Axon Tracing with Biocytin

1. Cut a gelfoam sponge into 2 mm^3 pieces.
2. Soak each 2 mm^3 gelfoam piece with 5 μL of a saturated Biocytin solution and use forceps to split each gelfoam piece into 4x 0.5 mm^3 gelfoam pieces.

3. For anterograde tracing, perform a secondary spinal cord transection 4 mm rostral to the primary lesion site, which corresponds to the brainstem level. To ensure complete transections, use a pair of forceps to confirm that the injured spinal cord stumps are fully retracted.
4. Place the Biocytin-soaked gelfoam piece between the transected stumps of the spinal cord.
5. Seal the wound with Vetbond and let it dry for 10–15 s. Place the fish in a recovery tank and use a transfer pipet to gently squirt zebrafish system water over the gills until regular gill movement is recovered.
6. Three hours after Biocytin labeling, euthanize the fish and proceed with histology.

3.5 Histology

1. Euthanize fish following an institutionally approved euthanization protocol.
2. Perform a rough dissection of the spinal cord with the surrounding bone and muscle tissues. Collect 3 mm of tissues on each side of the lesion and fix with 4% paraformaldehyde overnight.
3. For cryopreservation, perform 3× PBS washes at room temperature, then treat fixed tissues with 30% sucrose in PBS overnight.
4. Embed SC tissues in Optimal Cutting Temperature (OCT) compound for sectioning. For standard assessment of glial and axonal bridging, 16 μm serial transverse sections along the rostro-caudal axis of the fish are obtained. Section levels from 750 μm rostral to 1750 μm caudal to the lesion core are used for standard analysis.
5. For Gfap immunostaining and Biocytin labeling, dry slides for 30 min and draw a hydrophobic barrier around tissue sections.
6. Hydrate slides with 1 mL of PBS-Tween for 5 min.
7. Block with 250 μL Block solution for 1 h at room temperature.
8. Add 250 μL of primary Gfap antibody solution. Incubate overnight at 4 °C.
9. Wash slides 3× with PBS-Tween for 10 min.
10. Add 250 μL of secondary antibody solution. Incubate for 1 h at room temperature.
11. Wash slides 3× with PBS-Tween for 10 min.
12. For Biocytin labeling, incubate with Streptavidin for 2 h at room temperature.
13. Wash slides for 10 min with PBS-Tween.
14. Stain nuclei with Hoechst for 5 min.

15. Wash slides 3× with PBS-Tween for 10 min.
16. Mount and coverslip slides in Fluoromount G.

3.6 Imaging and Quantification of Glial and Axonal Bridging

1. To identify the lesion core, survey the extent of tissue damage in serial sections along the rostro-caudal axis of the fish. The section with the most severely damaged spinal cord tissue corresponds to the lesion core. Image the lesion core along with sections at 750 and 500 μm rostral and 500, 750, 1500, and 1750 μm caudal to the lesion.
2. The cross-sectional Gfap$^+$ area at the lesion core corresponds to the glial bridge. To calculate percent bridging, the cross-sectional area of the glial bridge Gfap$^+$ is normalized to the average cross-sectional area of the intact spinal cord at 500 μm and 750 μm rostral to the lesion.
3. For calculation of axon growth, Biocytin labeled axons are quantified using the "threshold" and "particle analysis" tools in Fiji. Axon growth is quantified proximal (500 μm and 750 μm caudal) and distal (1500 μm and 1750 μm caudal) to the lesion, and normalized to the efficiency of Biocytin labeling at 750 μm and 500 μm rostral to the lesion for each fish.

4 Notes

1. It is important to ensure the spinal cord is completely, but not too severely, transected. To ensure complete transections, use a pair of forceps to confirm that the injured spinal cord stumps are fully retracted. Complete spinal cord transections are performed without inducing any bleeding. Profuse bleeding is indicative of an overly severe injury. To prevent severe lesions, avoid deep injuries that transect the vertebral column causing bleeding and secondary internal injuries.
2. After a successful surgery, an injured fish is completely paralyzed caudal to the lesion and sits upside at the bottom of the recovery tank. Partial swimming indicates incomplete spinal cord transection.
3. Fish must be closely monitored twice a day for the first 3 days after injury. Partially or too severely injured fish are euthanized. Fish are monitored once a day after 3 days of injury and until the experimental end point.
4. We recommend using two pumps, referred to hereafter as flush pump and flow-through pump. To ensure water is circulating inside the swim tunnel, place the flush pump inside the buffer tank and connect it to the adjacent swim chamber with PVC tubes. To help refresh system water throughout the swim assay, place the flow-through pump between the additional flow-through tank and buffer tank.

Acknowledgments

This research was supported by grants from NIH (R01 NS113915), the Curators of the University of Missouri (Spinal Cord Injury and Disease Training Program), and funds from Washington University School of Medicine in St. Louis.

References

1. Becker T, Wullimann MF, Becker CG, Bernhardt RR, Schachner M (1997) Axonal regrowth after spinal cord transection in adult zebrafish. J Comp Neurol 377(4):577
2. Becker CG, Becker T (2008) Adult zebrafish as a model for successful central nervous system regeneration. Restor Neurol Neurosci 26(2–3):71
3. Reimer MM, Sorensen I, Kuscha V, Frank RE, Liu C, Becker CG, Becker T (2008) Motor neuron regeneration in adult zebrafish. J Neurosci 28(34):8510
4. Goldshmit Y, Sztal TE, Jusuf PR, Hall TE, Nguyen-Chi M, Currie PD (2012) Fgf-dependent glial cell bridges facilitate spinal cord regeneration in zebrafish. J Neurosci 32(22):7477
5. Mokalled MH, Patra C, Dickson AL, Endo T, Stainier DY, Poss KD (2016) Injury-induced ctgfa directs glial bridging and spinal cord regeneration in zebrafish. Science 354(6312):630
6. Cavone L, McCann T, Drake LK, Aguzzi EA, Oprisoreanu AM, Pedersen E, Sandi S, Selvarajah J, Tsarouchas TM, Wehner D, Keatinge M, Mysiak KS, Henderson BEP, Dobie R, Henderson NC, Becker T, Becker CG (2021) A unique macrophage subpopulation signals directly to progenitor cells to promote regenerative neurogenesis in the zebrafish spinal cord. Dev Cell 56(11):1617
7. Klatt Shaw D, Saraswathy VM, Zhou L, McAdow AR, Burris B, Butka E, Morris SA, Dietmann S, Mokalled MH (2021) Localized EMT reprograms glial progenitors to promote spinal cord repair. Dev Cell 56(5):613
8. Burris B, Jensen N, Mokalled MH (2021) Assessment of swim endurance and swim behavior in Adult Zebrafish. J Vis Exp 177

Chapter 15

Selective Cell Ablation Using an Improved Prodrug-Converting Nitroreductase

Timothy S. Mulligan and Jeff S. Mumm

Abstract

Selective cell ablation is an invaluable tool to investigate the function of cell types, the regeneration of cells, and the modeling of diseases associated with cell loss. The nitroreductase (NTR)-mediated cell ablation system is a simple method enabling the elimination of targeted cells through the expression of a nitroreductase enzyme and the application of a prodrug (such as metronidazole). The prodrug is reduced to a cytotoxic product by nitroreductase, thereby leading to DNA damage-induced cell death. In species with elevated regenerative capacity such as zebrafish, removing the prodrug allows endogenous tissue to replace the lost cells. Herein, we describe a method for the use of a markedly improved nitroreductase enzyme for spatially and temporally controlled targeted cell ablation in the zebrafish. Recently, we identified an NTR variant (NTR 2.0) that achieves effective targeted cell ablation at concentrations of metronidazole well below those causing toxic side effects. NTR 2.0 thereby enables the ablation of "resistant" cell types and novel cell ablation paradigms. These advances simplify investigations of cell function, enable interrogations of the effects of chronic inflammation on regenerative processes and facilitate modeling of degenerative diseases associated with chronic cell loss. Techniques for transgenic nitroreductase expression and prodrug application are discussed.

Key words Nitroreductase, NTR, NTR 1.1, NTR 2.0, NfsB, Metronidazole, MTZ, Ablation, Cell ablation, Zebrafish, Prodrug, Ronidazole, Nifurpirinol

1 Introduction

The ability to ablate specific cells can enable key insights in biological investigations of regenerative, developmental, and disease biology. Cell ablation can facilitate interrogations of cell function, investigations of mechanisms regulating cellular regeneration, and modeling diseases associated with cell death. The nitroreductase (NTR)-based cell ablation system is a simple method for killing cells through genetic expression of an NTR enzyme and application of a water-soluble prodrug. When used in conjunction with "no

James F. Amatruda et al. (eds.), *Zebrafish: Methods and Protocols*, Methods in Molecular Biology, vol. 2707,
https://doi.org/10.1007/978-1-0716-3401-1_15,

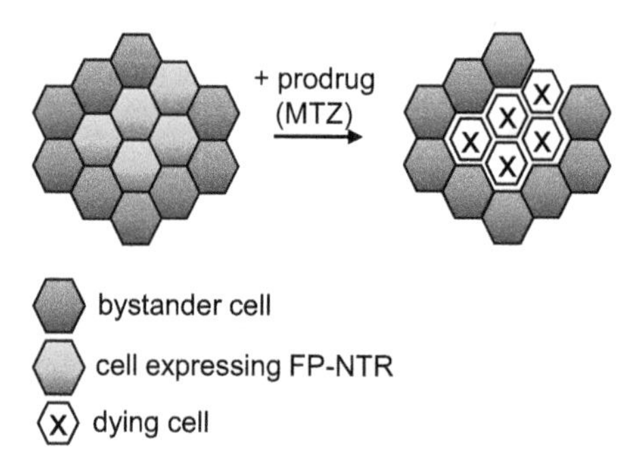

Fig. 1 NTR/prodrug selective cell ablation system. The nitroreductase/prodrug targeted cell ablation system comprises two components: a genetically encoded bacterial nitroreductase (NTR) enzyme and a prodrug that is reduced by NTR into a cytotoxic product. Exclusive expression of NTR in targeted cells and the use of specific prodrugs, such as metronidazole (MTZ), facilitates selective cell ablation with no loss of neighboring "bystander" cells. Co-expression of NTR with a fluorescent protein (FP-NTR) enables facile detection and quantification of the kinetics of cell loss and regeneration

bystander" prodrugs such as Metronidazole (MTZ), it enables an inducible, cell-specific, and cell cycle-independent method for ablating cells (Fig. 1).

A key feature of the NTR ablation system is spatiotemporal control of cell death. Spatial restriction is conferred by expressing NTR exclusively in a cell type or tissue of interest. This is typically accomplished by generating stable transgenic animals, but transient expression systems can also be used (*see* **Notes 1–5**). Temporal restriction is controlled by the timing of prodrug exposure. The extent of ablation can be controlled by the concentration and choice of prodrug and the duration of prodrug treatment (*see* **Note 6**). Removal of the prodrug results in cessation of ablation, allowing regenerative tissues to replace lost cells. The timing of NTR expression and cell loss/regeneration is typically visualized using co-expressed fluorescent reporters in the form of multicistronic expression systems (e.g., 2A viral peptides) or fusion proteins (e.g., NTR-mCherry) (*see* **Note 7**).

NTR-based reduction of prodrug substrates is believed to result in DNA damage through the formation of DNA adducts. Our recent work on rod photoreceptors suggests NTR/MTZ-induced ablation of retinal neurons involves a DNA-damage triggered form of cell death, termed parthanatos [1]. Prior studies ablating cells in the heart, pancreas, and skin suggest that the NTR/MTZ system elicits apoptotic cell death [2–4]. Thus, work remains to confirm the extent to which mechanisms of NTR-induced cell death are cell- and/or prodrug-specific. In addition to assaying loss of cell fluorescence linked to NTR, typical methods to assess cell death include visualizing nuclei fragmentation and the use of TUNEL and caspase 3 staining (*see* **Note 8**).

There are three common versions of NTR that have been used to ablate cells in zebrafish. These variants are referred to here as NTR 1.0, 1.1, and 2.0. These versions differ primarily in their efficacy to facilitate MTZ-induced cell loss. NTR 1.0 is the originally described enzyme isolated from *Escherichia coli* (*NfsB_Ec*). NTR 1.0 was initially developed as a potential cancer therapeutic in conjunction with the prodrug CB1954 to induce loss of targeted tumor tissue and surrounding cells [5]. NTR 1.0 can also induce cell-specific ablation in combination with prodrugs such as MTZ, Nifurpirinol (NFP), and Ronidazole [2, 6, 7]. Limitations of NTR 1.0 arose from its relatively weak prodrug reduction activity. For most cell types, this necessitates high concentration MTZ treatments (e.g., 10 mM), resulting in general prodrug toxicity after 24 h of exposure. NTR 1.1 (*NfsB_Ec T41Q/N71S/F124T*) incorporated several amino acid substitutions into NTR 1.0 that markedly enhanced CB1954-based cell ablation activity [8]. Despite NTR 1.1 also exhibiting a modest enhancement of MTZ-induced cell ablation activity in zebrafish [9], some cell types remained resistant to NTR/MTZ-mediated ablation. Recently, in collaboration with the lab of Dr. David Ackerley, we generated NTR 2.0 from a NfsB enzyme isolated from *Vibrio vulnificus* (*NfsB_Vv*) with enhanced MTZ-induced ablation efficiency. NTR 2.0 (*NfsB_Vv F70A/F108Y*) incorporates two amino acid substitutions that further increase its MTZ-induced ablation efficacy (~100-fold overall compared to previous versions) [10, 11]. Furthermore, zebrafish soaking in concentrations of MTZ suitable for ablation in the presence of NTR 2.0 ($\leq$1 mM) exhibited no general adverse effects over extended exposure times. Thus, NTR 2.0 enables novel sustained cell ablation paradigms that (1) enable facile ablation of previously resistant cell types, (2) temporally expand tests of cell function, (3) challenge regenerative capacities by prolonging cell loss and associated inflammation, and (4) can serve as inducible and titratable models of chronic degenerative diseases (Fig. 2). For these reasons, we recommend using the newly identified NTR 2.0 for targeted cell ablation paradigms in zebrafish (*see* **Note 9**).

2 Materials

2.1 Reagents

1. Zebrafish expressing nitroreductase in the cell type of interest.
2. (Optional 96-well plates) 96-well, black, U-shape plates that are compatible with downstream fluorescent plate reader screening such as a TECAN (Cat. #650209; Greiner bio-one) (*see* **Note 10**).

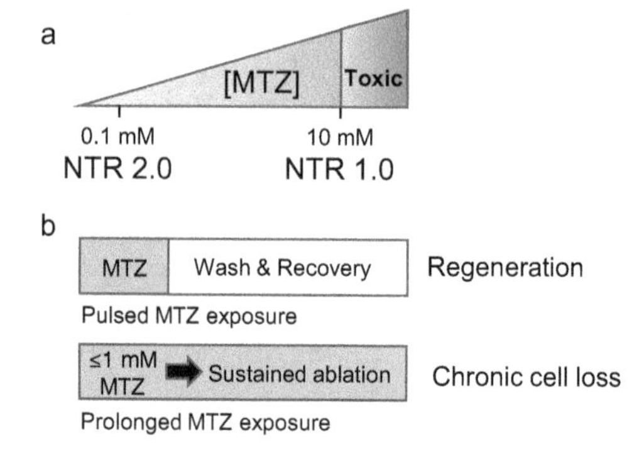

Fig. 2 NTR 2.0 enables novel cell ablation paradigms. (**a**) NTR 2.0 is ~100-fold more effective than NTR 1.0, facilitating targeted cell ablation at 0.1 mM versus 10 mM MTZ, respectively. Exposure to 10 mM MTZ for >24 h results in general toxicity while exposure to ≤1 mM is nontoxic and can be maintained indefinitely. (**b**) NTR 1.0 is thus limited to pulsed MTZ incubations of ~24 h to avoid general toxicity, with washout allowing recovery and regeneration. Conversely, NTR 2.0 enables both pulsed and sustained cell ablation with MTZ treatments of ≤1 mM. Sustained cell loss paradigms, in turn, will enhance tests of cell function, allow investigations of prolonged inflammation on regeneration, and can serve as inducible chronic degenerative disease models

2.2 Solutions

1. 1 mM MTZ: 17.116 mg Metronidazole (*see* **Note 11**), 100 mL E3 embryo media for larvae. (Protect solution from light.) Dilute desired concentration of MTZ (a typical range is 0.04–10 mM) in embryo media in a light-proof bottle (darkened or foil-wrapped) on a stirrer at least 30–60 min before use (*see* **Note 12**). For application to post-larval stage juveniles and adults maintained in standard aquaria, dilute MTZ in system water. (Optional) For assays that would be obscured by pigment production in the developing zebrafish, supplement embryo media with PTU (0.003% PTU final concentration).
2. 50× PTU: 750 mg 1-phenyl-2-thiourea, 500 mL dH_2O. (Stir overnight or until in solution).
3. E3 + 0.003% PTU: 20 mL 50× PTU, 980 mL Embryo Media.
4. 100× E3 Media: 292.2 g NaCl, 12.67 g KCl, 48.51 g $CaCl_2 \cdot 2H_2O$, 39.72 g $MgSO_4$ (anhydrous), 9 L H_2O. Adjust pH to 7.4 and total final volume to 10 L with H_2O.
5. E3 Embryo Media: 500 mL 100× E3 Media. Adjust with ~30 mL of 357 mM $NaHCO_3$ to pH 7.2–7.4, 49.47 L deionized H_2O.
6. System Water: Water collected from an aquaculture system.

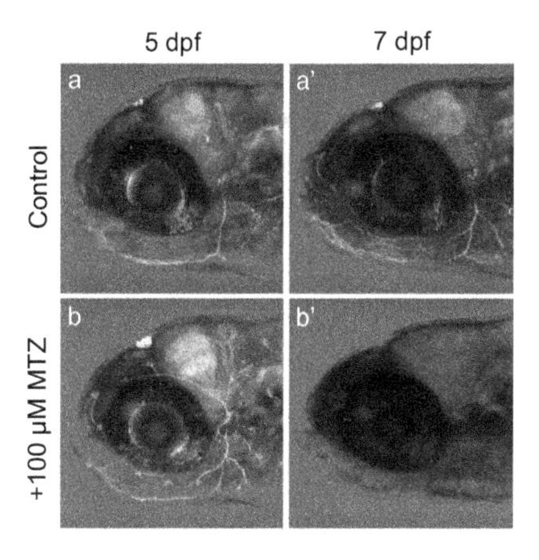

Fig. 3 NTR 2.0 enhanced cell ablation efficacy. Time series confocal in vivo imaging of transgenic zebrafish larvae co-expressing YFP and NTR 2.0 in neurons treated ±100 μM MTZ for 24 h (5–6 dpf), loss of YFP used as an indicator of cell ablation. (**a**-**a**') Representative control larva treated with embryo medium at 5 dpf (**a**) and 7 dpf (**a**'). (**b**-**b**') Representative experimental larva at 5 dpf, prior to MTZ exposure (**b**) and at 7 dpf, after MTZ exposure (**b**'); note near complete loss of YFP-expressing cells

3 Methods

3.1 Cell Ablation Protocol

At the desired stage of development:

1. Array individual larvae into single wells of a 96-well plate in 300 μL of embryo media; in groups of five larvae per 2 mL of a 6-well plate; or 75 larvae per 30–40 mL in a 30 mm dish. Treat adults at a density 1–6 fish per 1 L of system water.
2. Remove embryo media (or system water) and replace with the desired concentration of MTZ solution (*see* **Notes 13–16**).
   (Option for regenerative assays)
3. After allowing sufficient time for cell ablation (e.g., 24 h), wash out MTZ 2–3 times with embryo media/system water. Continue raising fish in embryo media/system water and feeding (if applicable) as usual until desired endpoint (*see* **Note 17**).
   (Option for modeling diseases with sustained cell loss)
4. Continue incubation in MTZ at concentrations of ≤1 mM for the desired length of time (*see* **Notes 18** and **19**).
5. Replace MTZ every 2–3 days to ensure effectiveness of prodrug and health of fish ≥12 days post fertilization (dpf). Feed larvae 5–11 dpf daily with dry food prior to replacing the prodrug until they are old enough to be fed brine shrimp. Zebrafish 12 dpf or older can be fed daily with brine shrimp (*see* **Note 20**).

3.2 Detecting Cell Loss

1. Monitor zebrafish embryos/larvae/adults to determine the timing of cell/fluorescence loss. This can be accomplished through several assays including stereo- or confocal microscopy (Fig. 3), fixation and immunofluorescence staining for cell specific markers, and quantification by fluorescent plate reader assays (*see* **Notes 21–23**).

4 Notes

1. Stable NTR-expressing transgenic lines can be generated with a variety of integration systems and thus tailored to individual expression needs. If the intent is to ablate an entire cell population, NTR should be expressed directly from a cell-type specific promoter or integrated into an endogenous locus exclusive to the targeted cells. Well-characterized promoters/enhancers will suffice, but BAC transgenesis or knock-in approaches can also be used if a defined promoter has yet to be identified. Integration at an endogenous locus can now be more readily accomplished with efficient knock-in methods involving CRISPR/Cas9, TALENs, or other double stranded break inducing tools [12–14] (*see* **Note 2**). In this case, co-expression systems, such as 2A peptides, are preferred unless the intent is to also disrupt the function of the targeted gene. For strategies containing a minimal promoter-driven NTR, Tol2 transposase-based transgenesis can be used for random integration into the genome (*see* **Note 3**). However, the integration site can result in variability of expression across lines due to position effect variegation. Integrases (e.g., PhiC31) can be used to insert transgenes at a defined "safe harbor" locus to address this issue [15, 16].
2. Weak expression from endogenous locus integration may be overcome by binary systems [12] (e.g., Gal4/UAS; QF/QUAS; *see* **Note 4**) or by encoding a single transgenic element that contains both the driver element and the NTR-linked reporter. This creates a positive feedback loop resulting in enhanced and continuous expression (e.g., "kalooping") [17]. Caution should be taken with such an approach to ensure that overexpression of the fluorescent reporter or NTR does not result in undesired mosaicism or cell lethality. Cell types may differ in their sensitivity to NTR 2.0 overexpression, so caution should be taken to avoid undesired cell lethality prior to addition of a prodrug. If this issue does arise, expression of NTR 1.1 may be better tolerated.
3. As with the generation of any transgenic animal with a random integration site, there exists the possibility of position site variegation affecting the level and specificity of transgene expression. The use of Tol2-based integration can also lead to

multiple integrations. This can make it challenging to identify an effective allele and maintain a well-characterized stable line. Caution should be taken to ensure proper co-expression of NTR and reporters through careful screening of transgenic families or integration into "safe harbor" integration sites.

4. The zebrafish community has developed a wealth of enhancer trap and promoter driven transgenic resources that utilize binary expression systems (e.g., Gal4/UAS, QF/QUAS, LexA/LexOP) [18, 19]. Transgenic fish expressing "driver" proteins (e.g., Gal4, QF) can be crossed to "reporter/effector" lines that co-express NTR 2.0 and a fluorescent protein (e.g., UAS:XFP-2A-NTR 2.0 or QUAS:XFP-2A-NTR2.0) in a target cell type of interest. This saves the trouble of having to create a unique NTR-XFP expressing transgenic line for each new cell type targeted or when an alternative reporter is desired. We recommend using variants of Gal4 and QF drivers that have been optimized for use in zebrafish [17, 20]. Some existing options of promoter-driven and enhancer-trap lines have been curated on zfin (GAL4: zfin.org/ZDB-EFG-080319-1; GAL4FF: zfin.org/ZDB-EFG-080516-1; KALTA4: zfin.org/ZDB-EFG-120217-2; and QF: zfin.org/ZDB-EFG-140721-1). Transgenerational silencing of UAS-based reporter/effector elements can be a concern when utilizing Gal4 systems and has led to attempts to decrease the CpG dinucleotide content and engineer varying numbers or nonrepetitive versions of repeats in UAS sequences [20–22].

5. In some cases, ablating only a subset of cells within the targeted population may be desired. Injecting NTR 2.0-expressing plasmids into developing embryos can facilitate this type of assay during early development. For later stages, we have found that when multiple UAS reporter transgenic lines are crossed to a single driver line, a mosaic pattern of expression often results, allowing random targeting of a subset of cells within the targeted cell population [10]. This can lead to opportunities to ablate a subset of cells within the larger targeted population and visualize the responses of the nontargeted neighboring cells.

6. NTR has also been used with the prodrug CB1954 due to its ability to induce a bystander effect, thereby killing both targeted and neighboring cells [5]. In addition to MTZ, "no bystander" prodrugs capable of specifically ablating NTR-expressing cells have been identified, including Nifurpirinol (NFP) and Ronidazole [6, 7]. When compared to MTZ, NFP and Ronidazole appear to offer comparable or improved cell-specific ablation efficacy. However, nonspecific toxicity remains an issue when using NFP and Ronidazole, as the

difference between their effective and toxic concentrations is minimal (i.e., similar to MTZ in this regard).

7. To facilitate visualization of cell loss in vivo, NTR is typically genetically engineered to be co-expressed with a fluorescent reporter protein either directly as a fusion protein or by a viral 2A peptide linker that is cleaved during translation. Several fluorescent reporter options exist that can be altered to suit individual needs, e.g., a specific fluorescent emission spectrum and/or a subcellular localized reporter. Existing options for transgenic lines and plasmids encoding fluorescently tagged NTR 2.0 include QUAS:YFP-2A-NTR2.0, UAS:YFP-2A-NTR2.0, UAS:GFP-2A-NTR2.0, and UAS:mCherry-2A-NTR2.0 [10] (addgene.org/browse/article/28211804/).
8. Assays such as terminal deoxynucleotidyl transferase dUTP nick end labeling (TUNEL) cannot delineate between parthanatos and apoptosis. Cell death mechanisms can often involve overlapping components and evolving definitions. Resources such as the Nomenclature Committee on Cell Death [23] can provide guidance when further characterization is desired of the type of cell death elicited by an NTR/prodrug combination in a specific cell type.
9. NTR 1.1 remains a viable option for cells that are known to be susceptible to NTR-based ablation or for assays that only require short-term prodrug exposures with MTZ, NFP, or Ronidazole. While NTR 1.1 has exhibited increased ablation efficacy when used with NFP and Ronidazole [7], NTR 2.0 has shown increased ablation efficacy with MTZ but not NFP [10]. NTR 2.0 efficacy with Ronidazole is currently being evaluated.
10. While not essential, these 96-well plates offer some protection for MTZ-treated larvae from the light and should be covered with a lid to ensure small volumes do not evaporate during treatment in an incubator.
11. MTZ powder should be stored at 4 °C for frequent use or −20 °C for long-term storage. Variation in ablation efficacy across providers and lots has been observed; thus, each new source of MTZ should be evaluated for ablation efficacy prior to large-scale use.
12. Prodrug solution is generally made fresh on the day it is used. MTZ stocks at concentrations of $\leq$1 mM can be stored at room temperature in the dark and used within a week of dilution without a significant decrease in effectiveness. Previously, we used DMSO (0.1%) to assist in getting high concentrations of MTZ (10 mM) into solution. The lower effective concentrations afforded with NTR 2.0-based cell ablations eliminate the need for DMSO. Since exposing larval zebrafish to DMSO has been reported to alter development, behavior, and to induce

stress-related signaling [24, 25], its use may confound experiments. Avoidance is therefore preferred whenever possible.

13. Alternatively, 2× MTZ stocks can be made up and diluted 1:1 into E3 media/system water when removing media is impractical, such as multi-well plates.

14. The concentration of MTZ needed for ablation will vary depending on the cell type, the version of NTR used, and the access of the prodrug to the targeted cells. Titrate prodrugs in initial experiments to determine an optimal concentration for each targeted cell type. The timing and duration of prodrug application will be dictated by the concentration of the prodrug, the cell type being ablated, the timing of NTR expression, and the desired outcome of the study. Prior NTR variants (NTR 1.0 and NTR 1.1) typically required a 10 mM MTZ exposure to induce effective levels of cell loss, which limited treatment to ~24 h before general toxicity issues arose. The improved NTR 2.0 variant allows ablation at nontoxic prodrug concentrations (e.g., <1 mM MTZ), enabling sustained ablation paradigms. Juvenile zebrafish have been incubated in up to 1 mM MTZ for over a month without detection of negative side effects [10]. Effects can vary but retinal neuronal cells (rod photoreceptor cells) typically begin to be lost around 12–16 h of 0.2 mM MTZ incubation. However, overt loss of detectable reporter often continues for up to 24 h after an acute prodrug exposure. Time-lapse imaging suggests this is likely due to the time it takes phagocytic cells to clear dead cell debris [26]. For example, a 24 h treatment with 0.2 mM MTZ is sufficient for near-total ablation of rod photoreceptor cells when analyzed at 48 h. For many cell types expressing NTR 2.0, a 24 h incubation with MTZ from 100 μM to 1 mM should be sufficient for total cell ablation. If cells persist, the concentration of MTZ can be increased or the incubation time can be extended.

15. We have noted that higher MTZ concentrations are not necessarily more effective when used in conjunction with NTR 2.0. In fact, MTZ titration assays often resulted in less effective ablation at concentrations above optimal concentrations (e.g., incubation with 5 or 10 mM MTZ typically shows diminished ablation efficacy compared to concentrations ranging between 40 μM and 1 mM). Whether this is due to prodrug-related stress or a compromised ability to efficiently clear the dying cells is unclear.

16. If prodrugs other than MTZ are to be used: Ronidazole shows general toxicity at >6 mM when used for >24 h, with concentrations at ~2 mM being effective in ablating cells expressing NTR 1.1 [7]. NFP shows general toxicity at 5–8 μM when used for >24 h, with concentrations at ~2.5 μM being effective in

ablating cells expressing NTR 1.1 [6, 7, 10]. We have noted variation in toxicity profiles of the alternative prodrugs, which could be due to differential sensitivity across strains. We therefore recommend defining toxic concentrations in the genetic background specific to each assay.

17. The time course of regeneration may vary depending on the tissue/cell type that has been ablated. For instance, larvae that have had their rod photoreceptors ablated for 24 h with MTZ (from days 5–6) have repopulated many of these cells 72 h after MTZ has been removed (at 9 dpf).
18. Zebrafish need not be kept in the dark during extended treatments with MTZ. We recommend maintaining MTZ-treated larvae/adults in the typical 14/10 h light/dark cycle to maintain normal circadian rhythmicity.
19. As the microbiome is now known to play key roles in many biological paradigms, caution is advised when using extended prodrug treatments; antibiotics, including MTZ, are known to kill bacteria at higher concentrations [27].
20. Take caution not to feed growing larvae with rotifers in the presence of MTZ as it may result in toxicity. Our experience suggests rotifers can reduce MTZ to its toxic by-product, leading to lethality upon consumption. Larvae that are too young to subsist on brine shrimp should be fed dry food.
21. Microscopy is generally a good indicator of overall cell loss but can also be misleading if simply looking at brightness of fluorescence at low magnification. Fragmented and dying cells may still contain strong fluorescent reporters, making it appear as though cells have not been ablated. Although time-consuming, the higher magnification and spatial resolution afforded by confocal microscopy provides a more definitive approach to discerning the morphologies of NTR targeted cells following MTZ exposure.
22. Visualizing the absence of co-markers or antibodies that normally label your desired cell type is another good confirmation of cell loss. If possible, confirmation by quantification of changes in cell nuclei (e.g., DAPI) is the most definitive means of confirming cell loss.
23. Methods for fluorescence-based plate reader analysis of live transgenic reporters have been previously described in detail [28]. The use of a plate reader for cell death analysis may require higher sample sizes due to variability in transgene expression and detection.

Acknowledgments

The authors would like to thank Jen Anderson and Dr. Meera Saxena for critical reading and comments on the manuscript.

References

1. Zhang L, Chen C, Fu J et al (2021) Large-scale phenotypic drug screen identifies neuroprotectants in zebrafish and mouse models of retinitis pigmentosa. elife 10. https://doi.org/10.7554/eLife.57245
2. Curado S, Anderson RM, Jungblut B et al (2007) Conditional targeted cell ablation in zebrafish: a new tool for regeneration studies. Dev Dyn 236:1025–1035. https://doi.org/10.1002/dvdy.21100
3. Chen C-F, Chu C-Y, Chen T-H et al (2011) Establishment of a transgenic zebrafish line for superficial skin ablation and functional validation of apoptosis modulators in vivo. PLoS One 6:e20654. https://doi.org/10.1371/journal.pone.0020654
4. Pisharath H, Rhee JM, Swanson MA et al (2007) Targeted ablation of beta cells in the embryonic zebrafish pancreas using E. coli nitroreductase. Mech Dev 124:218–229. https://doi.org/10.1016/j.mod.2006.11.005
5. Bridgewater JA, Knox RJ, Pitts JD et al (1997) The bystander effect of the nitroreductase/CB 1954 enzyme/prodrug system is due to a cell-permeable metabolite. Hum Gene Ther 8: 709–717. https://doi.org/10.1089/hum.1997.8.6-709
6. Bergemann D, Massoz L, Bourdouxhe J et al (2018) Nifurpirinol: a more potent and reliable substrate compared to metronidazole for nitroreductase-mediated cell ablations. Wound Repair Regen 26:238–244. https://doi.org/10.1111/wrr.12633
7. Lai S, Kumari A, Liu J et al (2021) Chemical screening reveals Ronidazole is a superior pro-drug to metronidazole for Nitroreductase-induced cell ablation system in zebrafish larvae. J Genet Genomics. https://doi.org/10.1016/j.jgg.2021.07.015
8. Jarrom D, Jaberipour M, Guise CP et al (2009) Steady-state and stopped-flow kinetic studies of three *Escherichia coli* NfsB mutants with enhanced activity for the prodrug CB1954. Biochemistry 48:7665–7672. https://doi.org/10.1021/bi900674m
9. Mathias JR, Zhang Z, Saxena MT, Mumm JS (2014) Enhanced cell-specific ablation in zebrafish using a triple mutant of Escherichia coli nitroreductase. Zebrafish 11:85–97. https://doi.org/10.1089/zeb.2013.0937
10. Sharrock AV, Mulligan TS, Hall KR, et al (2020) NTR 2.0: a rationally-engineered prodrug converting enzyme with substantially enhanced efficacy for targeted cell ablation. bioRxiv
11. Williams EM, Rich MH, Mowday AM et al (2019) Engineering *Escherichia coli* NfsB to activate a hypoxia-resistant analogue of the PET probe EF5 to enable non-invasive imaging during enzyme prodrug therapy. Biochemistry 58:3700–3710. https://doi.org/10.1021/acs.biochem.9b00376
12. Wierson WA, Welker JM, Almeida MP et al (2020) Efficient targeted integration directed by short homology in zebrafish and mammalian cells. elife 9. https://doi.org/10.7554/eLife.53968
13. Hoshijima K, Jurynec MJ, Grunwald DJ (2016) Precise editing of the zebrafish genome made simple and efficient. Dev Cell 36:654–667. https://doi.org/10.1016/j.devcel.2016.02.015
14. Shin J, Chen J, Solnica-Krezel L (2014) Efficient homologous recombination-mediated genome engineering in zebrafish using TALE nucleases. Development 141:3807–3818. https://doi.org/10.1242/dev.108019
15. Mosimann C, Puller A, Lawson KL et al (2013) Site-directed zebrafish transgenesis into single landing sites with the phiC31 integrase system. Dev Dyn 242:949–963. https://doi.org/10.1002/dvdy.23989
16. Roberts JA, Miguel-Escalada I, Slovik KJ et al (2014) Targeted transgene integration overcomes variability of position effects in zebrafish. Development 141:715–724. https://doi.org/10.1242/dev.100347
17. Distel M, Wullimann MF, Koster RW (2009) Optimized Gal4 genetics for permanent gene expression mapping in zebrafish. Proc Natl Acad Sci 106:13365–13370. https://doi.org/10.1073/pnas.0903060106

18. Kawakami K, Abe G, Asada T et al (2010) zTrap: zebrafish gene trap and enhancer trap database. BMC Dev Biol 10:105. https://doi.org/10.1186/1471-213X-10-105
19. Marquart GD, Tabor KM, Brown M et al (2015) A 3D searchable database of transgenic zebrafish Gal4 and Cre lines for functional neuroanatomy studies. Front Neural Circ 9:78. https://doi.org/10.3389/fncir.2015.00078
20. Burgess J, Burrows JT, Sadhak R et al (2020) An optimized QF-binary expression system for use in zebrafish. Dev Biol 465:144–156. https://doi.org/10.1016/j.ydbio.2020.07.007
21. Akitake CM, Macurak M, Halpern ME, Goll MG (2011) Transgenerational analysis of transcriptional silencing in zebrafish. Dev Biol 352:191–201. https://doi.org/10.1016/j.ydbio.2011.01.002
22. Goll MG, Anderson R, Stainier DYR et al (2009) Transcriptional silencing and reactivation in transgenic zebrafish. Genetics 182:747–755. https://doi.org/10.1534/genetics.109.102079
23. Galluzzi L, Vitale I, Aaronson SA et al (2018) Molecular mechanisms of cell death: recommendations of the nomenclature committee on cell death 2018. Cell Death Diff 25:486–541. https://doi.org/10.1038/s41418-017-0012-4
24. Xiong X, Luo S, Wu B, Wang J (2017) Comparative developmental toxicity and stress protein responses of dimethyl sulfoxide to rare minnow and zebrafish embryos/larvae. Zebrafish 14:60–68. https://doi.org/10.1089/zeb.2016.1287
25. Chen T-H, Wang Y-H, Wu Y-H (2011) Developmental exposures to ethanol or dimethylsulfoxide at low concentrations alter locomotor activity in larval zebrafish: implications for behavioral toxicity bioassays. Aquat Toxicol 102:162–166. https://doi.org/10.1016/j.aquatox.2011.01.010
26. White DT, Sengupta S, Saxena MT et al (2017) Immunomodulation-accelerated neuronal regeneration following selective rod photoreceptor cell ablation in the zebrafish retina. Proc Natl Acad Sci 114:E3719–E3728. https://doi.org/10.1073/pnas.1617721114
27. Sheng Y, Ren H, Limbu SM et al (2018) The presence or absence of intestinal microbiota affects lipid deposition and related genes expression in zebrafish (Danio rerio). Front Microbiol 9:1124. https://doi.org/10.3389/fmicb.2018.01124
28. Walker SL, Ariga J, Mathias JR et al (2012) Automated reporter quantification in vivo: high-throughput screening method for reporter-based assays in zebrafish. PLoS One 7:e29916–e29916. https://doi.org/10.1371/journal.pone.0029916

Chapter 16

Section Immunostaining for Protein Expression and Cell Proliferation Studies of Regenerating Fins

Scott Stewart and Kryn Stankunas

Abstract

Adult zebrafish fins fully regenerate after resection, providing a highly accessible and remarkable vertebrate model of organ regeneration. Fin injury triggers wound epidermis formation and the dedifferentiation of injury-adjacent mature cells to establish an organized blastema of progenitor cells. Balanced cell proliferation and redifferentiation along with cell movements then progressively reestablish patterned tissues and restore the fin to its original size and shape. A mechanistic understanding of these coordinated cell behaviors and transitions requires direct knowledge of proteins in their physiological context, including expression, subcellular localization, and activity. Antibody-based staining of sectioned fins facilitates such high-resolution analyses of specific, native proteins. Therefore, such methods are mainstays of comprehensive, hypothesis-driven fin regeneration studies. However, section immunostaining requires labor-intensive, empirical optimization. Here, we present detailed, multistep procedures for antibody staining and co-detecting proliferating cells using paraffin and frozen fin sections. We include suggestions to avoid common pitfalls and to streamline the development of optimized, validated protocols for new and challenging antibodies.

Key words Zebrafish, Fin regeneration, Blastema, Immunostaining, Antibodies, EdU labeling, Paraffin sections, Cryosections

1 Introduction

Adult zebrafish fins robustly regenerate to their original size and shape within a few weeks of injury [1]. The widely studied caudal fin is typified by 18 bony rays, or lepidotrichia, extending along the proximal–distal axis. Rays comprise highly segmented, semicylindrical, and apposed hemi-rays lined by osteoblasts that produce mineralized bone. Rays are tubular, encasing fibroblasts, vasculature, and sensory axons, and the entire fin skeleton is wrapped by a stratified epidermis. Fin amputation triggers immediate wound epidermis formation. Each ray then produces an organized mass of progenitor cells, the blastema, over an approximately 2-day "establishment phase." Blastema cells largely arise from the

James F. Amatruda et al. (eds.), *Zebrafish: Methods and Protocols*, Methods in Molecular Biology, vol. 2707,
https://doi.org/10.1007/978-1-0716-3401-1_16,

injury-induced dedifferentiation of mature cells into lineage-restricted progenitors. Spatiotemporally restricted cell signaling orchestrates the precise balance between self-renewal and redifferentiation to progressively re-form lost fin tissue, including the bony rays. Meanwhile, orthogonal regulatory networks restore skeletal pattern, including joints and ray branching. This "outgrowth phase" slows and then terminates as fin size is restored. Determining how cells change states during regeneration, including initial dedifferentiation and cell cycle entry, and how cells are positioned to restore pattern will inform how zebrafish and select other vertebrates readily regenerate limbs and other organs. Coordinated protein expression and regulation promote cell states and behaviors, and therefore hypothesis-driven studies of specific proteins remains central to resolving mechanisms.

In situ detection of mRNA in whole-mount regenerating fins is the most common approach to localize endogenous gene expression [2]. This technique is rapid and generally robust, specially requiring only synthesizing a probe using a cloned cDNA for the gene of interest [3]. While informative, whole-mount mRNA in situ hybridization techniques do not provide cell-level resolution of expression domains and can suffer from variable tissue penetration [2]. In situ hybridization using histological sections improves resolution but, of course, does not inform on the transcript's encoded protein. Transgenic reporters are also useful tools for gene expression studies, including increasingly common "knock-in" lines to monitor endogenous gene regulatory elements. Fluorescent and other reporters, however, can be misleading due to slow fluorescence maturation and prolonged protein perdurance. Even fusion protein "knock-ins," while enabling live imaging with fluorescent tags, can alter endogenous protein stability, localization, and activity. Therefore, rigorously resolving molecular mechanisms requires detailed knowledge of native proteins' expression patterns, relative abundances, subcellular and extracellular distributions, and, in some cases, activity.

Fluorescence immunostaining combined with high-resolution microscopy reveals expression patterns and subcellular localizations of protein targets in contexts of complex tissues containing an array of cell types. Whole mount staining can provide a three-dimensional perspective when coupled with confocal imaging and reconstructions. However, most antibodies do not perform well, light scattering and absorbance limit resolution, and image acquisition and analysis are time-consuming. Immunostaining using thin tissue sections is more reliable, higher throughput, and readily enables even super-resolution studies of protein localization and activity. Limitations include the inherent two-dimensional view and, relatedly, difficulty selecting well-matched sections between control and experimental samples. Other technical barriers include obtaining specific antibodies suitable for section immunostaining and optimizing staining protocols for each antigen–antibody pair. Formaldehyde fixation also causes protein–protein cross-links that obscure antigen–antibody binding [4, 5]. Fortunately, a variety of antigen retrieval techniques facilitate immunostaining on formaldehyde fixed tissue sections with excellent cellular resolution and signal:noise [6–8]. In our experience studying fin regeneration, section immunostaining has been instrumental to identify cell lineages for mosaic analyses [9, 10], osteoblast subtypes and proliferation states [11], and epidermis–osteoblast interactions underlying bony ray branching [12, 13].

Regrowing an adult fin in a few weeks requires widespread cell proliferation, most notably of blastema-populating osteoblast- and fibroblast-lineage cells during the outgrowth phase [1, 11]. Growth rates and extents are influenced by cell cycle control and/or overall numbers of proliferating cells. Accordingly, analyzing cell proliferation provides mechanistic insight into phenotypes associated with a genetic lesion, small molecule treatment, or other perturbation. Immunohistological techniques using tissue sections allow quantifying cell proliferation and simultaneously detecting co-stained proteins of interest.

Antibody-based approaches are often used to detect proliferating cells. For example, cells with high levels of proliferating cell nuclear antigen (PCNA) are considered proliferative [14, 15]. However, PCNA levels also can increase in nonproliferating cells responding to environmental insults, such as UV radiation [16, 17]. Therefore, it may be prudent to classify PCNA-expressing cells as having proliferation potential. Alternatively, mitotic cells can be distinguished using antibodies against phosphorylated Serine 10 Histone H3 present only at the onset of mitosis [18–20]. However, mitosis is brief and therefore usually few cycling cells are detected, challenging quantitative analyses.

Alternatively, monitoring incorporation of modified nucleotides into newly synthesized DNA directly identifies cells in S-phase. This is arguably the most straightforward and reliable way to detect cell proliferation in vivo, as evidenced by an extensive number of publications. Traditionally, this is carried out by delivering 5-bromo-2′-deoxyuridine (BrdU) to cells or to vertebrate animals in vivo by intraperitoneal (IP) injection [21–23]. Cells transiting S-phase incorporate BrdU into nascent DNA strands, allowing visualization using anti-BrdU antibodies [21, 22]. However, because BrdU antibody detection on sections requires acid treatment to nick DNA, co-staining with other antibody markers may be compromised [24]. “Click chemistry” to detect incorporation of the nucleotide analog 5-ethynyl-2′-deoxyuridine (EdU) in cells and animals overcomes this challenge [25–27]. This approach exploits the Cu(I)-catalyzed addition of fluorescent azides to EdU in DNA strands in situ [27], circumventing the need for acid treatment and antibodies for visualization.

We present methods to prepare paraffin and frozen fin sections and optimize antigen retrieval, blocking, and antibody staining. We describe how to integrate EdU labeling with section immunostaining to co-monitor cycling cells. We also emphasize the importance of primary antibody selection and validation when performing immunostaining and provide guidelines for assessing results in the context of two-dimensional fin sections. Together, we hope these suggestions will facilitate the incorporation of rigorous, cell-resolution studies of specific proteins toward understanding a most striking and accessible example of robust organ regeneration.

2 Materials

2.1 Equipment and Supplies

1. Plastic zebrafish crossing tanks. Aquaneering Inc., catalog ZHCT100T.
2. Hamilton syringe 702LT, luer tip, 25 μL volume.
3. Sterile 27 gauge, 1/2 inch length (27G ½″) needles.
4. Razor blades.
5. Paint brushes, small.
6. Glass vials with screw cap, 4 mL and 16 mL, e.g., Fisher Scientific., catalog 03-343-6C and 03-339-38F.
7. Coplin Jars with screw cap, 50 mL, e.g., Fisher Scientific Inc., catalog 19-4
8. Forceps, broad tip for slide handling and fine tip for embedding tissue.
9. Xylenes-safe histological wash containers, e.g., Fisher Scientific Inc., catalog 22-309-247
10. Consumer-grade electric pressure cooker, e.g., Cuisinart CPC-600 N1.
11. Microscope slides, e.g., Superfrost Plus, Fisher Scientific Inc., catalog 12-550-15.
12. Cover slips for microscope slides, 40 × 22 mm.
13. Paraffin embedding station or dispenser.
14. Embedding rings and base molds for paraffin embedding, Fisher Scientific Inc., catalog 22-038-197 and 22-363-552
15. Microtome.
16. Peel-A-Way Embedding Mold for frozen tissue, Polysciences Inc., catalog 18986-1
17. Cryostat.
18. Lab mixer, e.g., Fisher Scientific Inc., catalog 12-815-3Q.
19. Slide boxes with hinged lid.
20. Humidified chamber for slides. Prepared by layering damp paper towels on the bottom of a plastic slide box.

2.2 Required Buffers and Reagents

1. Prepare all solutions using molecular biology-grade water (resistivity of >18 MΩ·cm at 25 °C). Follow all safety procedures when working with hazardous materials and dispose of waste accordingly.
2. Zebrafish aquarium water. 60 mg Instant Ocean Sea Salt per liter H_2O. Instant Ocean Inc., catalog SS15-10.
3. Tricaine 100× stock solution (1.68%). Dissolve 1.68 g Tricaine (ethyl-3-aminobenzoate methanesulfonate, Sigma Aldrich Inc., E10521) in 100 mL H_2O. Store at 4 °C.
4. 10x concentrated phosphate buffered saline (PBS). 1.37 M NaCl, 27 mM KCl, 100 mM Na_2HPO_4, 18 mM KH_2PO_4. To prepare 1 L, dissolve 17.8 g $Na_2HPO_4 \cdot 2H_2O$, 2.4 g KH_2PO_4, 80 g NaCl, 2 g KCl in 500 mL H_2O. Once particles are dissolved, adjust to 1 L final volume using H_2O. Store at room temperature.
5. 4% paraformaldehyde (PFA) in 1x PBS. Combine 10 mL of 16% paraformaldehyde solution (Electron Microscopy Sciences Inc., catalog RT15700), 26 mL H_2O, and 4 mL 10× PBS. Store in aliquots at −20 °C for several months. Avoid repeated freeze/thawing. Paraformaldehyde is flammable and considered hazardous. Avoid inhalation and contact with skin.
6. PBS. Combine 100 mL 10× PBS stock solution and 900 mL H_2O. Store at room temperature.
7. Ethanol, 100% and 95%, molecular biology grade. Highly flammable and hazardous. Avoid inhalation and contact with skin.
8. Ethanol series. Prepare 25%, 50%, and 75% ethanol by diluting 95% ethanol with H_2O. Store at room temperature. Ethanol is flammable and hazardous. Avoid inhalation and contact with skin.
9. Xylenes. Histological grade, Sigma Aldrich Inc., catalog 534056. Hazardous. Use in fume hood, avoid inhalation and contact with skin.
10. 30% sucrose. For preparing frozen sections. Dissolve 30 g sucrose in 10 mL 10× PBS + 30 mL H_2O with vigorous mixing. Once in solution, adjust final volume to 100 mL with H_2O followed by vacuum sterilization using a 0.45 μm filter bottle. Store at 4 °C.
11. Tissue-Tek O.C.T. Compound. For frozen tissue embedding. Sakura Finetek USA Inc., catalog 4583.
12. 0.5 M EDTA, pH 8.0. Combine 186.1 g of disodium EDTA·$2H_2O$ with 800 mL of H_2O while stirring vigorously on a magnetic stirrer. Adjust the pH to 8.0 with NaOH pellets (roughly 20 g). Bring the final volume to 1 L using H_2O. Store at room temperature (*see* **Note 1**).

13. 5 M NaCl. To prepare 1 L, dissolve 292 g of NaCl in 700 mL of H_2O. Using a graduated cylinder, add H_2O to a final volume of 1 L. Store at room temperature.
14. 10% Tween-20. Combine 10 mL molecular biology grade Tween-20 with 90 mL H_2O and mix well. Store at room temperature.
15. 10% Triton X-100. Thoroughly mix 10 mL molecular biology grade Triton X-100 with 90 mL H_2O. Store at room temperature.
16. PBST. PBS containing 0.1% Tween-20. Combine 10 mL 10% Tween-20, 100 mL 10× PBS concentrate in a 1 L graduated cylinder. Adjust volume to 1 L using H_2O and store at room temperature.
17. EDTA-Tween antigen retrieval solution (1 mM EDTA, pH 8.0, 0.1% Tween-20). To prepare 50 mL of the above solution, mix 0.1 mL 0.5 M EDTA, pH 8.0, 0.5 mL 10% Tween-20 and bring volume to 50 mL using H_2O. Prepare freshly before each use.
18. Agilent-DAKO Target Retrieval Solution. 10× concentrate. Agilent Inc., catalog S169984-2. Prepare a working solution by diluting concentrate 1:10 in H_2O. Store at 4 °C (*see* **Note 2**).
19. ImmunoSaver Antigen Retriever. Electron Microscopy Sciences Inc., catalog 64142. Immediately prior to use, prepare a working solution by diluting ImmunoSaver 1:200 in H_2O. Contains highly toxic citraconic acid. Use in fume hood at all times. Avoid contact and inhalation. Waste is hazardous.
20. Trypsin-EDTA, 0.25% solution. ThermoFisher Inc., catalog 25200056.
21. 5-Ethynyl-2′-deoxyuridine (EdU) labeling and detection. Click-iT EdU Alexa Fluor 488 Imaging Kit, ThermoFisher Inc., catalog C10337, or Click-iT Plus EdU Alexa Fluor 488 Imaging Kit, ThermoFisher Inc., catalog C10637 (*see* **Note 3**).
22. EdU solution for intraperitoneal (IP) injection of adult zebrafish. Prepare a 1 mg/mL working stock by dissolving EdU powder from the Click-iT EdU kit in the appropriate volume of PBS. Warm to 37 °C until completely dissolved. Aliquot and store at −20 °C.
23. Primary Antibodies. Antigen dependent (*see* Subheading 3.10).
24. Fluorophore-conjugated secondary antibodies. Directed against the species of the primary antibody. For example, Rabbit anti-Mouse IgG (H + L) Secondary Antibody, Alexa Fluor 546 conjugate, ThermoFisher Inc., catalog A11060 (*see* **Note 4**).

25. Nonfat dry milk powder, blocking reagent, e.g., Carnation Instant Nonfat Dry Milk.
26. Bovine Serum Albumin (BSA) for blocking. Molecular biology grade.
27. Nonimmune serum for blocking. From the same species as the secondary antibody, typically goat or donkey, e.g., Rockland Inc., catalog B304 and D315-05.
28. Hoechst nuclear staining solution. Prepare a 10 mg/mL solution by dissolving 100 mg Hoechst 33342 (ThermoFisher Inc., catalog H1399) in 10 mL PBS. Prepare aliquots and freeze at −20 °C. Once thawed, solutions are stable at 4 °C for several months.
29. Mounting medium. Invitrogen SlowFade Diamond Antifade Mountant (ThermoFisher Inc., catalog S36963).
30. Clear nail polish, fast drying.

3 Methods

3.1 Intraperitoneal Injection of Adult Zebrafish with 5-Ethynyl-2′-Deoxyuridine (EdU)

To detect S-phase cells, EdU is delivered to zebrafish by intraperitoneal (IP) injection for incorporation into nascent DNA 2–4 h prior to tissue collection. The injection follows published guidelines with minor adaptations [11, 28, 29]. For efficiency, we employ Luer-tip Hamilton syringes, allowing use of standard disposable needles.

Carry out steps at room temperature or in a zebrafish facility (28 °C).

1. Thaw an aliquot of EdU 1 mg/mL stock solution at 37 °C. Vortex until all particulates are solubilized.
2. Attach a sterile 27G ½″ needle to a Hamilton syringe and flush with sterile PBS to remove any debris.
3. Prepare Tricaine by diluting 1.68% stock solution 1:100 in zebrafish aquarium water in a crossing tank or equivalent container.
4. Transfer three to six fish to diluted Tricaine for anesthesia.
5. Bathe fish in Tricaine until opercular movement slows (roughly 60 s). Transfer fish from Tricaine to the inverted petri dish lid.
6. Using a small paint brush, position fish on their backs and aligned against the lip of the petri dish lid.
7. Attach a fresh 27G ½″ needle to the Hamilton syringe and load 12 μL EdU stock solution (for a 1 g mature adult fish, adjust the EdU concentration or volumes proportionately). Use your nondominant hand and the brush to gently press the fish

against the lip of the petri dish lid. With the other hand, insert needle into the peritoneal space and inject the entire volume. The syringe/needle's void volume of ~2 μL results in an EdU dose of approximately 10 mg/kg body weight.

8. Transfer injected fish to crossing tanks (or equivalent) containing system water. Once swimming normally, transfer to housing tanks on flow-through water until fin collection.

3.2 Tissue Collection and Fixation

Perform **steps 1–6** at room temperature or in a zebrafish facility (28 °C). Complete **steps 8–10** at room temperature.

1. Thaw PFA-PBS, vortex, and then transfer required volume of fixative to a glass vial (4 mL vial for less than 5 fins; 16 mL vial for 6–20 fins). Use 1 mL PFA-PBS for 1 adult fin and scale linearly for multiple samples. Keep vial on ice.
2. Prepare Tricaine by diluting stock solution 1:100 in zebrafish water in a crossing tank or equivalent container.
3. Transfer fish to diluted Tricaine and anesthetize as described in Subheading 3.1.
4. Transfer fish to an inverted petri dish lid.
5. Using a fresh razor blade, amputate fin tissue and transfer fish to fresh fish water for recovery.
6. Wet the bristles of a small paint brush in PFA-PBS. Use the brush to gently pick up the resected tissue and transfer to the vial containing cold PFA-PBS. Immediately return vial to ice. Harvest tissue as quickly as possible and keep on ice to preserve biological integrity.
7. Once all tissue is collected, fix overnight at 4 °C with gentle agitation (*see* **Note 5**).
8. Carefully decant PFA-PBS and replace with 3 mL or 10 mL PBS (for 4 mL or 16 mL vials, respectively). Repeat three times to remove any residual PFA.
9. Incubate fins in for 1 h with gentle agitation. Decant and replace with the same volume of PBS. Repeat three more times. The total wash time is 4 h.
10. For paraffin sections, proceed to Subheading 3.3; for frozen sections, skip to Subheading 3.4. We recommend trying both paraffin and frozen sections when optimizing a new antibody for fin immunostaining (*see* **Note 6**).

3.3 Preparation of Paraffin Sections

Perform steps at room temperature unless noted otherwise.

1. Decant PBS, replace with an equal volume of 0.5 M EDTA, pH 8.0, and gently mix for 5 min. Decant and replace with fresh EDTA solution. Decalcify in EDTA for 4 days with gentle agitation.

2. Remove EDTA with six 1 h washes in 3 mL or 10 mL PBS.
3. Dehydrate tissue by sequential 30 min washes in 3 mL (10 mL) 25%, 50%, 75%, and 100% ethanol. Replace with fresh 100% ethanol and incubate overnight with gentle mixing.
4. Perform a final 30 min wash in 100% ethanol to ensure complete dehydration.
5. Remove ethanol and wash twice with 3 mL (10 mL) xylenes for 30 min. Xylenes are hazardous. Use a fume hood and wear gloves.
6. Carefully decant xylenes into a waste container and add 3 mL (10 mL) molten paraffin using a paraffin embedding station or dispenser. Keep vials at 65 °C.
7. Remove paraffin and replace it with an equal volume of fresh molten wax. Incubate 10 min at 65 °C. Repeat six times.
8. Incubate vial overnight at 65 °C.
9. Embed fins for either longitudinal or transverse sections. Transfer tissue from the glass vial to an embedding mold using a wide-bore transfer pipette. Fill mold with molten wax. Using pre-warmed (65 °C) fine forceps, quickly arrange the tissue while wax hardens. We recommend working under a stereomicroscope. For longitudinal sections, orient the tissue so that the dorsal–ventral axis is perpendicular to the bench top and the fin rays are parallel to the bench top. For transverse sections, position the fin with the dorsal–ventral axis parallel and the rays perpendicular to the bench top. Face the amputation site toward the bottom of the embedding mold.
10. Allow blocks to solidify at room temperature overnight. Transfer hardened blocks to −20 °C, where they can be stored indefinitely.
11. Collect 7 μm sections using a microtome. Transfer to glass slides pre-labeled with a pencil, preloaded with molecular biology-grade water, and placed on a slide warmer set to 37 °C. Section control and experimental blocks consecutively during the same session. Dry overnight at 37 °C.
12. Transfer dried slides to slide boxes and store at room temperature indefinitely.
13. Proceed to Subheading 3.5 when ready for immunostaining.

3.4 Preparation of Frozen Sections

Perform procedures at room temperature unless noted.

1. Decant PBS and replace with 3 mL or 10 mL (for 4 mL or 16 mL vials, respectively) 30% sucrose solution. Gently mix overnight at 4 °C.
2. Remove sucrose solution with a pipette and replace with O.C.T.

3. Using a wide-bore transfer pipette, transfer fin to an embedding mold containing O.C.T.
4. Place the mold on a flat block of dry ice. Quickly position the tissue using fine forceps for either longitudinal or transverse sections as described in Subheading 3.3.
5. Leave the mold on dry ice until the O.C.T. freezes. Store at − 80 °C.
6. Collect cryostat sections on slides and dry for 1 h. Store slides at −20 °C or −80 °C.
7. Proceed to Subheading 3.6 when ready for immunostaining.

3.5 Dewaxing and Hydration of Paraffin Sections

Handle slides with clean broad tip forceps while wearing gloves. Avoid touching the tissue sections. All steps are performed at room temperature.

1. Fill two organic solvent-resistant (e.g., glass) histology containers with xylenes. Using forceps, submerge slides in the first xylenes bath for 5 min. Transfer slides to the second xylenes-holding container for another 5 min.
2. Extract xylenes and rehydrate tissues by serial washes in pre-filled histology containers: 5 min each in 100%, 95%, 70%, 50%, and 25% ethanol. Transfer slides between containers using forceps.
3. Hydrate slides by transferring them to a screw-cap Coplin jar filled with PBS.
4. Proceed to antigen retrieval (Subheading 3.7).

3.6 Hydration of Frozen Sections

Wear gloves and use forceps when handling slides. Perform all steps at room temperature.

1. Bring slides to room temperature and use a pencil to label slides.
2. Place slides in a Coplin jar containing PBS.
3. Wash two times with PBS (5 min each) to remove all traces of O.C.T.
4. Proceed to Subheading 3.8 or the optional antigen retrieval step (Subheading 3.7).

3.7 Antigen Retrieval

Fixing experimental samples is essential to preserve cell and tissue morphology. Formaldehyde, widely employed as a fixative, reacts with primary amines, causing extensive protein cross-links [4]. While useful as a tissue preservative, formaldehyde notoriously masks antigens, precluding successful immunostaining [6, 7]. "Antigen retrieval" methods overcome this barrier [6, 7]. Antigen retrievals are typically used in conjunction with paraffin sections, but we have found they also can improve fixed frozen section immunostaining. Antigen retrieval is commonly carried out by heat or limited protease digestion. For heat induced antigen retrieval (HIAR), slides are immersed in various solutions and heated to <100 °C, typically in an electric pressure cooker designed for home kitchen use. Many HIAR buffers are available and must be empirically tested for each antigen–antibody pair. Here, we suggest select antigen retrieval solutions for immunostaining fin sections [6–8] (*see* **Note 7**). As a "last resort", we employ proteolytic digestion using a solution of cell culture grade trypsin–EDTA. Both heat and protease treatment can disrupt fin tissue adherence to slides. We recommend a relatively short HIAR (5 min) and room temperature protease digestion (5 min) as starting points.

Perform steps at room temperature unless noted. Use forceps and wear gloves to avoid touching slides.

1. Prepare 50 mL antigen retrieval solution and transfer to a screw cap Coplin jar. 1 mM EDTA, pH 8.0, 0.1% Tween-20 solution is an inexpensive and frequently effective HIAR solution to try first (*see* **Note 7**).
2. Using forceps, transfer de-waxed, hydrated slides to the Coplin jar containing antigen retrieval solution. Lightly screw cap in place. Do not fully tighten.
3. Add H_2O to the pressure cooker to a depth of 3–5 cm. Place the Coplin jar in the pressure cooker and steam for 5 min on high (*see* **Note 7**).
4. Slowly release steam from the cooker, remove the Coplin jar, and cool to room temperature for 1–2 h. Use caution as the Coplin jar will be extremely hot.
5. Proceed to Subheading 3.8 for blocking and permeabilization of tissue.

3.8 Blocking Nonspecific Antibody Binding

Prior to antibody addition, slides are incubated in a protein-rich solution to "block" nonspecific binding of antibodies to tissue sections. An array of blocking reagents are widely used [30], including proprietary, commercially available solutions. Successful immunostaining using temperamental antibodies often requires empirically determining the best blocking buffer.

Perform all steps at room temperature.

1. For paraffin sections, a reliable and economical option is 10% nonfat milk in PBS containing 0.1% Tween-20: 5 g nonfat dry milk (to block nonspecific binding of antibodies), 10 mL 10× PBS, 0.5 mL 10% Tween-20, add water to 50 mL. For frozen sections, use the same but substitute Triton X-100 (0.1% final concentration) for Tween-20 to increase tissue permeability. Combine 5 g nonfat milk, 5 mL 10× PBS, 0.5 mL 10% Triton X-100, then add water to 50 mL (*see* **Note 8**).

2. Transfer slides to a Coplin jar containing blocking reagent. Alternatively, and for more expensive blocking reagents, wipe dry the bottom and edges of the slide on a laboratory tissue and then place slides horizontally in a humidified slide box. Gently add 0.25 mL blocking buffer using a pipette and overlay with a cover slip. Incubate in blocking solution for 1–2 h.
3. Use forceps to pick up slides and hold vertically until the cover slip slides off into a waste container. Terminate blocking/permeabilization by rinsing slides in three changes of PBS in a Coplin jar.
4. If applicable, proceed to EdU detection.

3.9 EdU Detection

EdU detection generally is performed before immunostaining since the reaction is very robust and the fluorescent products are exceptionally stable.

Complete all steps at room temperature.

1. Using the Click-iT EdU kit (*see* **Note 3**) components and following manufacturer's guidelines, prepare: (a) Alexa Fluor azide working solution, (b) 1× Click-iT EdU reaction buffer, and (c) 10× Click-iT EdU buffer additive. Store the Alexa Fluor azide and the 10× buffer additive at −20 °C and the 1× reaction buffer at 4 °C.
2. Immediately prior to use, thaw the 10× Click-iT EdU buffer additive and Alexa Fluor azide. Prepare 1× Click-iT EdU buffer additive solution by diluting the stock solution in water 1:10. Next, prepare the Click-iT reaction cocktail according to kit guidelines. We find 150–200 μL of each solution per slide is sufficient. Scale accordingly. Use the reaction cocktail within 15 min of preparing it (*see* **Note 9**).
3. Use forceps to remove slides from PBS. Drain off excess PBS. Wipe the bottom and edges of the slide on a laboratory tissue. Arrange slides face-up within the humidified chamber. Work quickly so that the slides to dry out.
4. Pipette 150–200 μL of the Click-iT reaction cocktail on to each slide. To prevent evaporation, overlay slides with a cover slip and avoid air bubbles. Close lid of humidified chamber.
5. Incubate Click-iT reactions for 30 min.
6. To stop the reaction, use forceps to hold each slide vertically until the cover slip slides off into a waste container. Transfer slides to Coplin jars containing fresh PBS.
7. Wash three times in PBS (5 min each). Proceed with immunostaining.

3.10 Immunostaining

All antibodies, whether raised against zebrafish proteins or conserved antigens from other species, can produce specific and/or nonspecific staining. Multiple factors should be considered when selecting an unvalidated primary antibody and assessing signal specificity (*see* **Note 10**).

Handle slides with clean forceps. Perform all steps at room temperature unless noted otherwise.

1. Dilute antibodies in blocking buffer. Typically, 150 μL of diluted antibody is sufficient for a single slide. A starting range of 0.1–1.0 μg/mL is a good for purified IgGs (*see* **Note 11**). Multiple primary antibodies can be combined provided they are from different species or IgG subtypes.
2. Use forceps to remove slides from Coplin jars. Remove cover slips as previously described, wipe the slide bottom and edges on a laboratory tissue to remove excess buffer, and transfer slides face-up to a humidified chamber. Work quickly to avoid drying out slides.
3. Gently apply 150 μL antibody solution to each slide. Carefully lower a cover slip to avoid air bubbles.
4. Incubate overnight at 4 °C (*see* **Note 12**), ensuring slides remain fully horizontal.
5. After primary antibody incubation, hold slides vertically and allow cover slips to slide off. To remove stubborn cover slips, first immerse slides in a Coplin jar filled with PBST for 5 min. The cover slip should fall off into the jar for later disposal. Use forceps to gently dislodge the cover slip, if necessary.
6. Wash slides in Coplin jars with PBST for 5 min. Repeat four times (*see* **Note 13**).
7. Dilute secondary antibody in blocking buffer and apply to slides in a humidified chamber as for the primary antibody step. Apply a cover slip and incubate 1–2 h (*see* **Note 14**).
8. Wash slides in Coplin jars containing PBST for 5 min. Repeat four times.
9. Stain nuclei with 40 mL Hoechst (1:3000 of 10 mg/mL stock in PBST) for 5 min in a Coplin jar. Alternatively, place slides horizontally in a humidified slide box, add 0.25 mL of the same Hoechst solution, overlay with a cover slip, and incubate 5 min.
10. Transfer slides from Hoechst stain (remove cover slips if applicable) to a Coplin jar containing PBS and incubate 5 min. Replace with fresh PBS and wash for an additional 5 min. Slides are ready to mount.

11. To mount, drain off excess PBS, wipe dry slide bottom and edges, and lay the slide horizontally. Apply 50 μL mounting media (*see* **Note 15**) and carefully overlay with a cover slip while avoiding all air bubbles. Seal edges of cover slip with clear nail polish. Protect from light and air-dry. Store slides at −20 °C.

3.11 Imaging and Data Interpretation

Image slides within 24 h of mounting for best results. Store at −20 °C. Longer-term storage (>7 days) leads to signal loss. When applicable, control and experimental samples should be imaged in the same session with matching microscope and image acquisition settings. Interpreting immunostaining data requires flat, fully intact sections and an appreciation of the three-dimensional architecture of uninjured and regenerating teleost fins. Longitudinally embedded samples yield intra-ray (between hemi-rays) or inter-ray (between adjacent rays) sections. Landmarks include the apparent amputation site and the nascent regenerating bony ray tissue. Transverse sections through fin rays are characterized by repeating apposing semicircular bony hemi-rays separated by thin inter-ray tissue. Distal regenerating rays on transverse sections lack ossified bone but are still readily distinguished from inter-ray tissue by their thicker, oval appearance.

Judicious use of control antibodies is essential to conclusively classify regenerating fin sections, especially when comparing samples. The ideal intra-ray longitudinal section from a 2 day or later regenerate splits hemi-rays, yielding two discrete columns of Runx2- and/or sp7-expressing osteoblasts extending from the amputation site almost to the distal end of the blastema [11–13]. Fibroblast-lineage, blastema cells are medially adjacent, separating the two columns, and extend even further distally. Dachshund (Dach) transcription factors are convenient nuclear markers of the distal, growth factor-producing fibroblast-derived cells [31]. Figure 1 shows an example of anti-Dach and EdU double stained 4-day postamputation longitudinal fin section. Sections even slightly off the mid-line of each ray/blastema will not include the most distal blastema cells; frequently, consecutive sections must be stained to identify the optimal and most comparable sections. Thinner sections lacking osteoblasts are from inter-ray regions. Finally, rays themselves are not equivalent. For example, wider and longer regenerating fin rays possess correspondingly larger blastemas [31]. Therefore, sections analyzed from different samples (i.e., control vs. experimental) must be from comparable rays. Such rigor is essential to reduce variability for sensitive quantitative analyses.

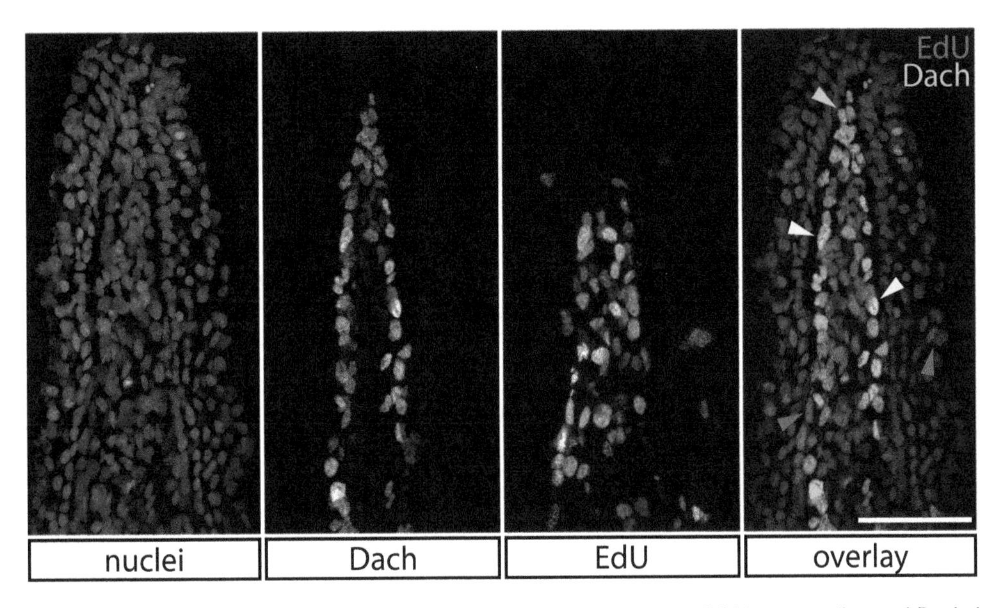

Fig. 1 Images of a distal regenerating zebrafish caudal fin section showing EdU incorporation and Dachshund (Dach) immunostaining. Methods: Fish at 4 days postamputation were injected with EdU and harvested 4 h later. Fins were collected, fixed in PFA, embedded in paraffin, and sectioned longitudinally. Slides were heated in 1 mM EDTA, pH 8.0, 0.1% Tween-20 for 5 min using a pressure cooker for antigen retrieval. After cooling, slides were blocked in 10% nonfat milk in PBST. EdU subsequently was detected following the manufacturer's recommendations (ThermoFisher Inc.). Slides were washed in PBS and incubated overnight with rabbit anti-Dach antibodies (Proteintech Inc., catalog 10914-1-AP). The next day, Alexa-conjugated goat anti-rabbit secondary antibodies were applied (ThermoFisher Inc.). After washing, nuclei were stained with Hoechst. Finally, slides were mounted and imaged with a Zeiss LSM 880 laser-scanning confocal microscope. Shown are maximum-intensity projections of each channel and a composite image. The overlay image shows EdU^+ nuclei in magenta, $Dach^+$ nuclei in green, and nuclei in gray. Magenta arrowheads point to EdU^+ nuclei lacking Dach expression. Yellow arrowheads highlight proliferating $EdU^+/Dach^+$ cells. The green arrowhead marks a representative $Dach^+$ cell lacking EdU. The scale bar indicates 50 μm

4 Notes

1. EDTA will not go into solution until the pH is adjusted to ~8.0.
2. We have successfully reused the Agilent-DAKO Target Retrieval Solution up to six times.
3. The Click-iT EdU kit (ThermoFisher Inc., catalog C10337) yields more robust EdU detection than the Click-iT Plus kit (ThermoFisher Inc., catalog C10637). However, the Click-iT Plus kit is compatible with fluorescent proteins, including EGFP and mCherry. A variety of Alexa conjugates are available in both Click-iT and Click-iT Plus formats.
4. Fluorescently conjugated secondary antibodies are available from a variety of vendors. Choosing appropriate fluorophores depends upon the number of antigens being distinguished,

matching excitation/emission spectra with available imaging capabilities, and whether fluorescent marker proteins are present in the tissue. Alexa-conjugated secondary antibodies work well, e.g., ThermoFisher Inc. Rabbit anti-Mouse IgG (H + L) Secondary Antibody, Alexa Fluor 546 conjugate, catalog A11060. Fin autofluorescence can be problematic when using conjugates emitting near 488 nm. As such, avoid the "green" channel or reserve it for the most robust antibodies. In contrast, fin tissue does not autofluoresce in "far red" channels (~647 nm emission). Therefore, use this channel for relatively weak immunostaining signals.

5. For frozen sections, a 2-h fixation at room temperature is usually sufficient. However, when optimizing immunostaining on frozen sections, consider shorter times (15 min at room temperature) or omitting fixation altogether. While more subject to tissue damage, unfixed tissue sections also retain protein fluorescence that is normally lost during fixation (e.g., mCherry and related red fluorescent proteins). This can be advantageous, but it also limits the fluorophore selection for secondary antibodies.

6. We suggest paraffin embedded tissues if attempting only one approach, since they are relatively easy to prepare and section. An experienced user can section a whole paraffin embedded fin and generate many usable slides in less than 15 min. Further, paraffin blocks and sections are very stable and can be stored indefinitely without negative effects on immunostaining. Finally, selecting matched sections across samples pre- and/or post-staining is relatively easy. A major drawback to paraffin sections is that some antigens are irreversibly damaged by the relatively harsh tissue preparation. In contrast, frozen tissue is processed without dehydration and organic solvents and therefore better preserves antigenicity. As another potential advantage, fluorescent proteins, including GFP variants, typically retain their fluorescence in frozen sections. However, collecting frozen sections takes considerably longer, and the sections have a finite storage lifespan (~1 year); freshly cut sections generally yield the best results.

7. Many antigen retrieval solutions, including proprietary commercial versions, are available. We generally start with 1 mM EDTA, pH 8.0, 0.1% Tween-20. If unsatisfying, we proceed with the commercial options: ImmunoSaver (citraconic acid-based solution) and/or Agilent-DAKO Target Retrieval Solution. The major drawbacks of these are toxicity and cost, respectively. Other commonly used solutions, including citrate, pH 6.0, should be explored [6, 7]. We typically perform antigen retrieval for 5 min in a pressure cooker; however, times up to 10 min can improve results. Longer incubations (>10 min)

usually strip sections off slides. Finally, protease-based antigen retrieval can determine success. We generally use the widely available trypsin/EDTA mix used for cell culture. However, take care to avoid overdigesting and detaching sections (typically, incubate <5 min at room temperature). Downstream immunostaining steps are identical.

8. We have encountered several temperamental antibodies that require a specific blocking condition; therefore, empirically determine the best blocking reagent for each antibody. For example, substitute 2% BSA or 5% nonimmune serum (matching the species source for secondary antibodies) in place of nonfat milk. Immunostaining using frozen fin sections is best if the tissue is first permeabilized to improve the penetration of relatively large antibody molecules. To do so, simply substitute Triton X-100 (final concentration of 0.1%) for Tween-20 in the blocking buffer used in Subheading 3.8.
9. We generally find that 50–60% of the recommended Alexa Fluor picolyl azide is sufficient.
10. Multiple criteria can help assess if a given antibody produces specific signal. The vast majority of commercially sourced antibodies are raised against mammalian proteins. Therefore, first, the investigator should ascertain the conservation of the immunogen in the zebrafish paralog(s). As a rule of thumb, 80% or greater amino acid identity conservation indicates the antibody likely will recognize the zebrafish form. Second, commercial validation and/or publication of an antibody for use on paraffin and/or frozen sections is encouraging. However, such supporting data should be scrutinized for rigor (e.g., negative controls) and reproducibility. Third, it is helpful to know a target's mRNA expression pattern, e.g., by in situ hybridization, to compare with the immunostaining results. Annotated single cell transcriptomics resources also can be helpful in this regard. Fourth, predicted subcellular protein localization (e.g., nuclear or plasma membrane) can also indicate that an observed pattern is likely specific. Finally, the gold standard for demonstrating antibody specificity is signal loss on tissue from a corresponding null mutant not expected to produce antigen-containing polypeptide. In practice, this is challenging since such null alleles are not always available and those for key genes acting during regeneration frequently are developmental lethal. In the latter case, antibodies can be validated using sectioned homozygous mutant embryos or larvae.
11. Optimal antibody concentrations must be determined empirically and should be tested over a wide range of concentrations when feasible. We follow the immunostainer's adage "dilute, dilute, dilute" to increase signal:noise ratio and reduce antibody expenses.

12. We usually perform the primary antibody incubation step overnight at 4 °C for convenience and consistency. However, 1–2 h incubations at room temperature are commonly used in other protocols and may work as well or better.
13. Consider performing a high salt wash to reduce background. Prepare high salt wash by diluting 5 M NaCl 1:10 in PBST and submerge slides in this buffer for 30 min. Rinse in PBST before continuing with staining.
14. In our experience, secondary antibodies are very robust and reliable, requiring little, if any, optimization. We routinely use 1:1000 of Alexa-conjugated antibodies, e.g., ThermoFisher Inc. Rabbit anti-Mouse IgG (H + L) Secondary Antibody, Alexa Fluor 546 conjugate, catalog A11060.
15. A variety of mounting media are commercially available and widely used for fluorescent microscopy. Invitrogen SlowFade Diamond Antifade Mountant (ThermoFisher Inc., catalog S36963) provides excellent photo-bleaching protection. However, more economical mounting media is sufficient for many applications, including larger scale optimization experiments.

Acknowledgments

We thank Astra Henner for technical expertise and edits. The National Institute of General Medical Sciences provided research funding (1R01GM127761).

References

1. Sehring I, Weidinger G (2021) Zebrafish fin: complex molecular interactions and cellular mechanisms guiding regeneration. Cold Spring Harb Perspect Biol:a040758. https://doi.org/10.1101/cshperspect.a040758
2. Smith A, Zhang J, Guay D, Quint E, Johnson A, Akimenko MA (2008) Gene expression analysis on sections of zebrafish regenerating fins reveals limitations in the whole-mount in situ hybridization method. Dev Dyn 237:417–425. https://doi.org/10.1002/dvdy.21417
3. Thisse C, Thisse B (2008) High-resolution in situ hybridization to whole-mount zebrafish embryos. Nat Protoc 3:59–69. https://doi.org/10.1038/nprot.2007.514
4. Fox CH, Johnson FB, Whiting J, Roller PP (1985) Formaldehyde fixation. J Histochem Cytochem 33:845–853. https://doi.org/10.1177/33.8.3894502
5. Howat WJ, Wilson BA (2014) Tissue fixation and the effect of molecular fixatives on downstream staining procedures. Methods 70:12–19. https://doi.org/10.1016/j.ymeth.2014.01.022
6. Krenacs L, Krenacs T, Raffeld M (1999) Antigen retrieval for immunohistochemical reactions in routinely processed paraffin sections. Methods Mol Biol 115:85–93. https://doi.org/10.1385/1-59259-213-9:85
7. Shi S-R, Shi Y, Taylor CR (2011) Antigen retrieval immunohistochemistry: review and future prospects in research and diagnosis over two decades. J Histochem Cytochem 59:13–32. https://doi.org/10.1369/jhc.2010.957191
8. Namimatsu S, Ghazizadeh M, Sugisaki Y (2005) Reversing the effects of formalin fixation with citraconic anhydride and heat: a universal antigen retrieval method. J Histochem

Cytochem 53:3–11. https://doi.org/10.1177/002215540505300102

9. Stewart S, Stankunas K (2012) Limited dedifferentiation provides replacement tissue during zebrafish fin regeneration. Dev Biol 365:339–349. https://doi.org/10.1016/j.ydbio.2012.02.031
10. Stewart S, Le Bleu HK, Yette GA, Henner AL, Robbins AE, Braunstein JA, Stankunas K (2021) Longfin causes cis-ectopic expression of the kcnh2a ether-a-go-go K+ channel to autonomously prolong fin outgrowth. Development 148. https://doi.org/10.1242/dev.199384
11. Stewart S, Gomez AW, Armstrong BE, Henner A, Stankunas K (2014) Sequential and opposing activities of Wnt and BMP coordinate zebrafish bone regeneration. Cell Rep 6: 482–498. https://doi.org/10.1016/j.celrep.2014.01.010
12. Armstrong BE, Henner A, Stewart S, Stankunas K (2017) Shh promotes direct interactions between epidermal cells and osteoblast progenitors to shape regenerated zebrafish bone. Development 144:1165–1176. https://doi.org/10.1242/dev.143792
13. Braunstein JA, Robbins AE, Stewart S, Stankunas K (2021) Basal epidermis collective migration and local sonic hedgehog signaling promote skeletal branching morphogenesis in zebrafish fins. Dev Biol 477:177–190. https://doi.org/10.1016/j.ydbio.2021.04.010
14. Galand P, Degraef C (1989) Cyclin/PCNA immunostaining as an alternative to tritiated thymidine pulse labelling for marking S phase cells in paraffin sections from animal and human tissues. Cell Tissue Kinet 22:383–392. https://doi.org/10.1111/j.1365-2184.1989.tb00223.x
15. Yu CC, Woods AL, Levison DA (1992) The assessment of cellular proliferation by immunohistochemistry: a review of currently available methods and their applications. Histochem J 24:121–131. https://doi.org/10.1007/BF01047461
16. Hall PA, Levison DA, Woods AL, Yu CC, Kellock DB, Watkins JA, Barnes DM, Gillett CE, Camplejohn R, Dover R (1990) Proliferating cell nuclear antigen (PCNA) immunolocalization in paraffin sections: an index of cell proliferation with evidence of deregulated expression in some neoplasms. J Pathol 162: 285–294. https://doi.org/10.1002/path.1711620403
17. Hall PA, McKee PH, Menage HD, Dover R, Lane DP (1993) High levels of p53 protein in UV-irradiated normal human skin. Oncogene 8:203–207
18. Hendzel MJ, Wei Y, Mancini MA, Van Hooser A, Ranalli T, Brinkley BR, Bazett-Jones DP, Allis CD (1997) Mitosis-specific phosphorylation of histone H3 initiates primarily within pericentromeric heterochromatin during G2 and spreads in an ordered fashion coincident with mitotic chromosome condensation. Chromosoma 106:348–360. https://doi.org/10.1007/s004120050256
19. Hsu JY, Sun ZW, Li X, Reuben M, Tatchell K, Bishop DK, Grushcow JM, Brame CJ, Caldwell JA, Hunt DF, Lin R, Smith MM, Allis CD (2000) Mitotic phosphorylation of histone H3 is governed by Ipl1/aurora kinase and Glc7/PP1 phosphatase in budding yeast and nematodes. Cell 102:279–291. https://doi.org/10.1016/s0092-8674(00)00034-9
20. Wei Y, Mizzen CA, Cook RG, Gorovsky MA, Allis CD (1998) Phosphorylation of histone H3 at serine 10 is correlated with chromosome condensation during mitosis and meiosis in Tetrahymena. Proc Natl Acad Sci U S A 95: 7480–7484. https://doi.org/10.1073/pnas.95.13.7480
21. Gratzner HG, Pollack A, Ingram DJ, Leif RC (1976) Deoxyribonucleic acid replication in single cells and chromosomes by immunologic techniques. J Histochem Cytochem 24:34–39. https://doi.org/10.1177/24.1.815428
22. Gratzner HG (1982) Monoclonal antibody to 5-bromo- and 5-iododeoxyuridine: a new reagent for detection of DNA replication. Science 218:474–475. https://doi.org/10.1126/science.7123245
23. Wynford-Thomas D, Williams ED (1986) Use of bromodeoxyuridine for cell kinetic studies in intact animals. Cell Tissue Kinet 19:179–182. https://doi.org/10.1111/j.1365-2184.1986.tb00728.x
24. Rakic P (2002) Neurogenesis in adult primate neocortex: an evaluation of the evidence. Nat Rev Neurosci 3:65–71. https://doi.org/10.1038/nrn700
25. Tornøe CW, Christensen C, Meldal M (2002) Peptidotriazoles on solid phase: [1,2,3]-triazoles by regiospecific copper(i)-catalyzed 1,3-dipolar cycloadditions of terminal alkynes to azides. J Org Chem 67:3057–3064. https://doi.org/10.1021/jo011148j
26. Rostovtsev VV, Green LG, Fokin VV, Sharpless KB (2002) A stepwise huisgen cycloaddition process: copper(I)-catalyzed regioselective “ligation” of azides and terminal alkynes. Angew Chem Int Ed Engl 41:2596–2599
27. Salic A, Mitchison TJ (2008) A chemical method for fast and sensitive detection of DNA synthesis in vivo. Proc Natl Acad Sci U

S A 105:2415–2420. https://doi.org/10.1073/pnas.0712168105
28. Kinkel MD, Eames SC, Philipson LH, Prince VE (2010) Intraperitoneal injection into adult zebrafish. J Vis Exp 2126. https://doi.org/10.3791/2126
29. Samaee S-M, Seyedin S, Varga ZM (2017) An affordable intraperitoneal injection setup for juvenile and adult zebrafish. Zebrafish 14:77–79. https://doi.org/10.1089/zeb.2016.1322
30. Buchwalow I, Samoilova V, Boecker W, Tiemann M (2011) Non-specific binding of antibodies in immunohistochemistry: fallacies and facts. Sci Rep 1:28. https://doi.org/10.1038/srep00028
31. Stewart S, Yette GA, Le Bleu HK, Henner AL, Braunstein JA, Chehab JW, Harms MJ, Stankunas K (2019) Skeletal geometry and niche transitions restore organ size and shape during zebrafish fin regeneration. bioRxiv https://doi.org/10.1101/606970

Part IV

Genetics and Genomics

Chapter 17

In Vivo Optogenetic Phase Transition of an Intrinsically Disordered Protein

Kazuhide Asakawa, Hiroshi Handa, and Koichi Kawakami

Abstract

Proteins containing intrinsically disordered regions (IDRs) control a wide variety of cellular processes by assembly of membrane-less organelles via IDR-mediated liquid–liquid phase separation. Dysregulated IDR-mediated phase transition has been implicated in the pathogenesis of diseases characterized by deposition of abnormal protein aggregates. Here, we describe a method to enhance interactions between the IDRs of the RNA/DNA-binding protein and TAR DNA-binding protein 43 (TDP-43) by light to drive its phase transition in the motor neurons of zebrafish. The optically controlled TDP-43 phase transition in motor neurons, in vivo, provides a unique opportunity to evaluate the impact of dysregulated TDP-43 phase transition on the physiology of motor neurons. This will help to address the etiology of neurodegenerative diseases associated with abnormal TDP-43 phase transition and aggregation, including amyotrophic lateral sclerosis (ALS).

Key words Membrane-less organelles, IDR, LLPS, Phase transition, TDP-43, Optogenetics, CRY2, Zebrafish, ALS

1 Introduction

Intrinsically disordered protein regions, or IDRs, typically contain amino acids with a low sequence complexity, lack a well-defined three-dimensional structure, and mediate multivalent interactions that facilitate liquid–liquid phase separation (LLPS) [1]. Macromolecular condensates that form via IDR-mediated LLPS, such as ribonucleoprotein (RNP) granules, have been extensively studied in vitro and in cellular models [2, 3], but relatively unexplored in vivo. To investigate LLPS occurring in multicellular systems, it is crucial to express the wild-type or mutant proteins with IDRs at physiologically relevant levels, because the mere exogenous expression of such a protein can alter its phase dynamics through modifications of its multivalent weak interactions.

James F. Amatruda et al. (eds.), *Zebrafish: Methods and Protocols*, Methods in Molecular Biology, vol. 2707,
https://doi.org/10.1007/978-1-0716-3401-1_17,

Trans-activation response element (TAR) DNA-binding protein 43 (TDP-43) is an evolutionarily conserved RNA/DNA-binding protein that bears an IDR at its C-terminus, and its phase transitions are implicated in multiple steps of gene expression ranging from transcription to translation [4, 5]. TDP-43 is also known for being a major constituent of cytoplasmic inclusions deposited in degenerating motor neurons in 97% of amyotrophic lateral sclerosis (ALS) [6, 7], the most common form of motor neuron disease. Furthermore, while the majority (>90%) of ALS cases are not linked to single gene mutation, the rare ALS-linked mutations occurring in the gene encoding TDP-43 (*TARDBP*) are mostly found in the IDR [8], suggesting that aberrant phase transition of TDP-43 underlies ALS pathogenesis. The phase behaviors of macromolecules in in vivo motor neurons, however, have been difficult to access microscopically as the neurons usually locate deep in the central nervous system and innervate the peripheral muscles through long and intricate nerve pathways.

In order to overcome these difficulties, we developed a protocol to control phase transition of TDP-43 through modulation of intermolecular interactions of a TDP-43 variant in the spinal motor neurons of zebrafish by external light illumination [9]. By exploiting the principle of the optoDroplet method [10], we first constructed a TDP-43 variant whose intermolecular interactions, likely including IDR–IDR interactions, can be enhanced by light-mediated clustering of a cryptochrome 2 (CRY2) tag. Second, we expressed the optogenetic TDP-43 variant in the *mnr2b*-positive spinal motor neurons by BAC transgenesis at the level where optogenetic TDP-43 display inherent nuclear-enriched localization under dark conditions. Third, by illuminating a blue light-emitting diode (LED) light onto the spinal motor neurons through a larval translucent body tissue, we induced phase transition of the optogenetic TDP-43, which resulted in deposition of cytoplasmic TDP-43 aggregates with some typical pathological signatures of the motor neurons in ALS patients. The light-mediated remote control of TDP-43 phase transition is applicable in analyzing its underlying mechanisms and cellular consequences in physiologically and pathologically relevant contexts.

2 Materials

2.1 Expression Construct for Optogenetic TDP-43

1. The hsp70lp-opTDP-43 h-polyA-Kan cassette [9] includes an optogenetic TDP-43 variant (opTDP-43 h) created by fusing the human wild-type TDP-43/*TARDBP* to mRFP1 and CRY2olig [11], which is a domain of *Arabidopsis* cryptochrome 2 carrying a mutation-enhancing light-dependent protein clustering at the N- and C-termini, respectively (Fig. 1a, b)

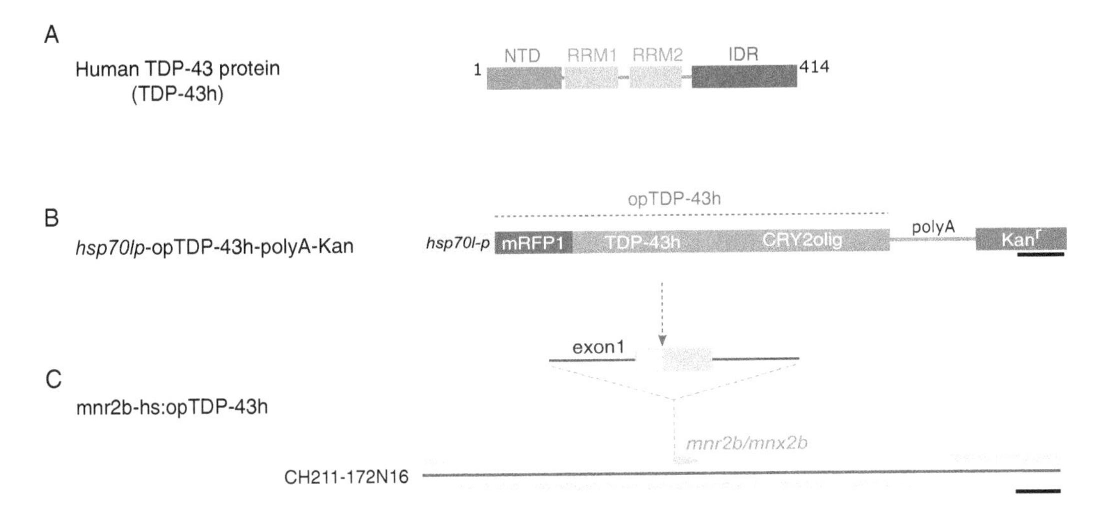

Fig. 1 BAC-mediated opTDP-43 h expression construct. (**a**) Structure of the human TDP-43 protein (414 amino acid residues). NTD *N-terminal domain*, RRM *RNA-recognition motif*, IDR *intrinsically disordered region*. (**b**) Structure of the opTDP-43 h expression cassette. (**c**) Zebrafish BAC containing *mnr2b* locus (CH211-172 N16). The opTDP-43 h expression cassette is inserted downstream of the 5′-untranslated region (UTR) of *mnr2b* (gray dashed arrow) in the first exon of *mnr2b*. The bars indicate 500 bp (**b**) and 10 k bp (**c**)

(*see* **Note 1**). It also includes the *hsp70l* promoter, SV40 polyA signal sequence, and kanamycin resistance gene encoding the aminoglycoside O-phosphotransferase APH(3′)-IIa (Fig. 1b).

2. The hsp70lp-opTDP-43 h-polyA-Kan cassette is amplified by PCR using a DNA polymerase with a low error rate (e.g., PrimeSTAR GXL DNA Polymerase, Takara) with primers that additionally contain 45-bp of the upstream and downstream sequences of the initiator codon of the *mnr2b* gene (mnr2b-hspGFF-Forward: 5′-tat cag cgc aat tac ctg caa ctc taa aca caa tca aag tgt tgc aGA ATT CAC TGG AGG CTT CCA GAA C-3′ and Km-r: 5′-ggt tct tca gct aaa agg gcg tcg atc ctg aag ttc ttt gac ttt tcc atC AAT TCA GAA GAA CTC GTC AAG AA-3′).
3. Obtain the zebrafish BAC clone CH211-172 N16 (BACPAC Genomics), which contains *mnr2b/mnx2b* locus encoding the Mnx-type homeobox protein-regulating motor neuron differentiation (Fig. 1c) [12, 13]. BAC is purified with the BAC Purification Kit (NucleoBond BAC100, MACHEREY-NAGEL).

2.2 Solutions and Reagents

1. E3 buffer: 5 mM NaCl, 0.17 mM KCl, 0.33 mM $CaCl_2$, 0.33 mM $MgSO_4$, 10^{-5}% Methylene Blue.
2. Injection solution: 40 mM KCl, phenol red (10%, v/v), 25 ng/μL of the mnr2b-hs:opTDP-43 h DNA, and 25 ng/μL Tol2 transposase mRNA [14].

3. E3-PTU: E3 buffer containing 0.003% (w/v) *N*-phenylthiourea.
4. E3-Tricane: E3 buffer containing 250 μg/mL of ethyl-3-aminobenzoate methanesulfonate salt (Tricane).
5. E3-LMA-Tricane: E3-Tricane containing 1% low melting temperature agarose (NuSieve GTG Agarose, Lonza). Frozen aliquots can be melted each time.

2.3 Equipment for Optogenetics

Fish are incubated in the 6-well dish (353046, Falcon) and illuminated from below with the LED panel (Nanoleaf AURORA smarter kit, Nanoleaf Limited). The wavelength and intensity of the LED light are calibrated with the spectrometer probe (VIS-50, BLUE-Wave, StellerNet Inc.) and the optical power meter (3664, Hioki) equipped with the optical sensor (9742-10, HIOKI), respectively.

2.4 Microscopy

1. Fish expressing RFP fluorescence are selected by the epifluorescence microscope (Axioimager Z1, Zeiss) equipped with the objective lens Plan-Neofluar 5×/0.15 (Zeiss).
2. Confocal microscope (FV1200, Olympus) equipped with the objective lens XLUMPlanFL N 20×/1.00 (Olympus) is used to scan a fish embedded in E3-LMA-Tricane on a glass base dish.

3 Methods

3.1 BAC-Mediated Expression of opTDP-43 in the Spinal Motor Neurons

1. Purify and introduce the BAC DNA of CH211-172 N16 into SW102 *Escherichia coli* (*E. coli*) cells by electroporation, as described by Warming et al. [15] (*see* **Note 2**).
2. Integrate iTol2-amp cassette [14] into the backbone of CH211-172 N16 by homologous recombination for *Tol2* transposon-mediated BAC transgenesis (CH211-172 N16-iTol2A), as described by Asakawa et al. [16].
3. Amplify the *hsp70l*-opTDP-43 h-polyA-Kan cassette by PCR with primers mnr2b-hspGFF-Forward and Km-r for homologous recombination-mediated integration to obtain a PCR product of ~5.7 bp.
4. Purify the amplified *hsp70l*p-opTDP-43 h-polyA-Kan DNA by agarose gel electrophoresis. Introduce the purified *hsp70l*p-opTDP-43 h-polyA-Kan DNA into CH211-172 N16-iTol2A by electroporation as described by Warming et al. [15]. Select ampicillin- and kanamycin-resistant transformants on Luria-Bertani agar plates.
5. Purify the CH211-172 N16-iTol2A carrying the *hsp70l*p-opTDP-43 h-polyA-Kan cassette (Fig. 1c) (hereafter, mnr2b-hs:opTDP-43 h) and dissolve it at 250 ng/μL in TE buffer after phenol/chloroform extraction.

6. Prepare the injection solution for the mnr2b-hs:opTDP-43 h DNA. Inject 1 nL of the injection solution into the cytosol of wild-type zebrafish embryos at the one-cell stage with a microinjector.
7. Obtain F1 offspring of the injected fish by crossing with the wild-type fish. Examine the F1 fish at 3 dpf for RFP fluorescence in the spinal cord using an epifluorescence microscope. On average, a founder fish is identified from 10 to 20 injected fish by examining ~50 F1 offspring for each injected fish (the germline transmission rate is 5–10%).

3.2 Light-Induced Phase Transition of opTDP-43h

1. Place an empty 6-well dish above the LED panel by keeping the dish and LED panel 5 mm apart with a spacer (e.g., with five slide glasses stacked) (Fig. 2a). Turn on the LED light through the application installed on a tablet/phone and adjust the light to the wavelength peaking at ~456 nm by putting the spectrometer probe (Fig. 2b, c). Then, place the optical sensor in

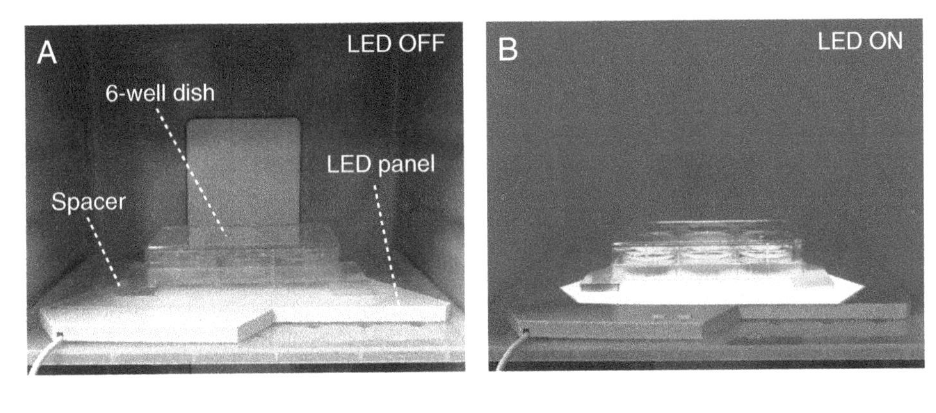

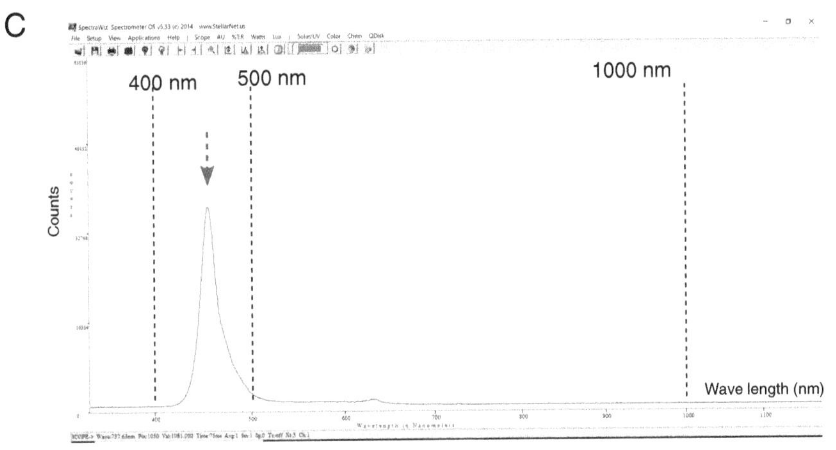

Fig. 2 LED illumination setup. (**a**, **b**) A 6-well dish is placed above the LED panel with the spacers in an incubator. (**c**) Measurement of LED light spectrum with a spectrometer. The dashed blue arrow indicates the peak of the LED light

the empty well and adjust the power of the LED light (~0.61 mW/cm^2). Place the LED panel in an incubator. Complete these steps before the following **step 4**. The LED light setting can be saved and is retrievable in the application.

2. Cross Tg[mnr2b-hs:opTDP-43 h] fish with Tg[mnr2b-hs:EGFP-TDP-43z] fish, which expresses EGFP-tagged zebrafish TDP-43 in the *mnr2b*-positive cells [9] (*see* **Note 3**), and raise the resulting embryos in a plastic dish containing 30 mL of E3 buffer. Replace the E3 buffer with E3-PTU at 8–10 h post-fertilization (hpf) to inhibit melanogenesis. Cover the plastic dish with aluminum foil after 30 hpf.
3. Dechorionate the embryos by 47 hpf and select Tg[mnr2b-hs:opTDP-43 h] Tg[mnr2b-hs:EGFP-TDP-43z] double transgenic fish by examining fish for RFP and GFP signals using the epifluorescence microscope as described above.
4. Briefly anesthetize Tg[mnr2b-hs:opTDP-43 h] Tg[mnr2b-hs:EGFP-TDP-43z] fish at 48 hpf in E3-Tricane.
5. Preheat E3-LMA-Tricane at 42 °C and put a drop of the E3-LMA-Tricane on the glass base dish at the room temperature with the diameter of the dome-shaped agarose drop on the glass dish being 8–10 mm. Add the anesthetized fish to the low melting temperature agarose, and mix by pipetting.
6. During the solidification of E3-LMA-Tricane, maintain the fish on its side by using a syringe needle to keep the spinal cord in an appropriate horizontal position. Put a few drops of E3 buffer onto the dome-shaped agarose-mounted fish after the solidification.
7. Obtain serial confocal z-sections of the spinal cord by scanning with a confocal microscope, using a step size of 1.0 μm per slice for the objective and excitation/emission wavelengths 473/510 nm for EGFP and 559/583 nm for mRFP1 (Fig. 3a). Remove the fish from the agarose by carefully cracking the agarose with a syringe needle as soon as the imaging is complete.
8. Put the imaged fish into the 6-well fish containing 6–7.5 mL of E3 buffer, place the dish on the LED panel in the incubator set at 28 °C and illuminate the dish with the blue light as shown in Fig. 2b.
9. Keep some of the Tg[mnr2b-hs:opTDP-43 h] Tg[mnr2b-hs:EGFP-TDP-43z] fish in a separate 6-well dish in the dark by covering the dish with aluminum foil as unilluminated control fish.
10. After the illumination (e.g., for 72 h at 28 °C), image the spinal cord of the illuminated fish by repeating **steps 4–8** (*see* **Note 4**) (Fig. 3a, b).

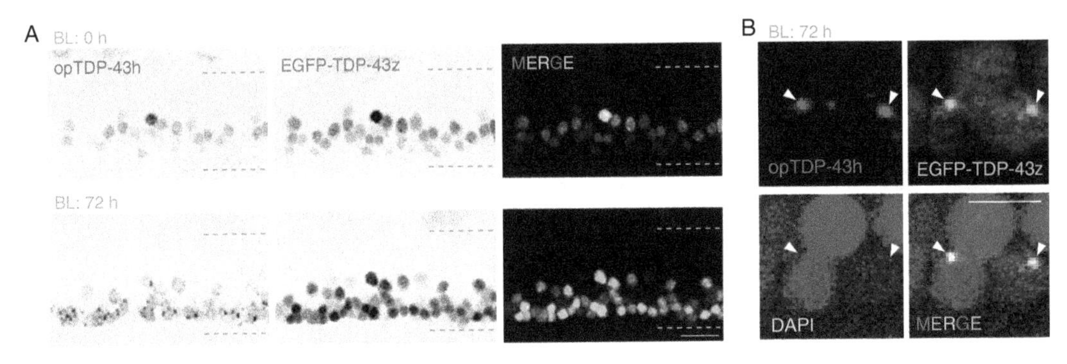

Fig. 3 Live imaging of opTDP-43 h before and after LED light stimulation. (**a**) Max intensity projection images of the z-series confocal images of the ventral spinal cord of Tg[mnr2b-hs:opTDP-43 h] Tg[mnr2b-hs:EGFP-TDP-43z] double transgenic fish before (top, 48 hpf) and after (bottom, 120 hpf) the LED light stimulation. The dashed lines demarcate the dorsal and ventral limits of the spinal cord. (Figures are adapted from Asakawa et al. [9]). (**b**) Cytoplasmic opTDP-43 mislocalization after the 72-h blue LED light illumination. Arrowheads indicate cytoplasmic opTDP-43 aggregates that include non-optogenetic EGFP-TDP-43z. The bars indicate 20 μm (**a**) and 5 μm (**b**)

4 Notes

1. The N-terminus tagging of $CRY2_{PHR}$ [17] and CRY2olig [18] and C-terminal tagging of CRY2olig [9] have been demonstrated to induce light-dependent TDP-43 phase transition. This implies that position of these CRY2 tags may not be critical in promoting TDP-43 phase transition.
2. Freshly prepared BAC DNA from *E. coli* on the same day of electroporation gives a higher success rate of BAC transformation.
3. Use the wild-type fish, instead of Tg[mnr2b-hs:EGFP-TDP-43z] fish, when non-optogenetic TDP-43 marker (EGFP-TDP-43z) is not necessary.
4. For finer spatiotemporal control of opTDP-43 h, blue light illumination can be performed by scanning an agarose-embedded fish with a confocal laser [9].

Acknowledgments

This work was supported by SERIKA FUND (K.A.), KAKENHI Grant numbers JP19K06933 (K.A.) and JP20H05345 (K.A.).

References

1. Tompa P, Fuxreiter M, Oldfield CJ, Simon I, Dunker AK, Uversky VN (2009) Close encounters of the third kind: disordered domains and the interactions of proteins. BioEssays 31(3):328–335. https://doi.org/10.1002/bies.200800151
2. Brangwynne CP (2013) Phase transitions and size scaling of membrane-less organelles. J Cell Biol 203(6):875–881. https://doi.org/10.1083/jcb.201308087
3. Hyman AA, Weber CA, Julicher F (2014) Liquid-liquid phase separation in biology. Annu Rev Cell Dev Biol 30:39–58. https://doi.org/10.1146/annurev-cellbio-100913-013325
4. Prasad A, Bharathi V, Sivalingam V, Girdhar A, Patel BK (2019) Molecular mechanisms of TDP-43 Misfolding and pathology in amyotrophic lateral sclerosis. Front Mol Neurosci 12:25. https://doi.org/10.3389/fnmol.2019.00025
5. Asakawa K, Handa H, Kawakami K (2021) Multi-phaseted problems of TDP-43 in selective neuronal vulnerability in ALS. Cell Mol Life Sci 78(10):4453–4465. https://doi.org/10.1007/s00018-021-03792-z
6. Arai T, Hasegawa M, Akiyama H, Ikeda K, Nonaka T, Mori H et al (2006) TDP-43 is a component of ubiquitin-positive tau-negative inclusions in frontotemporal lobar degeneration and amyotrophic lateral sclerosis. Biochem Biophys Res Commun 351(3):602–611. https://doi.org/10.1016/j.bbrc.2006.10.093
7. Neumann M, Sampathu DM, Kwong LK, Truax AC, Micsenyi MC, Chou TT et al (2006) Ubiquitinated TDP-43 in frontotemporal lobar degeneration and amyotrophic lateral sclerosis. Science 314(5796):130–133. https://doi.org/10.1126/science.1134108
8. Lagier-Tourenne C, Cleveland DW (2009) Rethinking ALS: the FUS about TDP-43. Cell 136(6):1001–1004. https://doi.org/10.1016/j.cell.2009.03.006
9. Asakawa K, Handa H, Kawakami K (2020) Optogenetic modulation of TDP-43 oligomerization accelerates ALS-related pathologies in the spinal motor neurons. Nat Commun 11(1):1004. https://doi.org/10.1038/s41467-020-14815-x
10. Shin Y, Berry J, Pannucci N, Haataja MP, Toettcher JE, Brangwynne CP (2017) Spatiotemporal control of intracellular phase transitions using light-activated optoDroplets. Cell 168(1–2):159–71 e14. https://doi.org/10.1016/j.cell.2016.11.054
11. Taslimi A, Vrana JD, Chen D, Borinskaya S, Mayer BJ, Kennedy MJ et al (2014) An optimized optogenetic clustering tool for probing protein interaction and function. Nat Commun 5:4925. https://doi.org/10.1038/ncomms5925
12. Wendik B, Maier E, Meyer D (2004) Zebrafish mnx genes in endocrine and exocrine pancreas formation. Dev Biol 268(2):372–383. https://doi.org/10.1016/j.ydbio.2003.12.026
13. Seredick SD, Van Ryswyk L, Hutchinson SA, Eisen JS (2012) Zebrafish Mnx proteins specify one motoneuron subtype and suppress acquisition of interneuron characteristics. Neural Dev 7:35. https://doi.org/10.1186/1749-8104-7-35
14. Suster ML, Abe G, Schouw A, Kawakami K (2011) Transposon-mediated BAC transgenesis in zebrafish. Nat Protoc 6(12):1998–2021. https://doi.org/10.1038/nprot.2011.416
15. Warming S, Costantino N, Court DL, Jenkins NA, Copeland NG (2005) Simple and highly efficient BAC recombineering using galK selection. Nucleic Acids Res 33(4):e36. https://doi.org/10.1093/nar/gni035
16. Asakawa K, Abe G, Kawakami K (2013) Cellular dissection of the spinal cord motor column by BAC transgenesis and gene trapping in zebrafish. Front Neural Circ 7:100. https://doi.org/10.3389/fncir.2013.00100
17. Zhang P, Fan B, Yang P, Temirov J, Messing J, Kim HJ et al (2019) Chronic optogenetic induction of stress granules is cytotoxic and reveals the evolution of ALS-FTD pathology. elife 8:8. https://doi.org/10.7554/eLife.39578
18. Mann JR, Gleixner AM, Mauna JC, Gomes E, DeChellis-Marks MR, Needham PG et al (2019) RNA binding antagonizes neurotoxic phase transitions of TDP-43. Neuron 102(2):321–38 e8. https://doi.org/10.1016/j.neuron.2019.01.048

Chapter 18

Colorimetric Barcoding to Track, Isolate, and Analyze Hematopoietic Stem Cell Clones

Dorothee Bornhorst, Brandon Gheller, and Leonard I. Zon

Abstract

In zebrafish, hematopoietic stem cells (HSCs) are born in the developing aorta during embryogenesis. From the definitive wave of hematopoiesis onward, blood homeostasis relies on self-renewal and differentiation of progeny of existing HSCs, or clones, rather than de novo generation. Here, we describe an approach to quantify the number and size of HSC clones at various times throughout the lifespan of the animal using a fluorescent, multicolor labeling strategy. The system is based on combining the multicolor Zebrabow system with an inducible, early lateral plate mesoderm and hematopoietic lineage specific *cre* driver (draculin (*drl*)). The *cre* driver can be temporally controlled and activated in early hematopoiesis to introduce a color barcoding unique to each HSC and subsequently inherited by their daughter cells. Clonal diversity and dominance can be investigated in normal development and blood disease progression, such as blood cancers. This adoptable method allows researchers to obtain quantitative insight into clonality-defining events and their contribution to adult hematopoiesis.

Key words Hematopoietic stem cells (HSCs), Zebrabow system, Colorimetric barcoding, Cre-loxP recombination, Lineage analyses, In vivo multicolor imaging

1 Introduction

The hematopoietic system is populated from a limited number of hematopoietic stem cells (HSCs) that are seeded in the adult marrow early in development. These HSCs replenish multilineage progenitors, which give rise to all differentiated blood cells. Normal function and output of HSCs ensure a balanced production of all peripheral blood cell types throughout the lifespan. Because there is no additional de novo generation of HSCs after early development [1], tools can be developed that label individual HSCs, or clones, early in development and allow for assessment of clones in the adult hematopoietic system [2]. Production in the number and type of

Dorothee Bornhorst and Brandon Gheller contributed equally to this work

James F. Amatruda et al. (eds.), *Zebrafish: Methods and Protocols*, Methods in Molecular Biology, vol. 2707,
https://doi.org/10.1007/978-1-0716-3401-1_18,

daughter cells from individual HSC clones fluctuates throughout the lifespan. Progressive clonal expansion of an HSC that harbors a mutation is the basis for diverse human cancers and has been linked to a more than ten-fold increase in the risk of developing a blood cancer [3, 4]. An overrepresentation of progeny from a limited number of HSCs also arises under a number of adverse conditions such as after irradiation [2] and with advanced aging [3–5]. Hence, understanding the nature of clonal dynamics in normal and malignant cell populations is essential for the rational development of therapeutics to prevent HSC clonal dominance-related diseases.

To quantify clonal dynamics and the number of HSCs present in the zebrafish over time, the Zon laboratory adapted the Zebrabow methodology [6], specifically to the hematopoietic system. In the Zebrabow system, three fluorescent proteins (dTomato, cyan fluorescent protein (CFP), and yellow fluorescent protein (YFP)) are driven by a ubiquitous promoter and flanked by lox sites. The Zebrabow transgene expresses dTomato in its default state and upon stochastic recombination by *cre* or *creERT2*, either CFP or YFP is exclusively expressed. To generate a more diverse repertoire of colors, multiple independent integrations of the Zebrabow construct in the genetic background were established, which leads to different expression levels of each fluorophore and therefore multiple, unique color barcodes. The resulting unique color barcodes are inherited by all descendants of the labeled cell [2]. This system allows for clonal tracking of HSCs without the need for additional perturbation such as transplantation. Hence, the clonal dynamics of each HSC can be analyzed and its trajectory can be reconstructed. The Zon laboratory combines the Zebrabow system [6] with the draculin (*drl*) promoter [7] to drive Cre recombinase expression specifically for tracking HSCs and their progeny.

Using this robust method has led to several discoveries about HSC biology. For example, Henninger et al. revealed that the entire adult hematopoietic system in zebrafish is populated from approximately 21 HSCs [2]. Additionally, this study demonstrated that stressors, including sublethal irradiation and transplantation, reduce clonal diversity and lead to clonal selection [2]. More recently, the Zon laboratory has paired this system with mosaic mutagenesis using the CRISPR/Cas9 system to examine the effect of genetic perturbations on clonal hematopoiesis in adult zebrafish [8]. Furthermore, this system can be combined with fluorescence-activated cell sorting to prospectively isolate different clonal populations for downstream analysis such as single-cell RNA sequencing and DNA sequencing [8].

The strength and versatility of multicolor genetic labeling in zebrafish provides a well-established tool that can be readily adapted. The system is a powerful method to evaluate clonal diversity in both native and malignant settings and to better understand the genetics underlying these conditions. Using different Cre combinations and mosaic gene knockouts, experiments can be designed covering a wide array of approaches to suit the biological question. Thus, the Zebrabow system provides a resource for systematic anatomical and lineage studies during zebrafish blood development.

2 Materials

2.1 *Genotyping of* Tg (ubi:Zebrabow-M)

1. gDNA from fin clip (as previously described https://zfin.atlassian.net/wiki/spaces/prot/pages/356155929/Fin+Amputations).
2. Qiagen blood and tissue kit for gDNA extraction (Qiagen, Cat. #69504).
3. QX200 AutoDG Droplet Digital PCR System (automated droplet generation) or QX200 Droplet Generator (manual droplet generation).
4. Manual droplet generation: DG8 Cartridges for QX100/QX200 Droplet Generator (Biorad, Cat. # 186-4008).
5. Manual droplet generation: Droplet Generator DG8 Gaskets (Biorad, Cat. # 186-3009).
6. Manual droplet generation: Droplet Generation Oil for Probes (Biorad, Cat. # 186-3005).
7. Automated droplet generation: Pipet Tips for Automated Droplet Generator (Biorad, Cat. # 186-4120).
8. Automated droplet generation: DG32 Cartridge for Automated Droplet Generator (Biorad, Cat. # 186-4108).
9. Automated droplet generation: Automated Droplet Generation Oil for Probes (Biorad, Cat. # 186-4110).
10. twin.tec PCR Plate 96-well LoBind semi-skirted (Eppendorf, Cat. # 30129504).
11. ddPCR Supermix for Probes (no dUTP) (Biorad, Cat # 186-3024).
12. Pierceable Foil Heat Seal (Biorad, Cat. # 181-4040).
13. ddPCR Reader Oil (Biorad, Cat # 186-3004).
14. 96-well generator tip boxes.
15. Green droplet cassettes.
16. Chilled 96-well adaptor for destination plate.
17. Ice bucket + chilled 96-well heat block.
18. FAM Probe standard diluted to 20× using nuclease free water (*gapdh*).
19. HEX Probe standard diluted to 20× (*tdTomato* for ubi:zebrabow) (*see* **Note 1**).

2.2 Zebrafish

1. *Tg(ubi:Zebrabow-M)* [6] and *Tg(drl:creERT2)* [7] zebrafish lines are crossed to produce the experimental zebrafish.
2. E3 embryo medium (5 mM NaCl, 0.17 mM KCl, 0.25 mM $CaCl_2$, and 0.15 mM $MgSO_4$).
3. 28.5 °C incubator.

2.3 creERT2-Mediated Recombination

1. 4-Hydroxytamoxifen (4-OHT) solution (3.86 mg/mL in 100% ethanol, protect from light and vortex at RT for 20 min to produce a 10 mM stock. Store aliquots for single use at −20 °C) (Sigma, H7904) (*see* **Note 2**).
2. Pronase (Roche, Cat. # 10165921001; 50 mg/mL).
3. 6-well plastic cell culture dishes.
4. 15 or 50 mL conical tubes.
5. Aluminum foil.

2.4 Cardiac Bleed and Kidney Marrow Collection

1. Iced water.
2. Fish net.
3. Heparin (heparin sodium salt from porcine intestinal mucosa, Sigma, Cat. # 2106).
4. 1.5 mL microtubes.
5. Blood buffer (for 250 mL blood buffer: 225 mL 1% PBS, 5 mL FBS (fetal bovine serum), sterile filter, and store at 4 °C. Immediately add heparin prior to use to 1 USP units/mL).
6. Surgical tools (forceps, needles, and scissors).
7. P10 pipette and tips.
8. Dissecting microscope.
9. Styrofoam board with 30g needles.

2.5 Analysis of the Recombined Kidney Marrow

1. Blood buffer (0.9% PBS, 5% FBS, 1000× heparin, as described in Subheading 2.4).
2. Collected kidney marrow samples suspended in blood buffer.
3. FACS tubes (Falcon 5 mL Round Bottom Polystyrene Test Tube, with Cell Strainer Snap Cap; Falcon, Cat. # 352235).
4. DRAQ7 live cell stain (a far-red fluorescent dye that only stains the nuclei in dead and permeabilized cells; Abcam, Cat. # ab109202).
5. P1000 pipette and P10 pipette.
6. Flow cytometer with 440, 488, 561, and 640 laser lines (e.g., BD LSR *Fortessa*).

7. FlowJo software (Becton, Dickinson & Company).
8. Zebrabow Color Analysis via Zebrabow App or MATLAB code (see Github link: https://github.com/jehenninger/2020_Avagyan_Henninger_et_al).

3 Methods

3.1 Genotyping of Adult Tg(ubi:Zebrabow-M)

1. Fin clip the adult *Tg(ubi:Zebrabow-M)* zebrafish [6] (Fig. 1a; as previously described). As a control, include a zebrafish with a single insertion and one with no insert (wild type).
2. Perform the gDNA extraction using the Qiagen blood and tissue kit (Qiagen, Cat. #69504). Follow the manual instructions using 180 μL ALT buffer plus 20 μL proteinase K solution per fin clip and elute in 50 μL nuclease-free water.
3. For optimal ddPCR results, dilute the gDNA solution to 10–15 ng/μL with nuclease-free water.
4. Thaw 2× Supermix on ice (to avoid freezing and thawing cycles aliquot to 1 mL and freeze).

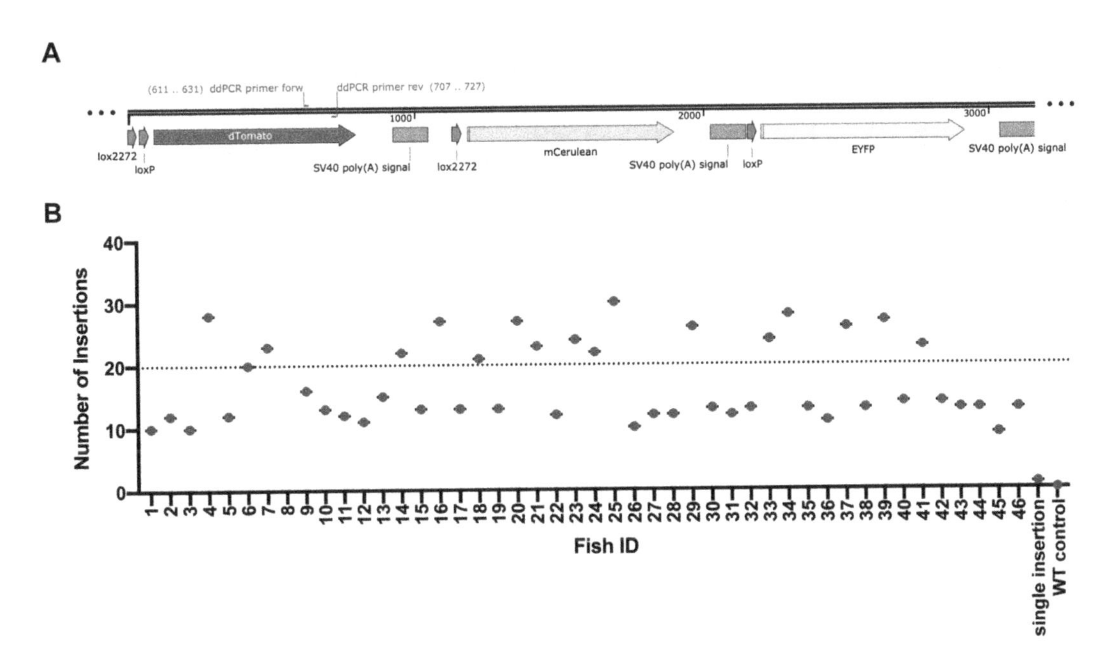

Fig. 1 Genotyping strategy for *Tg(ubi:Zebrabow-M)* zebrafish. (**a**) Schematic of the Zebrabow color cassette depicting the location of the ddPCR primer location within the *tdTomato* sequence. (**b**) Number of insertions are calculated from the event reads for each sample. Insertions vary and zebrafish with more than 20 insertions should be kept. The single insertion has one insertion detected and the wild-type control has no insertions, which indicated a successful genotyping procedure

Table 1
ddPCR reaction mix for *Tg(ubi:Zebrabow-M)* genotyping

	Volume per well/reaction (μL)	Mix for a 96-well plate (+10%) (μL)
Super mix	12.5	1320
Probe 1 (20×; gapdh)	1.25	132
Probe 2 (20×; tdTomato)	1.25	132
DNA sample (15–25 ng/μL)	2.5	–
Water	2.5	792
Total	25	2376

5. Pipette the PCR reaction with the gDNA samples into one 96-well Eppendorf plate placed in a 96-well heat block on ice using the volumes described in Table 1.
6. Pipette 22.5 μL master mix into each well and add 2.5 μL gDNA into the respective well. Mix via pipetting up and down five times. **Steps 7–12** describe the automated droplet generation and **steps 13–19** the manual droplet generation.
7. Automated droplet generation:

   Place the sample plate into the correct position (see manual) and make sure the plate is trapped within the clamps.
8. Place a clean destination plate into a chilled 96-well adaptor and place it into the correct position (green light under the block appears).
9. Place up to two generator tip boxes into the correct positions.
10. Place up to three sets of droplet generator cassettes into the correct positions. The green side of the cassette faces the left side (green light appears).
11. Close the automated droplet generator. On the front panel, select the correct number of samples and wait for the initialization to determine the time to finish the droplet generation.
12. Heat seal the destination plate using 180 °C for 5 s and place it into a thermocycler. Make sure that the 96-well plate will fit into the thermocycler.
13. Manual droplet generation:

    Put cartridge into the holder with the flat end to the right and close it properly.
14. Transfer 20 μL reaction into the middle wells of the cartridge (touch the tip to the bottom and pull up as dispensing the solution to avoid bubbles). Be careful that no bubbles are in the center bottom of the well over the port since droplets will not be generated. Samples in all eight wells are needed to induce the generation.

Table 2
PCR program for *Tg(ubi:Zebrabow-M)* genotyping

95 °C	10 min	
94 °C 60 °C	30 s 60 s	40 cycles
98 °C	10 min	
4 °C	Forever	

15. Transfer 70 μL oil into the oil wells.
16. Put gasket over four posts.
17. Open the QX200 door, put cartridge in (unidirectional), and close the door (droplet generation begins automatically and takes 2 min per 8 samples).
18. Remove cartridge, discard gasket, and gently transfer the 40 μL of emulsion into the 96-well twin-tec plate. Pipette slowly and at an angle so as not to shear droplets (air bubbles are fine at this point). Make sure that the 96-well plate will fit into the thermocycler.
19. Heat seal the destination plate using 180 °C for 5 s and place it into the thermocycler.
20. Run the following PCR program in a thermocycler (Table 2). After the PCR, the samples are stable at 4 °C for up to 24 h.
21. Read the plate on QX200 plate reader and in the QuantaSoft software setup Ch1 as *gapdh* and Ch2 as *dTomato* with dye set FAM/HEX.
22. Choose CNV2 (indicates wild-type copy) and setup a template labeling all parameters.
23. Export a CNV file for a later analysis using Excel or analyze within QuantaSoft (*see* **Note 3**).
24. Analyze copy number: Divide the concentration of *tdTomato* with the *gapdh* concentration times 2 (tdTomato conc./gapdh conc. × 2 = copy number).
25. Keep zebrafish with more than 20 insertions/copies (Fig. 1b).

3.2 Genotyping of Tg (drl:creERT2) [7]

1. Cross a potential *Tg(−6.3drl:creERT2,alpha-crystallin:YFP)* adult zebrafish with a wild-type zebrafish.
2. Let the embryos develop until *alpha-crystallin:YFP* is strongly expressed in the eye from approximately 24 h postfertilization (hpf) (green eyes indicate a positive embryo).
3. Select all embryos with green eyes and grow them to adulthood. Adult *Tg(−6.3drl:creERT2,alpha-crystallin:YFP)* zebrafish can be identified using a fluorescent binocular with a GFP filter.

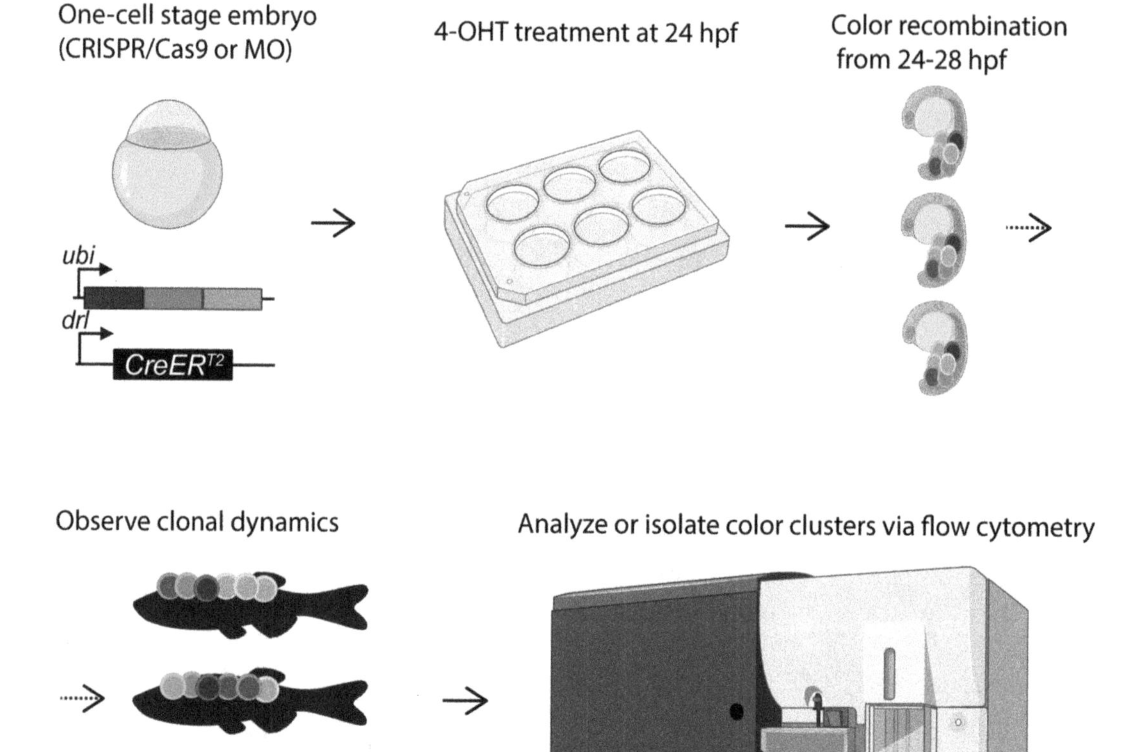

Fig. 2 Experimental workflow for a Zebrabow analysis. Double transgenic *Tg(ubi:Zebrabow-M);Tg(drl:creERT2)* single-cell embryos are collected and a genetic manipulation could be performed. At 24 hpf, 4-OHT treatment is performed for 4 h to induce the color recombination in *drl-CreERT2* positive cells. After growing the zebrafish to adulthood, the color barcodes can be analyzed via flow cytometry and clones of interest could be sorted for further analysis using RNA- or DNAseq techniques

3.3 Zebrabow Fish Breeding and Color Induction

1. To generate the experimental zebrafish, cross *Tg(ubi:Zebrabow-M)* [6] zebrafish with *Tg(drl:creERT2)* [7] zebrafish (Fig. 2).
2. Collect the fertilized embryos in a 10 cm dish containing E3 medium (if desired, a genetic intervention may be introduced at the one-cell stage following standard protocols such as CRISPR-induced mutations [9] or morpholino knockdown [10]).
3. At 24 hpf, embryos can be either dechorionated manually using forceps or using a Pronase treatment (Use 1 mL of the Pronase stock solution for a 10 mL working solution of 1 mg/mL. Pour the working solution into a small petri dish and place the embryos still in their chorion into the solution. Swirl the embryos until the chorion becomes soft and embryos start to dechorionate, typically after 3–5 min at 24 hpf. Immediately transfer the embryos into a petri dish with fresh E3 medium. Wash three times in fresh E3 medium.).

4. The dechorionated embryos should be placed at a density of 25–35 embryos per well into a 6-well dish.
5. Based on the number of embryos to be treated, create a 15 μM mixture of 4-OHT (rehydrated in ethanol) in E3 medium in a tin foil-covered conical tube (protected from light).
6. Using a thin-tipped pipette, remove all E3 media from the 6-well dish without disturbing embryos and immediately replace with 6 mL per well of the 15 μM 4-OHT working solution.
7. Wrap the 6-well plate in aluminum foil and incubate at 28.5 °C for 4 h.
8. After 4 h, remove the 6-well plate from the incubator and transfer the 4-OHT treated embryos to a 10 cm dish in fresh E3 media.
9. Sort the embryos using a fluorescent binocular for green expression of *cryaa:GFP* in the eyes (*drl:creERT2* genotyping) and a red ubiquitous expression of tdTomato (*ubi:Zebrabow-M* genotyping).
10. Keep the embryos at 28.5 °C in an incubator until analysis or transfer the zebrafish to your facility's nursery.

3.4 Determine Hematopoietic Clone Number and Clone Size

1. Isolate the zebrafish kidney marrow by first, euthanizing the zebrafish via ice-cold water submersion; second, making a lateral cut along the length of the underside of the zebrafish using a micro-scissor. Third, make two parallel incisions to the initial cut so that the skin of the zebrafish can be fixed to a Styrofoam board with 30g needles exposing the animal's innards.
2. Remove internal organs leaving only the kidney marrow exposed using fine tip forceps. Collect the kidney marrow using fine tip forceps and place into 300 μL blood buffer on ice until all samples are ready for the analysis via flow cytometry. A total of 300 μL volume is recommended for a 3-month postfertilization (mpf) zebrafish, and it should be modified depending on the size of the kidney marrow.
3. Immediately prior to analysis, pipette the isolated marrow in blood buffer with a 1000 μL pipette 30 s up and down to mechanically dissociate the kidney marrow into a single cell suspension and to release the hematopoietic cells. Slowly pipette the suspended kidney marrow through the cell strainer attached to a FACS tube.
4. Add 1 μL DRAQ7 into the mixture prior to analysis via flow cytometry. DRAQ7 is used as a viability stain and detected in far-red channel, thereby not interfering with the Zebrabow channels/colors.

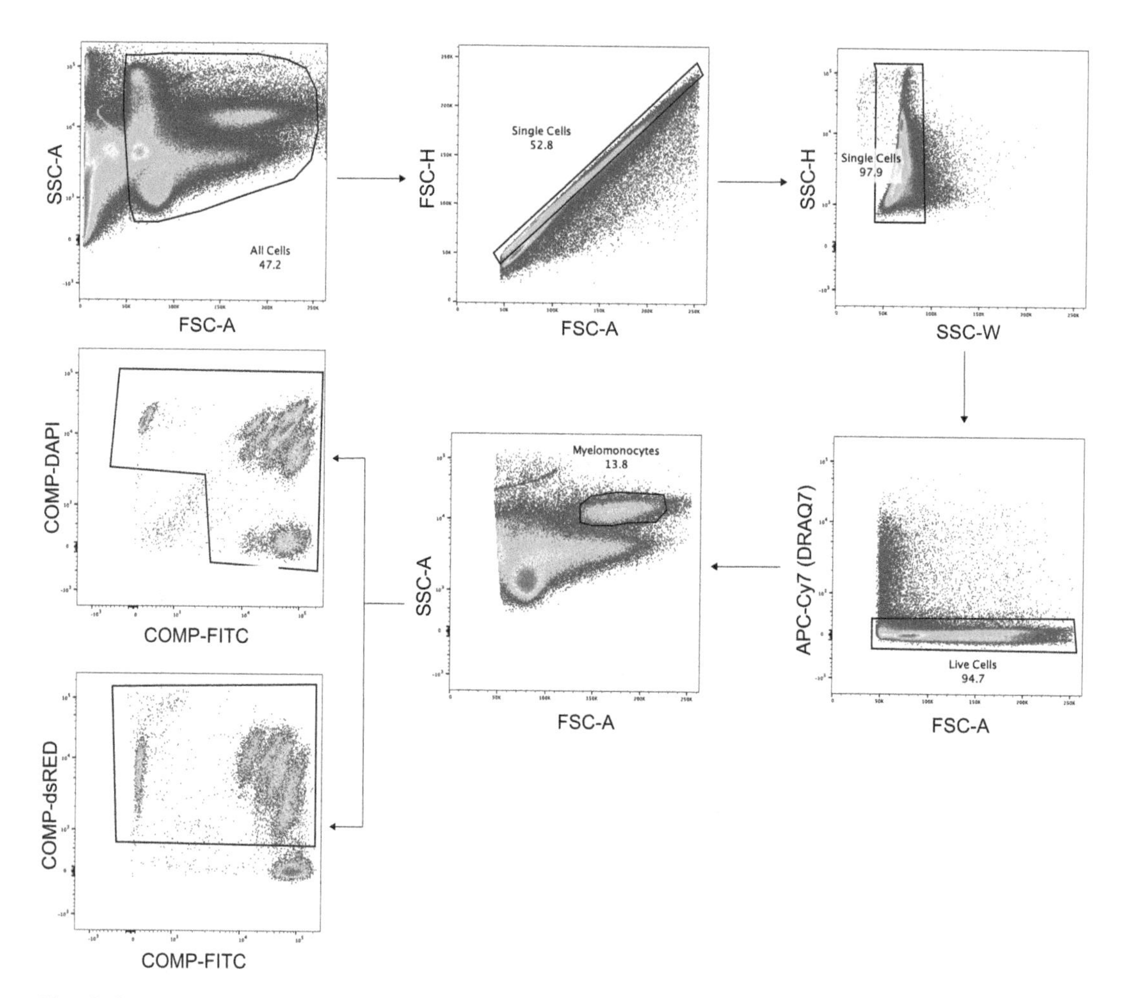

Fig. 3 Representative flow cytometry gating strategy to prepare data for the Zebrabow Color Analysis Application. An example of flow cytometry data obtained from a single, dissociated kidney marrow of a *Tg*(*ubi:Zebrabow-M*);*Tg*(*drl:creERT2*) zebrafish treated with 4-OHT at 24 hpf. The gating strategy is depicted from left to right as indicated by the arrows. Single cells are identified via FSC-A and FSC-H and further by SSC-W and SSC-H, live cells are identified by the exclusion of DRAQ7 staining in the APC-Cy7 channel, and myelomonocytes are identified by their characteristic FSC-A and SSC-A properties. From the myelomonocyte gate, tdTomato, CFP, and YFP are used to identify all cells for analysis in the Zebrabow Color Analysis Application. Two gates are separated, either cells expressing CFP detected in the DAPI channel and/or YFP detected in the FITC and cells expressing tdTomato detected in the dsRED channel. From these gates a Boolean gate is generated, indicating all cells that will be analyzed by the Zebrabow Color Analysis Application. FSC-A *Forward scatter area*, FSC-H *forward scatter height*, SSC-A *side scatter area*, SSC-H *side scatter height*, SSC-W *side scatter width*, Comp *compensation*

5. Samples should be run through a flow cytometer equipped with an 85 μm filter and lasers detecting dTomato, YFP, and CFP protein. An example of the gating strategies is presented in Fig. 3 (*see* **Note 4**).
6. Record a minimum of 20,000 myelomonocyte events for a robust analysis of the color barcoding. Import the .fcs files from the flow cytometer into the FlowJo software.

7. Add a compensation matrix containing each color to be analyzed to the sample.
8. Process the data using FlowJo as depicted in Fig. 3. Using a "OR GATE" or Boolean gate function, create a population that includes cells expressing "blue and green" or "red" and export the compensated colors exclusively (*see* **Note 5**).
9. Download the Zebrabow Color Analysis Application here: https://github.com/jehenninger/2020_Avagyan_Henninger_et_al. Blind the files for an unbiased analysis of color clusters.
10. Load the exported .fcs file into the Zebrabow Color Analysis Application and observe the clustering solutions provided. A full description of the software can be found here: https://github.com/jehenninger/2020_Avagyan_Henninger_et_al. Representative plots obtained by the program can be observed in Fig. 4.

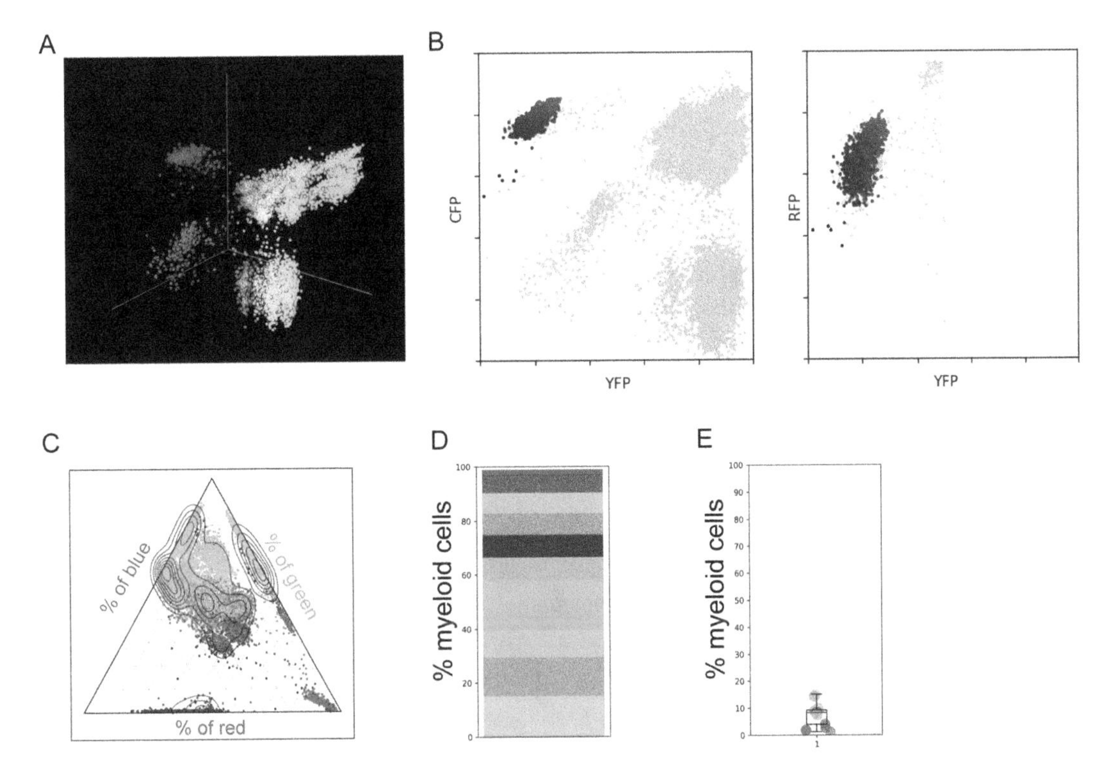

Fig. 4 Example graphs obtained using the Zebrabow Color Analysis Application. (**a**) A three-dimensional rendering of myelomonocytes from the kidney marrow of a 3 mpf *Tg(ubi:Zebrabow-M);Tg(drl:creERT2)* zebrafish, 4-OHT treated at 24 hpf. (**b**) An individual cluster is highlighted in two-dimensional CFP/YFP and RFP/YFP plots. These plots can be used to inform gating strategies to sort specific clones using a flow cytometer. (**c**) The expression level of tdTomato, CFP, and YFP of each myelomonocyte is plotted in the ternary plot, respectively. (**d/e**) Two different methods of graphical representation are depicted, representing the percentage of analyzed cells belonging to each annotated color cluster

11. Evaluate clustering solutions produced by the Zebrabow Color Analysis Application and fuse or split clusters that were falsely labeled using the embedded functions, respectively.
12. Create a destination folder and save the files. Saving the files will generate folders with all analyzed data including images (Fig. 4) and clone size information.
13. Using the backgating images provided by the Zebrabow Color Analysis Application, identified clusters can be isolated for downstream applications such as transplantation, DNA sequencing, RNA sequencing, or metabolic assays.

4 Notes

1. Probe information sheet, see Table 3.
2. 4-OHT is generally unstable and light sensitive (work quickly and keep the solution protected from light).
3. Data will only be collected for wells marked as containing samples. It is normal to see "no call" initially. Positive signals for FAM/HEX ≈12,000 and negative ≈2000.
   1. If the droplet reader shows so-called "rain" between the positive and negative droplets on the *Y* axis, the annealing temperature should be increased.
4. In some instances, a 405 or 488 nm laser may be used for CFP detection. However, for optimal detection of CFP signal and subsequent cluster analysis, optimal detection using a 440 or 458 nm laser is recommended.
5. For the analysis, the compensation step is mandatory for Zebrabow Color Analysis Application to accept the data. Exclude samples that have ≥25% of cells only expressing *tdTomato* as this is an indication that the *cre* recombination was inefficient.

Table 3
Probe information for *Tg(ubi:Zebrabow-M)* genotyping

Gene	Probe seq	5′ Reporter	Int	3′ Quencher	Primer 1	Primer 2
dTomato	CTACGTGGA [ZEN]CACC AAGCTGG ACAT	HEX	ZEN	Iowa Black FQ	GATGGTGTAG TCCTCGTTG TG	CACTACCTGG TGGAG TTCAAG
zf-gapdh	AAGGGTGAG [ZEN]GTTAA GGCAGAAGGC	6-FAM	ZEN	Iowa Black FQ	GCATGACCA TCAA TGACCAG TTTG	AGTGCTTG TTTC TTCACAGG TTTAC

Acknowledgments

We would like to thank Serine Avagyan and Jonathan Henninger for technical support and fruitful discussions. Figures were created with BioRender.com under an academic license. D.B. was supported by the German National Academy of Sciences Leopoldina project LPDS 2021-01. B.G. was supported by a Canadian Institutes for Health Research Postdoctoral Fellowship. Research reported in this publication was supported by the National Institute of Diabetes and Digestive and Kidney Diseases of the National Institutes of Health under Award Number U24DK126127. The content is solely the responsibility of the authors and does not necessarily represent the official views of the National Institutes of Health.

References

1. Bertrand JY, Chi NC, Santoso B, Teng S, Stainier DYR, Traver D (2010) Haematopoietic stem cells derive directly from aortic endothelium during development. Nature 464(7285): 108–111
2. Henninger J, Santoso B, Hans S, Durand E, Moore J, Mosimann C, Brand M, Traver D, Zon L (2017) Clonal fate mapping quantifies the number of haematopoietic stem cells that arise during development. Nat Cell Biol 19: 17–27
3. Sperling AS, Gibson CJ, Ebert BL (2016) The genetics of myelodysplastic syndrome: from clonal haematopoiesis to secondary leukaemia. Nat Rev Cancer 171(17):5–19
4. Bowman RL, Busque L, Levine RL (2018) Clonal hematopoiesis and evolution to hematopoietic malignancies. Cell Stem Cell 22:157–170
5. Jaiswal S, Fontanillas P, Flannick J, Manning A, Grauman PV, Mar BG, Lindsley RC, Mermel CH, Burtt N, Chavez A, Higgins JM, Moltchanov V, Kuo FC, Kluk MJ, Henderson B, Kinnunen L, Koistinen HA, Ladenvall C, Getz G, Correa A, Banahan BF, Gabriel S, Kathiresan S, Stringham HM, McCarthy MI, Boehnke M, Tuomilehto J, Haiman C, Groop L, Atzmon G, Wilson JG, Neuberg D, Altshuler D, Ebert BL (2014) Age-related clonal hematopoiesis associated with adverse outcomes. N Engl J Med 371: 2488–2498
6. Albert Pan Y, Freundlich T, Weissman TA, Schoppik D, Cindy Wang X, Zimmerman S, Ciruna B, Sanes JR, Lichtman JW, Schier AF (2013) Zebrabow: multispectral cell labeling for cell tracing and lineage analysis in zebrafish. Development 140:2835–2846
7. Mosimann C, Panáková D, Werdich AA, Musso G, Burger A, Lawson KL, Carr LA, Nevis KR, Sabeh MK, Zhou Y, Davidson AJ, Dibiase A, Burns CE, Burns CG, Macrae CA, Zon LI (2015) Chamber identity programs drive early functional partitioning of the heart. Nat Commun 6:8146
8. Avagyan S, Henninger JE, Mannherz WP, Mistry M, Yoon J, Yang S, Weber MC, Moore JL, Zon LI(2021) Resistance to inflammation underlies enhanced fitness in clonal hematopoiesis. Colorful Clones in the Blood Science 374(6568):768–772. https://doi.org/10.1126/science.aba9304
9. Ablain J, Durand EM, Yang S, Zhou Y, Zon LI (2015) A CRISPR/Cas9 vector system for tissue-specific gene disruption in zebrafish. Dev Cell 32:756–764
10. Brent LJ, Drapeau P (2002) Targeted 'knockdown' of channel expression in vivo with an antisense morpholino oligonucleotide. Neuroscience 114:275–278

Chapter 19

Mutation Knock-in Methods Using Single-Stranded DNA and Gene Editing Tools in Zebrafish

Sergey V. Prykhozhij and Jason N. Berman

Abstract

Introduction or knock-in of precise genomic modifications remains one of the most important applications of CRISPR/Cas9 in all model systems including zebrafish. The most widely used type of donor template containing the desired modification is single-stranded DNA (ssDNA), either in the form of single-stranded oligodeoxynucleotides (ssODN) (<150 nucleotides (nt)) or as long ssDNA (lssDNA) molecules (up to about 2000 nt). Despite the challenges posed by DNA repair after DNA double-strand breaks, knock-in of precise mutations is relatively straightforward in zebrafish. Knock-in efficiency can be enhanced by careful donor template design, using lssDNA as template or tethering the donor template DNA to the Cas9-guide RNA complex. Other point mutation methods such as base editing and prime editing are starting to be applied in zebrafish and many other model systems. However, these methods may not always be sufficiently accessible or may have limited capacity to perform all desired mutation knock-ins which are possible with ssDNA-based knock-in methods. Thus, it is likely that there will be complementarity in the technologies used for generating precise mutants. Here, we review and describe a suite of CRISPR/Cas9 knock-in procedures utilizing ssDNA as the donor template in zebrafish, point out the potential challenges and suggest possible approaches for their solution ultimately leading to successful generation of precise mutant lines.

Key words CRISPR/Cas9, Knock-in, ssODN, Long ssDNA, Point mutation, Disease model, Zebrafish

1 Introduction

Generating precise mutations efficiently is one of the most important applications of genome editing technologies, and this task has driven innovation and optimization of gene editing. There are currently five methods utilizing clustered regularly interspaced short palindromic repeats (CRISPR), Cas9, as well as other CRISPR nucleases and their derivatives: ssODN-based knock-ins, lssDNA-based knock-ins, Donorguide knock-in, prime editing, and base editing (Fig. 1). Due to the limited scope of this chapter, we will focus on the first three. However, prime editing has been

James F. Amatruda et al. (eds.), *Zebrafish: Methods and Protocols*, Methods in Molecular Biology, vol. 2707, https://doi.org/10.1007/978-1-0716-3401-1_19,

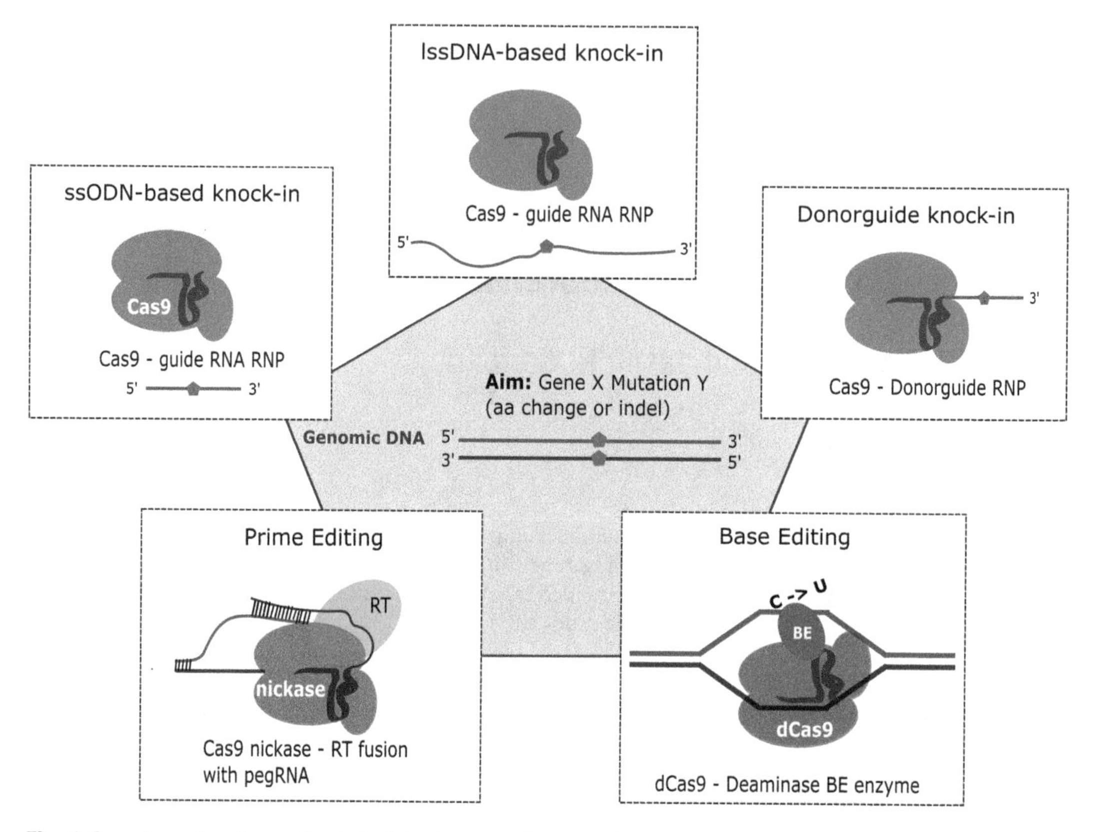

Fig. 1 Overview of major methods to introduce precise mutations in zebrafish. In the center of the figure, a genomic DNA site of Gene X is shown with a precise mutation Y indicated by a red pentagon (amino acid (aa) change, insertion or deletion (indel)). Introducing the mutation knock-ins using *ssODN* requires an oligonucleotide (ssODN) indicated by a blue line with the red pentagon, Cas9 protein, and a guide RNA (purple ribbon). *lssDNA*-based knock-ins is basically the same but a longer ssDNA molecule is used. Sometimes two guide RNAs are used to generate a DNA substrate for lssDNA interaction. In the knock-in with *Donorguides*, the cut is programmed by the same molecule that provides the template for repair by means of a covalent linkage between the tracrRNA and a DNA oligonucleotide. This hybrid molecule binds to a CRISPR RNA (crRNA) molecule to make a full chimeric synthetic guide RNA. *Prime editing* is a flexible gene editing mechanism based on a Cas9 nickase–reverse transcriptase (RT) protein fusion complexed with a prime editing guide RNA (pegRNA) that encodes the target modification. *Base editing* (BE) involves targeting of a catalytically inactive Cas9 (dCas9) fusion with a deaminase BE enzyme. Several different deamination events can occur and then result in mutations

demonstrated in zebrafish [1], and despite some initial accessibility issues, this method is likely to be very important in the future as it is very flexible, can be more efficient than CRISPR/Cas9, and is less prone to induce undesirable mutations. Base editing involves guide RNA-directed binding of the catalytically inactive dCas9 fusion with either a cytidine deaminase or an Adenine Base Editor enzyme leading to the respective deaminations and transition mutations [2–4]. Base editors have a significant track record in zebrafish since 2017 [5], so if your target mutation is a transition at an overlapping guide RNA site, base editing could be a great approach to try.

The use of CRISPR/Cas9 and chemically synthesized oligonucleotides has been the earliest and primary approach to introduce precise mutations into the zebrafish genome. Multiple studies emerged in the last 7–8 years that have explored the applications, advantages, and drawbacks of this technique [6–12]. These studies demonstrated that oligonucleotides from 50 to about 130 in length can serve as templates in mutation knock-ins after the cuts induced by Cas9 injected either as protein or mRNA. Guide RNAs can be provided as enzymatically or chemically synthesized single guide RNA (sgRNA) or as a two-component synthetic crRNA (CRISPR RNA): tracrRNA complex, but we are not aware of the systematic studies of the relative efficiencies of different formulations for the same guide RNA in zebrafish. The main advantage of the oligonucleotide approach is the significant flexibility in how you can define a mutation knock-in. The limitation of this technology is that the double-strand breaks and DNA repair machinery are still required for mutant generation, thus limiting efficiency to a maximum of <10%.

The zLOST (zebrafish long single-stranded DNA template) method or, more precisely, CRISPR/Cas9-mediated point mutation knock-ins with enzymatically-synthesized lssDNA of 300–500 nucleotides as templates have recently been shown to have higher efficiencies in zebrafish compared to shorter oligonucleotides or double-stranded DNA molecules [13]. Another zebrafish study successfully knocked in multiple epitope tags in several genes using an lssDNA production method based on a single-strand excision from a plasmid, Cas9 protein and the two-component synthetic guide RNA [14].

Attaching the donor template oligonucleotide to the Cas9 nuclease can be intuitively expected to increase the knock-in efficiency. This idea has been demonstrated by a fusion of Porcine Cyclovirus 2 Rep protein to Cas9 (PCV-Cas9), which is able to covalently bind an oligonucleotide and stimulate knock-in efficiency in cell culture systems [15]. Donorguide is a chimeric crRNA:tracrRNA-oligonucleotide complex that binds to the Cas9 protein and increases knock-in efficiency in both zebrafish and cell lines [16].

In the following protocol, we have summarized design and experimental considerations needed to perform precise mutation knock-in experiments using single-stranded DNA in zebrafish.

2 Materials

2.1 General Molecular Biology Reagents

1. Agarose.
2. 10× TBE (Tris/boric acid/EDTA) buffer (1 M Tris, 1 M boric acid, 20 mM EDTA, pH 8.3).
3. Gel Loading Dye, purple (6×).

4. Nuclease-free water.
5. SYBR Green I nucleic acid gel stain (10,000× in DMSO).
6. Tris–EDTA (TE) buffer (10 mM Tris–HCl, 1 mM EDTA, pH 8.0).
7. Taq DNA polymerase.
8. dNTP mix, 100 mM stock.
9. Q5® High-Fidelity 2× Master Mix (New England Biolabs).
10. MEGAshortscript T7 kit (Thermo Fisher).
11. MEGAscript T7 kit (Thermo Fisher).
12. mMessage mMachine T3 (Thermo Fisher).
13. QIAquick Gel Extraction kit (QIAGEN).
14. QIAquick PCR purification kit (QIAGEN).
15. NucAway kit (Thermo Fisher).
16. SuperScript™ III First-Strand Synthesis System (Thermo Fisher).
17. Lambda exonuclease (New England Biolabs).
18. Exonuclease III (New England Biolabs).
19. MinElute kit (QIAGEN).
20. Alt-R crRNAs (Integrated DNA Technologies).

    Chimeric CRISPR-Cas9 tracrRNAs from IDT as RNA Ultramers (https://www.idtdna.com/site/order/oligoentry/index/UltraRNA) with the general sequence (for Subheading 3.5): "rArGrCrArUrArGrCrArArGrUrUrArArArArUrArArGrGrCrUrArGrUrCrCrGrUrUrArUrCrArArCrUrUrGrArArArArArGrUrGrGrCrArCrCrGrArGrUrCrGrGrUrGrCrUrUrUrUrUrUrUr UNNNNNNNNNNNNNNNNNNNNNNNNNN" (where rN = RNA bases comprising the tracrRNA sequence and N = DNA bases comprising the DNA donor portion).
21. Duplex Buffer (30 mM HEPES, pH 7.5; 100 mM potassium acetate) (Integrated DNA Technologies).
22. Alt-R® S.p. Cas9 Nuclease V3, 100 μg (Integrated DNA Technologies).
23. Alt-R® S.p. HiFi Cas9 Nuclease V3, 100 μg (Integrated DNA Technologies).

2.2 Growing and Handling Zebrafish

1. E3 medium: 5 mM NaCl, 0.17 mM KCl, 0.33 mM $CaCl_2$, 0.33 mM $MgSO_4$, 0.0001% Methylene Blue.
2. Tricaine stock: 0.4% water solution of Tricaine buffered to pH 7.2 with 1 M Tris–HCl, pH 9.0 (~4 mL per 100 mL of solution).

3 Methods

In this section, we will focus on the precise knock-in of short modifications by means of CRISPR/Cas9 (*see* **Note 1**) and single-stranded DNA (ssDNA) molecules. These modifications are typically <10 nucleotides (nt) and are most frequently amino acid replacements, because the majority of naturally occurring mutations are at the single-nucleotide level. ssDNA molecules used in knock-in methods can be divided into ssODN, which are <200 nt in length and are made synthetically and long ssDNA (lssDNA) up to 3000 nt and prepared enzymatically. Megamers are commercial synthetic ssDNA molecules of 201–2000 nt. Knock-in of larger DNA cassettes using lssDNA has been recently covered [14, 17, 18] and we focus here on the knock-in of short modifications. In general, short modifications are used for modeling human diseases, whereas larger cassette knock-ins are more relevant for generating experimental tools.

In the following, we will assume that an active guide RNA or more rarely a pair of guide RNAs is determined and available, either prepared enzymatically or synthetic, and that the primers used for analysis of its activity (site assay primers) are available.

3.1 Sequencing of the Target Region in the Zebrafish Strain(s) Used for Experiments

Proceed with the following steps to determine polymorphisms in the knock-in region compared to the reference (UCSC or Ensembl) DNA sequences.

1. Cut fin clips with a scalpel from 20 to 30 adult zebrafish from your desired strain(s) by anesthetizing them briefly in 0.02% Tricaine in small batches of (4–8 fish) and put the clips into 40 μL of 50 mM NaOH.
2. Heat the samples at 95 °C for 10 min in a thermocycler and then vortex them for a few seconds.
3. Incubate the samples on ice for 5 min and neutralize them with 4 μL of 1 M Tris–HCl, pH 8.0. Make a pooled sample by taking an aliquot from every sample.
4. Amplify the site assay region with the following Taq-based procedure (*see* **Note 2**):

   Use 3 μL of pooled DNA solution per 20 μL reaction and perform the touchdown PCR method: 94 °C for 3 min; 10 cycles (94 °C for 30 s, T_{anneal} + 10 [with 1 °C decrease every cycle] for 30 s, 72 °C for 30 s), 25 cycles (94 °C for 30 s, T_{anneal} for 30 s, 72 °C for 30 s).
5. Run the samples on a 1.5 or 2% agarose gel in 0.5× TBE and gel-extract a well-defined band of the correct size using QIAquick Gel Extraction kit according to standard instructions.

6. Sanger sequence the PCR product from each side of the amplicon.
7. Analyze the sequencing reads at the level of final sequences after base calling and at the level of chromatograms to check if there are positions with significant polymorphism in your zebrafish population. Deep sequencing ($<10^4$–10^5 reads) can also be used for more precise polymorphism determination, but the overall amplicon size and sequencing settings need to be compatible, e.g., 250 bp amplicon and paired reads of 150-nt each are a suitable combination.
8. Create a consensus sequence for the amplicon and check for any differences with the reference.

3.2 Design of ssODN for Introducing Short Indel Modifications by CRISPR/Cas9

1. The strategies for indel class mutations can be designed manually using a DNA sequence editor, simple text editor, or word processing software. Introduce the desired modification into the reference or the consensus sequence from Subheading 3.1 manually.
2. Using the modified sequence, determine the binding or target (T) strand of your DNA molecule and locate the cut site based on your guide RNA and enzyme. If your cut site is inside a deletion, you should consider the cut site to be located between the deletion edges.
3. To find the optimal oligo, it is helpful to design two larger asymmetric oligos and two smaller symmetric ones (ca 60 nt) (Fig. 2). If your gRNA site PAM sequence is in the exonic strand sequence, the following designations will apply; otherwise, the design process is the same, but the oligo naming will have to switch (Fig. 2). T 126 A left (Target Asymmetric left): position your cursor at the cut site, move 36 nt to the left (5′), select 126 nt +/− insertion or deletion size to 3′ (to the right) and reverse-complement the selected sequence. NT 126 A right (Non-target Asymmetric right): move 36 nt to the right (3') of the cut site and select 126 nt +/− insertion or deletion size to the left (5′). Make sure the short arm length is not affected by the modification. For the symmetric oligos, use 30 nt as the homology arm length on both sides. In case of deletions, your homology arms will be positioned on each side of the deletion and for insertions you will need to increase the size of your oligos by the insertion length. Select homology arms and the modification symmetrically around the cut site from the NT strand and reverse-complement this sequence to obtain the T strand short oligo sequence. The availability of these four oligos is generally sufficient to explore their relative efficiency and to find the best one by the genotyping assays and high-throughput sequencing approaches (*see* **Note 3**).

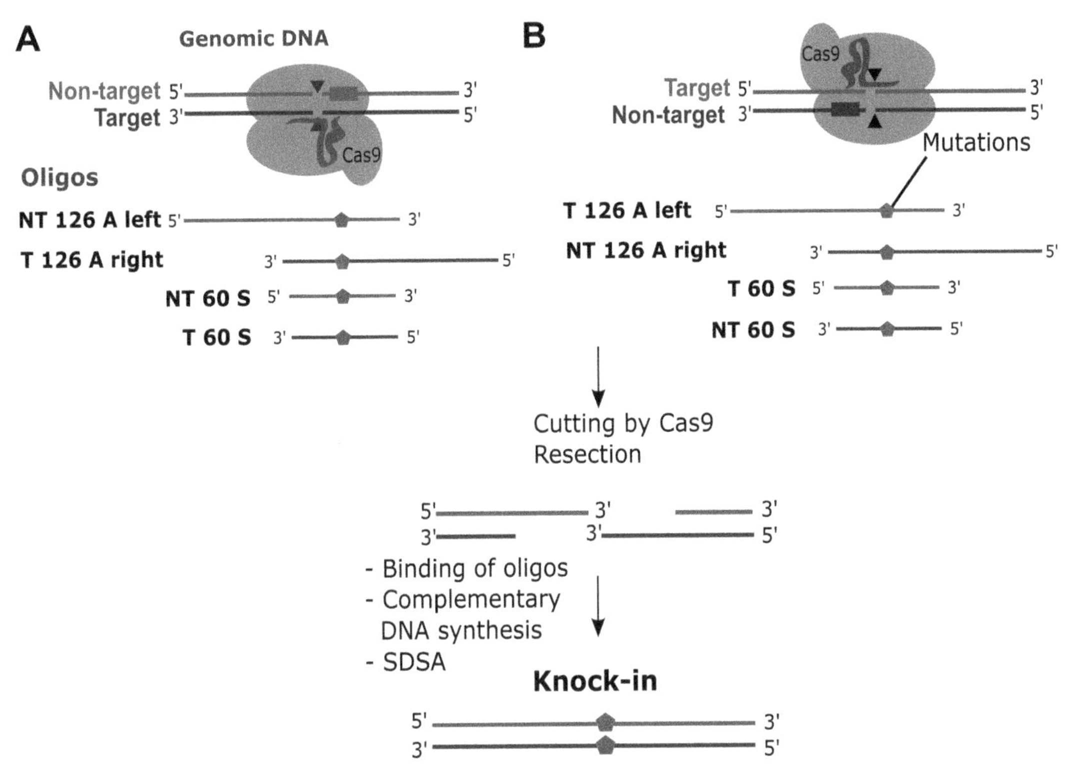

Fig. 2 Design of longer asymmetric (A) and short symmetric (S) oligos for CRISPR/Cas9 of defined mutations for both guide RNA orientations. If the top strand corresponds to the exonic or coding sequence, guide RNAs can bind either the complementary strand (**a**) or the top strand itself (**b**). We therefore designate strands as either Target (T) or Non-target (NT) and the corresponding oligonucleotides for the two guide RNA binding cases are nearly equivalent in structure but have different designations. However, these oligos will likely engage similar gene editing mechanisms after a Cas9-mediated cut and resection of DNA ends albeit with different efficiencies. The current model is that the oligos bind to the complementary ssDNA regions and induce synthesis-dependent strand annealing (SDSA) pathway resulting in a precise knock-in in a certain percentage of cases

4. We also recommend phosphorothioate linkages between the last three nucleotides on all oligo ends. Integrated DNA Technologies (IDT) uses "*" for these bonds (such as T*C*G). Check with your provider for the convention. Although we have not tested this, IDT also offers proprietary Alt-R™ HDR Donor Oligos at a similar cost to Ultramer oligos that we usually order. The shorter (ca 60 nt) ssODN can be ordered as desalted oligos.
5. AS-PCR for indel strategies is possible but you may experience some problems (*see* **Note 4**). For insertion strategies, the knock-in primer must end inside the insertion (6 or fewer nt), and the wild-type primer will typically start at the same position and span the site of the insertion such that in the insertion-homozygous sample its 3′-end will be strongly

mismatched. The deletion-specific primer should have most of its sequence before the deletion but the last five 3′ nucleotides should bind after the deletion in such a way that these nucleotides are strongly mismatched when aligned to the wild-type sequence, whereas the wild-type primer should bind the sequence within the deletion.

3.3 Design of ssODN for Introducing Amino Acid Changes by CRISPR/Cas9

Given the known coding sequence, any mutation that leads to an amino acid change can be specified at the genomic DNA level, coding DNA level, and ultimately at the protein level such as with the simple A123B code, where A and B are amino acids and 123 is the residue number. Such precise mapping of mutations to codons enables easier and more automatable design of CRISPR donor templates. If we are modeling a human variant leading to an amino acid change, we may determine that the conservation at the amino acid level does not always extend to the codon conservation. However, the mutant codon can be the same or different than in the mutant human cDNA. We will start with a hypothetical situation when a mutation in a human gene is known, and we want to create a corresponding precise point mutant zebrafish model.

3.3.1 Manual Design of Oligos for Amino Acid Changes

1. Identify the correspondence between the human protein and the orthologous zebrafish one by BLAST (https://blast.ncbi.nlm.nih.gov/Blast.cgi). Verify that the mutated residue is conserved. Otherwise, this exact model may not yield the desired results depending on the importance and similarity of the zebrafish residue to the human one. In case of conservation, note the exact residue number in the zebrafish protein and a short stretch of neighboring amino acids.
2. From the RefSeq protein record found in the BLAST search, look up the DBSOURCE line and follow the link to the mRNA RefSeq NCBI record. This record can be downloaded as a Genbank file and then opened in any DNA sequence editor such as Vector NTI. From this cDNA file, isolate the coding sequence either as another sequence file in the same software or as text in another text editor software.
3. Identify the mutant codon location in the coding sequence based on the alignment obtained by BLAST or from a focused human–zebrafish alignment (Fig. 3). To do this, select the triple of the residue number nucleotides and the last three nucleotides will be the desired codon. Verify the correctness of the codon selection.
4. Download the gene file from Ensembl in the Export Data section in the correct orientation and open it in the DNA sequence editor.

1. Check the correspondence between the human mutation and the zebrafish one.

2. Identify mutation location in cDNA:

```
ATGGCGCAAAACGACAGCCAAGAGTTCGCGGAGCTCTGGGAGAAGAATTTGATAAGTATTCAGCCCCCAGGTGGTGGCTCTTGCTG
GGACATCATTAATGATGAGGAGTACTTGCCGGGATCGTTTGACCCCAATTTTTTTGAAAATGTGCTTGAAGAACAGCCTCAGCCAT
CCACTCTCCCACCAACATCCACTGTTCCGGAGACAAGCGACTATCCCGGCGATCATGGATTTAGGCTCAGGTTCCCGCAGTCTGGC
ACAGCAAAATCTGTAACTTGCACTTATTCACCGGACCTGAATAAACTCTTCTGTCAGCTGGCAAAAACTTGCCCCGTTCAAATGGT
GGTGGACGTTGCCCCTCCACAGGGCTCCGTGGTTCGAGCCACTGCCATCTATAAGAAGTCCGAGCATGTGGCTGAAGTGGTCCGCa
gatgcccccatcatgagcga
```

3. Take a short stretch (10-15 nt) on either side of the target residue and search it inside the genomic sequence to find the exon.

4. Mark relevant regions in the exon in a program or manually, split or mark codons

```
TATTCACCGGACCTGAATAAACTCTTCTGTCAGCTGGCAAAAACTTGCCCCGTTCAAATGGTGGTGGACGTTGCCCCTCCACAGGGCTCCG
TGGTTCGAGCCACTGCCATCTATAAGAAGTCCGAGCATGTGGCT GAA GTG GTC CGC AGA TGC CCC
CATCATGAGCGAACCCCGGATGGAGATAgtacagacatttttttttccatatccattcttgcatcattctaggc
sgRNA   R143  intron
```

5. Introduce PAM and target mutations

```
TATTCACCGGACCTGAATAAACTCTTCTGTCAGCTGGCAAAAACTTGCCCCGTTCAAATGGTGGTGGACGTTGCCCCTCCACAGGGCTCCG
TGGTTCGAGCCACTGCCATCTATAAGAAGTCCGAGCATGTGGCT GAA GTC GTC CAC AGA TGC CCC
CATCATGAGCGAACCCCGGATGGAGATAgtacagacatttttttttccatatccattcttgcatcattctaggc
sgRNA   R143  R143H  PAM mutation  intron
```

6. If restriction sites are desired, apply the program WatCut
http://watcut.uwaterloo.ca/template.php?act=silent_new
to about 40 bp around the mutation site. Here is the result for R143H:

```
TATTCACCGGACCTGAATAAACTCTTCTGTCAGCTGGCAAAAACTTGCCCCGTTCAAATGGTGGTGGACGTTGCCCCTCCACAGGGCTCCG
TGGTTCGAGCCACTGCCATCTATAAGAAGTCCGAGCATGTGGCT GAA GTC GTC CACAGGTGCC CC
CATCATGAGCGAACCCCGGATGGAGATAgtacagacatttttttttccatatccattcttgcatcattctaggc
sgRNA   R143  R143H  PAM mutation  intron  BanI mutation  BanI site
```

7. Select the relevant size and orientation from the above sequence and order as an oligo. Homology arms should be counted from the cut site and not from any of the mutations.

Fig. 3 Manual design process for amino acid change donor template oligonucleotides. The basic steps of the manual design process to generate oligonucleotides to knock-in amino acid change mutations. The example shown is a *tp53* R143H mutation in zebrafish. Some steps such as **5** and **6** have multiple steps within them, which are explained in the text. The steps shown here do not directly correspond to the steps in Subheading 3.3.1

5. Copy a short stretch (10–15 nt) on either side of the target residue codon and search it inside the genomic sequence to find the corresponding exon. Once the genomic location of the target codon is found, isolate about 600 nt of the sequence around the codon to two identical files (DNA sequence editor, or word processing software), one of which will be mutated in silico. If the experimentally determined consensus sequence is available and different from the reference, replace the differing part of the reference sequence with the consensus one.
6. Label the gRNA region and the target codon in the smaller genomic region sequence (Fig. 3). In the DNA sequence editor, it is also helpful to translate nearby codons on each side to their amino acids and compare the results to the human–zebrafish protein alignment.
7. Introduce the target mutation and then determine if the target mutation is likely to inactivate the guide RNA binding site by mutating the PAM or the gRNA region next to it. If the target mutation does not inactivate gRNA, you need to check if the PAM or gRNA spacer overlapping codons can be mutated to prevent possible recutting. Introduce one or more of these synonymous changes into the sequence.
8. For genotyping purposes, it may be helpful to identify any new restriction sites introduced by mutations. Make new files from both the wild-type and virtually mutated sequence files by limiting the sequence to the amplicon used to verify gRNA activity using primer sequences and the search tool in the DNA sequence editor. Identify unique restriction sites in each file and determine if there are extra sites in the mutated file compared to the wild-type file.
9. If no sites have been identified, it is possible to introduce restriction sites by synonymous codon mutations using WatCut http://watcut.uwaterloo.ca/template.php?act=silent_new). Select up to four codons before and up to four codons after the target codon, and submit to WatCut. Next, submit the full amplicon-mutated sequence to NEBCutter v2 (http://nc2.neb.com/NEBcutter2/) and identify non-cutter enzymes. You can then cross-reference the table of the resulting candidate restriction enzymes and codon mutations in WatCut and non-cutter enzymes. The enzymes appearing in both sets are of interest; select the enzyme that is readily available commercially and is located most closely to the target codon. Make the corresponding synonymous mutation in the sequence file. Label the introduced restriction site and record the enzyme information for ordering purposes.
10. Select the oligos as in **steps 3** and **4** of Subheading 3.2.

11. Manually design allele-specific primers by choosing the strand maximizing the divergence between the wild-type and knock-in alleles. They can be paired with a newly designed primer or with one of the site assay amplicon primers. NEB Tm Calculator (https://tmcalculator.neb.com/#!/main) helps with optimizing allele-specific primer length when combining them with one of the existing primers.

3.3.2 Design of Amino Acid Change Point Mutations Using CRISPR Knock-in Designer

The full series of steps for the manual amino acid change strategy design in Subheading 3.3.1 is rather involved. Therefore, we aimed to automate the entire process and published the CRISPR Knock-in Design tool (https://crisprtools.shinyapps.io/knockinDesigner/) in a recent paper [19], which can be consulted for an in-depth description of the website algorithms. Here, we will provide a basic practical guide of how the website can be used and what input data are necessary (Fig. 4) and how the input interface appears (Fig. 5).

1. *Provide mutation and sequence data.* GENE MUTATION input tab should be used to provide Gene name and Mutation. GENE SEQUENCE DATA input tab should be used to provide the sequence data. You should choose either MANUAL DATA INPUT or ID-BASED INPUT tab for your input. One advantage of the MANUAL DATA INPUT is that you can use the experimentally determined consensus sequence throughout the design process.

   For the MANUAL DATA INPUT, you need to enter the Coding DNA sequence, Mutation site exon sequence (required), 5′ flanking fragment, and 3′ flanking fragment (optional if they do not contain other input sequences). Flanking fragments are typically introns but can be other genomic sequences when mutations are close to the beginning or end of the gene. Which flanking sequences are required is determined by the mutation and primer locations.

   For ID-BASED INPUT, you need to provide the correct Ensembl Transcript ID input, which will automatically allow the program to infer the species and retrieve the reference sequences.

2. *Provide primer sequences.* The PCR PRIMERS section allows for input of primers needed to produce a PCR amplicon around the knock-in mutation site, which we will also call the site assay. Each of these primers should be ≥100 bp and ≤ 300 bp (600 bp is the maximum distance) from the mutation site. It is best to use the primers that you used initially for gRNA activity analysis.

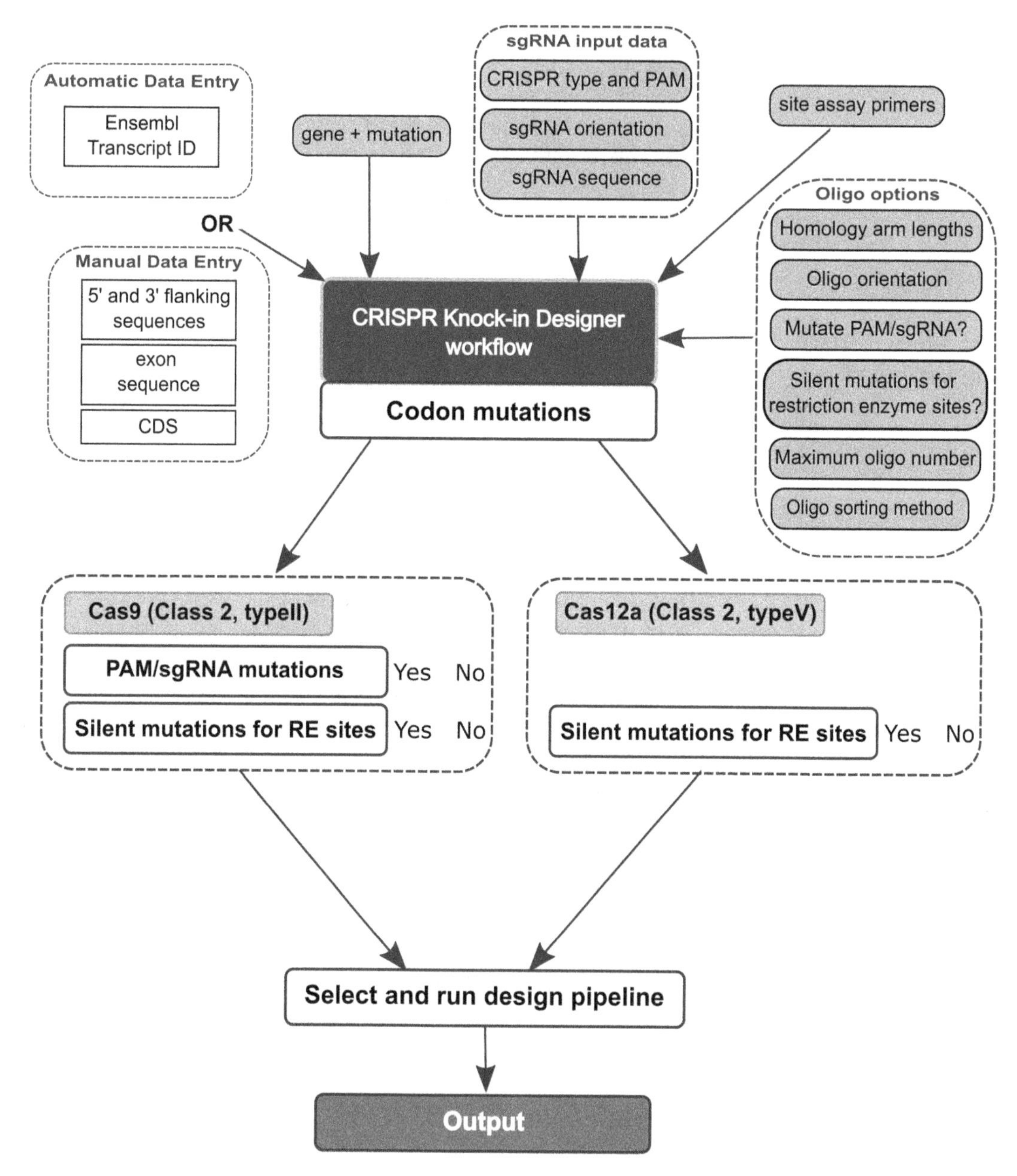

Fig. 4 An algorithm outline of CRISPR Knock-in Designer. The protein-coding gene mutation is defined here by the gene symbol and a mutation string such as X123Y (gene + mutation) in single-letter amino acid code. The relevant sequences are either retrieved automatically using the Ensembl Transcript ID input (Automatic Data Entry) or provided manually. The Manual Data Entry option requires coding DNA sequence (CDS) and exon sequence of the target residue codon, the flanking sequences being optional depending on the primer binding sites. Another important input is site assay (amplicon around the target site) primers that define the genomic region for the genotyping assay designs. Single guide RNA (sgRNA) input data consist of the CRISPR type and PAM parameter, sgRNA orientation and sequence. The Oligo options include homology arm length, oligo orientation, "Mutate PAM/sgRNA?" (whether to mutate the PAM or sgRNA sequence), "Silent mutations for restriction enzyme sites?" (whether to introduce restriction sites by synonymous mutations near the target codon), Maximum oligo number, and Oligo sorting method. The program then performs the basic target codon

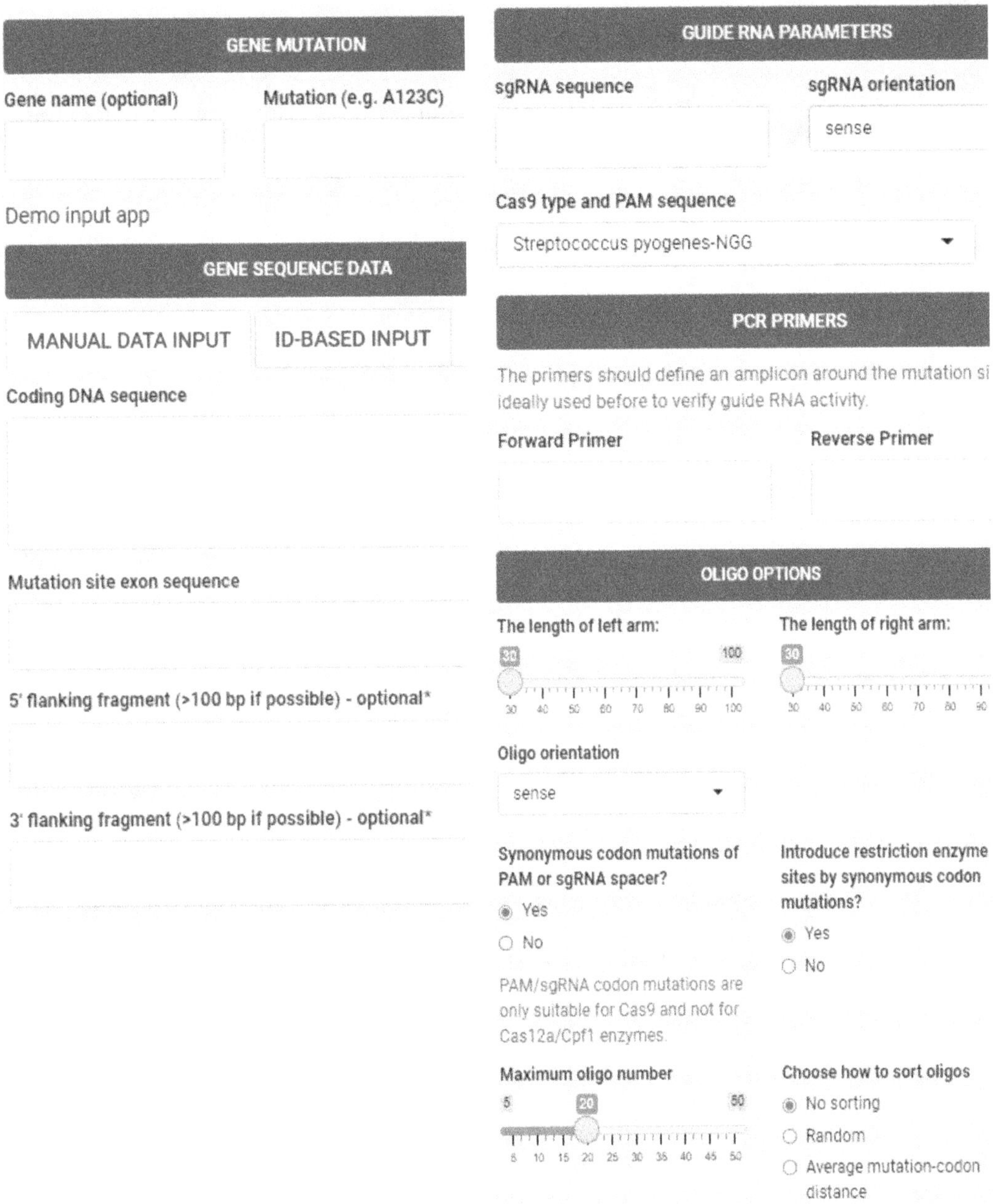

GENE MUTATION

Gene name (optional)

Mutation (e.g. A123C)

Demo input app

GENE SEQUENCE DATA

MANUAL DATA INPUT

ID-BASED INPUT

Coding DNA sequence

Mutation site exon sequence

5' flanking fragment (>100 bp if possible) - optional*

3' flanking fragment (>100 bp if possible) - optional*

GUIDE RNA PARAMETERS

sgRNA sequence

sgRNA orientation

sense

Cas9 type and PAM sequence

Streptococcus pyogenes-NGG

PCR PRIMERS

The primers should define an amplicon around the mutation si

ideally used before to verify guide RNA activity.

Forward Primer

Reverse Primer

OLIGO OPTIONS

The length of left arm:

30 100

30 40 50 60 70 80 90 100

The length of right arm:

30

30 40 50 60 70 80 90

Oligo orientation

sense

Synonymous codon mutations of PAM or sgRNA spacer?

Yes

No

PAM/sgRNA codon mutations are only suitable for Cas9 and not for Cas12a/Cpf1 enzymes.

Introduce restriction enzyme sites by synonymous codon mutations?

Yes

No

Maximum oligo number

5 20 50

5 10 15 20 25 30 35 40 45 50

Choose how to sort oligos

No sorting

Random

Average mutation-codon distance

Fig. 5 CRISPR Knock-in Designer input panels

Fig. 4 (continued) mutation procedure followed by additional mutation procedures that are chosen based on the oligo options and the CRISPR type. All paths of program execution are shown and those that can be chosen based on the CRISPR type are overlaid with the CRISPR type label. However, in each run, only a single design pipeline can be chosen, which results in the output of the designed oligonucleotides and the corresponding genotyping strategies

Table 1
Long asymmetric oligo recommended options

	PAM within exonic (sense) strand Oligo options			PAM in the opposite (anti-sense) strand Oligo settings		
	Left arm	**Right arm**	**Oriented**	**Left arm**	**Right arm**	**Oriented**
T-AS oligo	35	90	Anti-sense	90	35	Sense
NT-AS oligo	90	35	Sense	35	90	Anti-sense

3. *Enter guide RNA parameters.* The GUIDE RNA PARAMETERS section is essential for specifying how your target site will be cut by your CRISPR strategy. For your gRNA, you need to enter its sequence, orientation, and CRISPR type/PAM sequence.
4. *Specify oligo options.* The OLIGO OPTIONS section allows the user to specify the structure of the oligos and extra mutations to be designed by the program. The orientation, left arm and right arm length options define the homology arm structure of the oligos (Table 1).
   - The next option is "*Synonymous codon mutations of PAM or sgRNA spacer?*". This option only applies to class 2, type II (Cas9) enzymes because the PAM is close to the cut (3 bp), whereas the Cas12a/Cpf1 enzymes have 18 bp distance between the PAM and cut site.
   - The next option is "*Introduce restriction enzyme sites by synonymous codon mutations?*". This option enables restriction site introduction by a single synonymous codon mutation per oligo in one of the three codons closest to the target mutation on either side.
   - The "*Maximum oligo number*" limits the number of output oligos, and the "*Choose how to sort oligos*" option set allows the user to do no sorting, random sorting, or sorting by the average of absolute distances between the target codon mutation and additional mutations. The random sorting option is mainly useful to reorder oligos with different replacement codons. The "average mutation–codon distance" option sorts oligos in such a way that oligos with more mutation clustering will occupy higher positions in the list thus increasing the likelihood that all the designed mutations will be introduced into the genome together and improving the specificity of their detection.
5. *Design output.* The program will generate normal web page outputs as well as downloadable text files each time it runs. The user must decide which exact modification will be chosen. The

variables are the exact replacement codon, restriction site(s), and gRNA site mutation(s). The last two variables are optional. The design of oligos with different homology arms lengths and orientations can be designed in the same session by dynamically adjusting homology arms lengths and orientations. Choosing oligos with the same number for different settings will ensure that they all correspond to the same modification. AS-PCR primers and the genotyping assays are designed automatically for each oligo, which reduces knock-in strategy design workload.

3.4 Design and Preparation of lssDNA Donor Template Oligos

Recent advances in methods to generate lssDNA molecules (>200 nt) have enabled researchers in multiple model systems to demonstrate their effectiveness at introducing point mutation and large gene cassette knock-ins. Although the usage of lssDNA for replacing one or a few nucleotides in the genome may seem surprising and is not always necessary, there have been case studies such as the zLOST (zebrafish long single-stranded DNA template) study where performance of 300-nt lssDNAs was superior compared to ~100-nt oligonucleotides [13]. We also envision that lssDNA can be useful for cases when no usable gRNAs can be found at the target site and the researchers need to resort to deletions with a pair of gRNAs (Fig. 6a). lssDNAs with homologies at each side of such deletions can help restore the deleted region and introduce the desired modification. In the following, we will describe the design process of lssDNAs for point mutation knock-ins and methods for their enzymatic preparation.

1. Sequence the genomic region of the future donor template in your zebrafish strain if you have not done so as described in Subheading 3.1. If you already have the sequence or want to use the reference one, proceed to the next step.
2. Design your intended set of knock-in mutations using one of the methods in Subheading 3.2 or 3.3 and place them into the genomic reference or sequenced consensus DNA context, for example, by replacing the wild-type sequence with its mutated version.
3. Design primers in your favorite primer design software to amplify the region to be converted to lssDNA. Amplicon should be about 300-bp in length or longer for targeting knock-in mutations into deletions (increase your lssDNA by the deletion size). In case you identify repetitive regions, you may also consider adjusting the amplicon region to avoid them. There are then at least three methods to generate the final lssDNA.

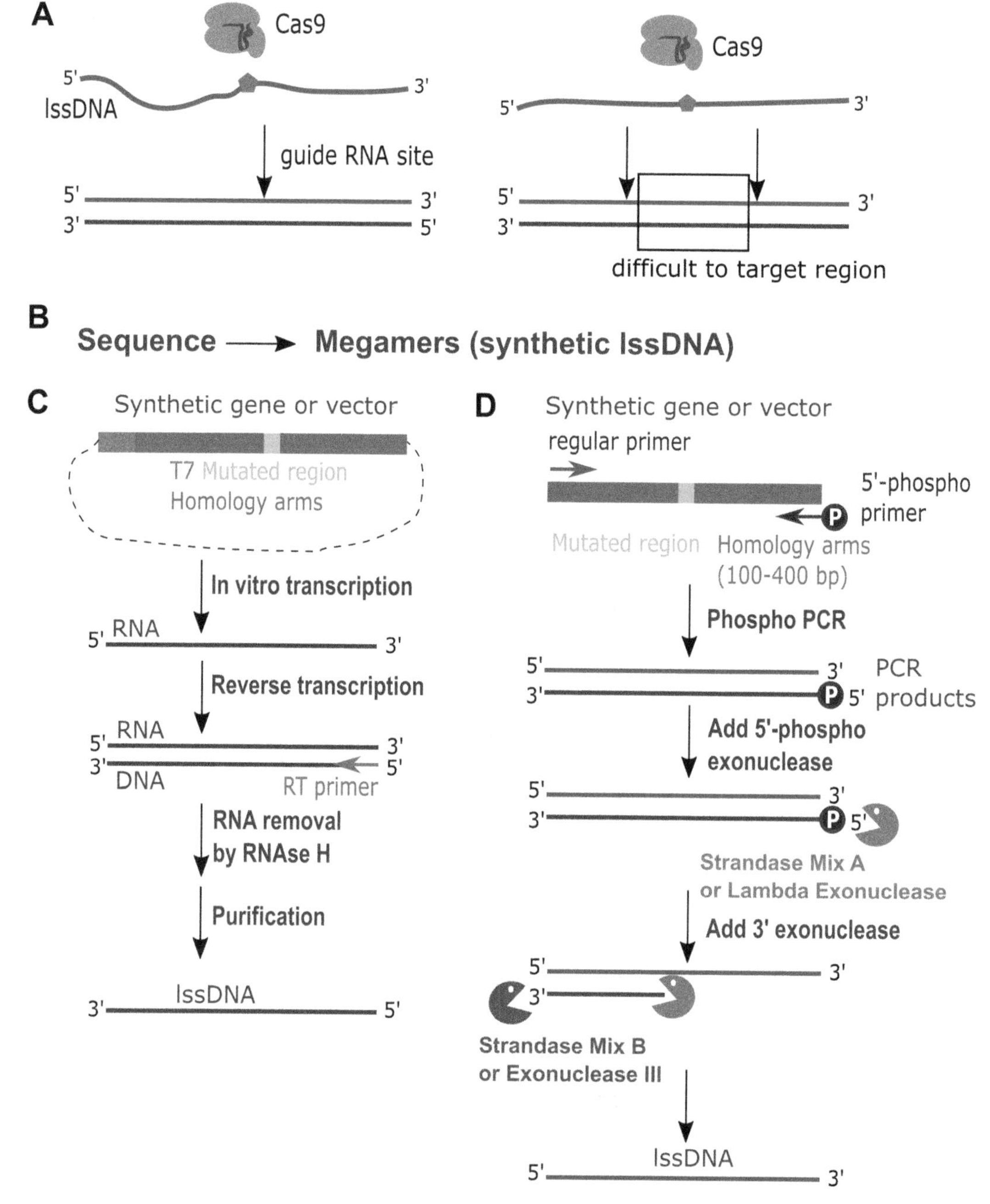

Fig. 6 lssDNA targeting strategies and synthesis methods. (**a**) lssDNA can be used for CRISPR/Cas9-based knock-in when either a specific guide RNA is used (left) or in case a region is difficult to target, a deletion can be generated by two guide RNAs, and lssDNA can both repair a deletion and introduce the desired modification. (**b**) Based on a specific sequence, it is possible to order synthetic Megamers, although they are expensive compared to other molecules and the waiting period is currently about 20 business days (IDT). (**c**) lssDNA synthesis based on an in vitro transcription and reverse transcription starting from either a linear or vector DNA molecule. The other strand lssDNA can be generated by placing T7 promoter on the other strand at its 5′ end. (**d**) Exonuclease-based generation of lssDNA from a PCR product, where one of the primers contains a phosphate at the 5′-end. Putting the phosphate on the other primer will allow generation of lssDNA from the complementary strand

3.4.1 Synthetic lssDNA

1. Using the genomic reference or sequenced consensus wild-type and in silico mutated DNA sequence, select the amplicon region from **step 3** above for consistency.
2. Order Megamers of either one or both strands of the chosen sequence (Fig. 6b). This step may not be for routine use since they are quite expensive and take a long time to synthesize.

3.4.2 Reverse Transcription-Based Synthesis of lssDNA (Fig. 6c)

1. Order Gblocks (IDT) with the previously defined mutated sequence and a T7 promoter located at the start of either sense or anti-sense strand. Include the primers designed in **step 3** above in the same order as well as a T7 promoter primer. If you are about to run out of these Gblocks, you can perform a PCR on a Gblock with T7 and a primer binding the opposite strand. However, this step may have some issues such as additional incorrect products requiring further optimization.
2. Set up 10 μL MEGAscript T7 reaction(s) with 400 ng of Gblocks: 1 μL of each NTP, 1 μL of 10× Reaction Buffer, 4 μL of a DNA template (100 ng/μL), and 1 μL of T7 Enzyme mix.
3. Incubate at 37 °C for 4 h.
4. Add 1 μL of TurboDNAse and incubate at 37 °C for 15 min.
5. Dilute the sample with 30 μL of Nuclease-free water.
6. Purify synthesized RNA samples using NucAway kit according to the kit instructions and measure the concentration.
7. Analyze RNA on an agarose gel to determine its integrity. Mix 2 μL of each RNA sample with 2 μL of 2× Loading Dye from the kit and 1 μL of 100× SYBR Green I and run the sample on a 1.5% agarose/TBE gel together with a 100 bp ladder sample (5 μL of ladder with 1 μL 100× SYBR Green I).
8. Set up a reverse transcription reaction according to the Superscript III Reverse Transcriptase protocol. Add the following components to a PCR tube (Tube A):

   RNA template from **step 3** (~5 μg).

   1.5 μL of RT primer* (100 μM)

   3 μL of 10 mM dNTP mix

   Water to 30 μL.

   *RNA strand–RT primer correspondence: sense RNA will require reverse primer as RT primer and anti-sense RNA will require forward primer.
9. Mix by pipetting and centrifuge at maximum speed at room temperature for 30 s. RT primers are the primers ordered in **step 1**, which can bind to Gblock strands. All incubations ≥37 °C are carried out in a thermocycler.

10. Incubate the tube at 65 °C for 5 min.
11. Put the sample on ice and combine the reagents for cDNA synthesis (Tube B) according to the following list:

    6 μL of 10× RT buffer

    3 μL of 0.1 M DTT

    12 μL of 25 mM $MgCl_2$

    3 μL of RNase Out

    3 μL of SuperScript III Reverse Transcriptase

    3 μL of Nuclease-free water
12. Combine the tubes A and B, mix by pipetting, and incubate at 50 °C for 50 min. Stop the reaction by heat-inactivation at 85 °C for 5 min. Cool the reaction to room temperature.
13. Add 3 μL of RNase H to the tube, mix by pipetting, and incubate the tube at 37 °C for 20 min.
14. Add 6 μL of 100× SYBR Green I and 10 μL of 6× Loading Dye to the sample and run the mixture in multiple wells of a 1.5% agarose gel.
15. Gel-extract the prominent band of the correct size and measure its concentration.

3.4.3 Exonuclease-Based Production of lssDNA (Fig. 6d)

The protocol described in Subheading 3.4.2 is somewhat complex and error-prone since reverse transcription generates not only a well-defined lssDNA product but also other molecules requiring additional gel extraction. On the other hand, Subheading 3.4.2 utilizes well-established and readily available kits and methods, thus helping in its straightforward adoption. However, the need for the more precise lssDNA production methods has resulted in some improved techniques such as Guide-it™ Long ssDNA Production System that relies on a PCR, where one primer contains 5′ phosphate, and a subsequent application of two exonucleases to remove the strand containing this phosphorylated primer; this method has been already applied to the mouse [17, 20] and *Xenopus* species [18]. We describe the procedure based on the work by Inoue et al. [20] that utilizes specific exonuclease enzymes and aims to generate lssDNA of both strands.

For both lssDNA primers, order their 5′-phosphorylated versions as well as a Gblock with introduced mutations, but without any sequence additions at the end. This mutated Gblock can be the same size as the amplicon of lssDNA primers or somewhat longer.

1. Perform 100 μL PCRs for both lssDNA strands with Q5 polymerase 2× mix with one 5′-P-primer and the other regular primer under optimized conditions (single PCR product) such as sufficiently low template (needs to be determined empirically by dilutions) and 0.2 μM primers, as well as an optimized annealing temperature to ensure a single specific PCR product.

2. Purify the PCR products using QIAquick PCR purification kit according to the kit instructions and measure their concentration. The goal is to obtain 10 μg of each PCR product. This may require a repeat and scaling of **steps 2** and **3**.
3. Set up a 50–100 μL Lambda exonuclease reaction with 10 units of enzyme, 10 μg of dsDNA substrates in 1X NEBuffer 1 to digest 5′-phospholyrated chains of dsDNAs.
4. Incubate at 37 °C for 30 min and then inactivate the enzyme by heating at 75 °C for 10 min.
5. Add 90 units of Exonuclease III to the reaction mix and incubate for 2 min. Adjust the incubation depending on the starting dsDNA's length. As a standard, incubate the reaction mixture for 2 min per 300 bases at 37 °C.
6. Inactivate the enzyme at 65 °C for 5 min and check a small portion of the reaction mixture on the agarose gel electrophoresis to confirm the ssDNA purity.
7. Purify the remaining mixture by MinElute kit to obtain ssDNAs. For microinjections, elute ssDNAs with 20 μL of 0.1× TE buffer. To increase the ssDNA recovery, preheat 0.1× TE buffer to 70 °C before use and incubate the column for 5 min before the spin down.
8. Measure the ssDNA concentration on NanoDrop. The concentration could be 100 ng/μL.
9. Preserve 5 μL aliquots at −80 °C until use.

3.5 Design and Preparation of Chimeric RNA:DNA Donorguide Molecules

The Donorguide concept has recently been described in a biorxiv preprint [16] as a molecular complex consisting of a crisprRNA (crRNA) and a modified tracrRNA that contains a covalently attached DNA oligonucleotide. Both components are synthetic and must be assembled into a Donorguide complex upon arrival. Functionally, the Donorguide acts somewhat analogously to the pegRNA in prime editing except that the intended modification can be copied from a DNA molecule and reverse transcription is therefore not required. The following is a protocol of how to design Donorguides and use them in a gene-editing experiment.

1. Design your precise mutational strategy and a short *Non-Target* symmetric oligonucleotide containing the modification. Trim the oligo to contain at least 25 nt with homology arms of at least 10–15 nt. It is possible to have longer homology arms (the authors tested up to 24 nt on each side) at a somewhat greater cost.
2. Order Alt-R crRNAs from IDT (www.idtdna.com/CRISPR-Cas9) and resuspend them at 20 μM in Duplex Buffer (30 mM HEPES, pH 7.5, 200 mM potassium acetate).

3. Resuspend chimeric CRISPR-Cas9 tracrRNAs at 20 μM in Duplex Buffer.
4. Combine equal volumes of the crRNA and the chimeric tracrRNA solutions, e.g., 10 μL each and heat the mix at 95 °C for 5 min and then let the sample cool to room temperature.
5. Dilute Cas9 protein in Cas9 working buffer (20 mM HEPES–KOH, 150 mM KCl, pH 7.5) to 6.1 μM (1 μg/μL). For maximum efficiency, use the wild-type Cas9, but the HiFi Cas9 can also be tested if it works with the guide RNA used in the experiment.
6. On the morning of the injection, mix 2 μL of Cas9 solution with 2 μL of the 10 μM Donorguide and incubate at 37 °C for 10 min to assemble the RNP complex at 3 μM. Add 0.4 μL of 0.5% Phenol Red solution and perform injections.

3.6 Reagent Preparation, Microinjection of Zebrafish Eggs and Sample Collection

In a previous protocol in this book series, we described in detail our procedure for producing sgRNAs and several versions of zebrafish codon-optimized Cas9 mRNA [21], including the high-fidelity ones, all of which contained vector-encoded poly-A tails (A71) [22]. Gagnon and colleagues provided one of the earliest protocols to synthesize sgRNAs, Cas9 mRNA, and to produce the Cas9 protein [9]. Many zebrafish labs have successful sgRNA and Cas9 mRNA synthesis protocols of their own.

1. Set up adult zebrafish in breeding tanks with dividers the day before the injection.
2. The next morning, prepare the injection mixes. Here, there are several options according to the reagents used and the knock-in method employed.
3. Prepare the mixture of Cas9 mRNA and sgRNAs:
   (a) 625 ng of nCas9n-A71 mRNA or 1500 ng of standard nCas9n-SV40 mRNA
   (b) 500 ng of sgRNA or crRNA:tracrRNA complex
   (c) 0.5 μL of 10 μM template oligo or 1 μL of 100 ng/μL lssDNA solution
   (d) Add water to 5 μL and 0.55 μL of 0.5% Phenol Red.
4. For Cas9 protein injections, mix the following:
   (a) 2.5 μL of 1 μg/μL Cas9 protein solution
   (b) 500 ng of sgRNA or crRNA:tracrRNA complex
   (c) 0.7 μL of 2 M KCl
   (d) water to 4.5 μL,
   (e) Incubate the mix at 37 °C for 5 min.

(f) Add 0.5 μL water or 0.5 μL of 10 μM template oligo or 100–200 ng/μL lssDNA.

(g) Finally, add 0.55 μL of 0.5% Phenol Red.

5. The same morning after injection mix preparation, breed the fish for 10–15 min and collect the eggs in E3 medium.
6. Array the one cell-stage fish eggs in troughs of an agarose injection plate and orient them in such a way that the cell of each egg faces the injection needle.
7. Inject zebrafish eggs with the injection mix. The bolus diameter should not exceed 20% of the egg diameter (*see* **Note 5**).
8. Grow the injected fish to at least 48 hours postfertilization (hpf) before genotyping.
9. Before starting with sample preparation, treat embryos (48–72 hpf) with 0.02% Tricaine in fish medium for several minutes.
10. For knock-in injected embryo genotyping, transfer them into PCR tubes and replace the fish medium with 40 μL of 50 mM NaOH.
11. Prepare eight single-embryo extracts from uninjected embryos (wild-type control) and 16 samples from each knock-in strategy injected embryo clutch.
12. Heat the samples at 95 °C for 10 min. Place the single-embryo samples into the thermocycler.
13. Vortex the single-embryo samples for 3 s.
14. Cool samples on ice for 5 min and neutralize with 4 μL of 1 M Tris–HCl, pH 8.0.

3.7 Detection of Precise and Imprecise Knock-in Edits

The approaches described in this chapter are aimed at introducing precise genomic modifications. Before proceeding with growing up any modified zebrafish, you should confirm that the strategy is successful and prove that the modification occurred in a sufficient proportion of embryos and cells of individual embryos. Typically, an efficiency of 5–10% in point mutation knock-ins is considered high, but we have recovered successful knock-in founders from strategies with 1–2% efficiency. We usually detect the knock-ins using allele-specific PCR. Efficiency determination should then be undertaken with deep sequencing. Restriction digestion of the sites introduced by the knock-in oligo may be a valid method in some cases but is much less sensitive, so we reserve it for screening zebrafish in the subsequent generations. The challenge with all these methods is that in addition to the wild-type and the correct knock-in modification, there will be many other indel and partially correct knock-in modifications (*see* **Note 6**).

3.7.1 Allele-Specific PCR

1. If you have not designed the primers focused on the target modification and the corresponding wild-type region, design them before proceeding as described in Subheadings 3.2 and 3.3. For an additional method of primer design when single-nucleotide precision is required, *see* **Note 7**.
2. Calculate the annealing temperature (T_{anneal}) of your PCR using the NEB Tm Calculator online tool (http://tmcalculator.neb.com/#!/) and/or determine the optimal temperature empirically using gradient PCR. We typically add 2 °C to the predicted temperature because it is usually much lower than predicted by other tools.
3. Use the touchdown PCR method: 94 °C for 3 min; 10 cycles (94 °C for 30 s, T_{anneal} + 10 [with 1 °C decrease every cycle] for 30 s, 72 °C for 30 s), 25 cycles (94 °C for 30 s, T_{anneal} for 30 s, 72 °C for 30 s). Use this PCR protocol for pooled samples and 8–16 single-embryo samples for knock-in strategies.
4. For knock-in strategies, run the wild-type assay on 4–8 uninjected (wild-type) embryos (PCR size control) and the knock-in assay on both 4–8 uninjected and 8–16 injected embryo samples. The knock-in assays are expected to fail or produce different nonspecific bands in uninjected samples that serve as controls for the knock-in PCR assay.
5. Run the resulting PCR products on a 2% agarose gel to assess whether amplification occurred. This step visualizes specificity of the knock-in assay, whether it worked and the apparent relative efficiencies of different strategies if >1 knock-in strategy is employed in parallel.

3.7.2 Preparation of Samples for Deep Sequencing

1. Collect pooled samples (pool single-embryo samples in Subheading 3.6) from all the knock-in strategies employed as well as a wild-type sample. In case of a need for statistical analysis, collect at least three samples for each strategy from independent injections.
2. Find out the read length from your deep sequencing provider and adjust your sequencing amplicon size (design new primers for a shorter amplicon) to make sure that the reads from the two sides have >50 nucleotides overlap. In some cases, you may want to introduce short random 8-nucleotide barcodes into your primers to increase multiplexing and decrease the number of samples, but this is optional.
3. Amplify the deep-sequencing amplicon from all samples with a proof-reading Q5 polymerase in 40-μL reactions.
4. Run all the PCR products on a 2% agarose gel and gel-extract them.

5. Submit the purified PCR products according to your sequencing provider's instructions.
6. Analyze the data. We do not provide the detailed instructions here since this is a conceptual protocol chapter, but we list the suitable resources in **Note 6**.

4 Notes

1. Any programmable nuclease can be used in principle if it is compatible with oligonucleotides as donor templates. Cas9 serves as the currently available most robust system.
2. Any polymerase and almost any DNA extraction method can be used at this step. It may also be advantageous to amplify a longer region if you want to perform knock-ins with a long ssDNA.
3. We recommend performing deep sequencing of knock-in samples generated using different donor templates regardless of apparent efficiency determined by other genotyping methods since deep sequencing allows one to determine the correct knock-in percentage.
4. Unfortunately, allele-specific primers to detect indel mutations are not always reliable and specific likely due to bulges of the target genomic DNA with primers. For example, wild-type specific primers may still work in a homozygous indel mutant sample or deletion-specific primers may still work in the wild-type sample. A recently published protocol describes a two-allele genotyping procedure where the size of the mutant product is different from the size of the wild-type product [23]. With appropriate controls, this should alleviate specificity concerns about AS-PCR. For established mutant lines, we also routinely use and recommend the heteroduplex mobility assay (HMA) for indel genotyping as we described previously [24] or a more advanced version of HMA named PRIMA that can distinguish mutations down to 1-bp insertion or deletion [25].
5. Many protocols give precise injection volumes such as "1 nL." It can be quite difficult to estimate such small volumes. Therefore, we prefer to inject different concentrations of active molecular ingredients such as Cas9/sgRNA mixes or morpholino oligonucleotides and use visual judgement to make sure that we do not inject too much. Embryo viability and developmental abnormalities following injection should also be used to guide the highest nontoxic dosages of the injected material.
6. Complexity of the mutations observed after CRISPR/Cas9-based knock-in requires application of specialized software to determine the mutational outcomes and, most importantly, the

precise knock-in rate. We recommend the following software tools for knock-in sequencing analysis: CRIS.py [26], CrispR-Variants [27], BATCH-GE [28].

7. AS-PCR knock-in primers for precise point mutations are typically sensitive and specific if there are at minimum two or three mismatches compared to the wild-type allele. However, single-nucleotide differences may be more challenging to genotype. PCR using SuperSelective primer assays may help in this situation [29].

References

1. Petri K, Zhang W, Ma J et al (2021) CRISPR prime editing with ribonucleoprotein complexes in zebrafish and primary human cells. Nat Biotechnol. https://doi.org/10.1038/s41587-021-00901-y
2. Qin W, Lu X, Liu Y et al (2018) Precise a•T to G•C base editing in the zebrafish genome. BMC Biol 16:139. https://doi.org/10.1186/s12915-018-0609-1
3. Zhao Y, Shang D, Ying R et al (2020) An optimized base editor with efficient C-to-T base editing in zebrafish. BMC Biol 18:190. https://doi.org/10.1186/s12915-020-00923-z
4. Rosello M, Vougny J, Czarny F et al (2021) Precise base editing for the in vivo study of developmental signaling and human pathologies in zebrafish. elife 10:e65552. https://doi.org/10.7554/eLife.65552
5. Zhang Y, Qin W, Lu X et al (2017) Programmable base editing of zebrafish genome using a modified CRISPR-Cas9 system. Nat Commun 8:118. https://doi.org/10.1038/s41467-017-00175-6
6. Armstrong GAB, Liao M, You Z et al (2016) Homology directed knockin of point mutations in the zebrafish tardbp and fus genes in ALS using the CRISPR/Cas9 system. PLoS One 11:1–10. https://doi.org/10.1371/journal.pone.0150188
7. Boel A, De Saffel H, Steyaert W et al (2018) CRISPR/Cas9-mediated homology-directed repair by ssODNs in zebrafish induces complex mutational patterns resulting from genomic integration of repair-template fragments. Dis Model Mech 11:dmm035352. https://doi.org/10.1242/dmm.035352
8. Prykhozhij SV, Fuller C, Steele SL et al (2018) Optimized knock-in of point mutations in zebrafish using CRISPR/Cas9. Nucleic Acids Res 46:e102. https://doi.org/10.1093/nar/gky512
9. Gagnon JA, Valen E, Thyme SB et al (2014) Efficient mutagenesis by Cas9 protein-mediated oligonucleotide insertion and large-scale assessment of single-guide RNAs. PLoS One 9:5–12. https://doi.org/10.1371/journal.pone.0098186
10. Farr GH III, Imani K, Pouv D, Maves L (2018) Functional testing of a human PBX3 variant in zebrafish reveals a potential modifier role in congenital heart defects. Dis Model Mech 11: dmm035972. https://doi.org/10.1242/dmm.035972
11. Tessadori F, Roessler HI, Savelberg SMC et al (2018) Effective CRISPR/Cas9-based nucleotide editing in zebrafish to model human genetic cardiovascular disorders. Dis Model Mech 11:dmm035469. https://doi.org/10.1242/dmm.035469
12. de Vrieze E, de Bruijn SE, Reurink J et al (2021) Efficient generation of knock-in zebrafish models for inherited disorders using crispr-cas9 ribonucleoprotein complexes. Int J Mol Sci 22:9429. https://doi.org/10.3390/ijms22179429
13. Bai H, Liu L, An K et al (2020) CRISPR/Cas9-mediated precise genome modification by a long ssDNA template in zebrafish. BMC Genomics 21:1–12. https://doi.org/10.1186/s12864-020-6493-4
14. Ranawakage DC, Okada K, Sugio K et al (2021) Efficient CRISPR-Cas9-mediated knock-in of composite tags in zebrafish using long ssDNA as a donor. Front Cell Dev Biol 8: 1–20. https://doi.org/10.3389/fcell.2020.598634
15. Aird EJ, Lovendahl KN, St. Martin A et al (2018) Increasing Cas9-mediated homology-directed repair efficiency through covalent tethering of DNA repair template. Commun Biol 1:54. https://doi.org/10.1038/s42003-018-0054-2

16. Simone BW, Lee HB, Daby CL, et al (2021) Chimeric RNA:DNA donorguide improves HDR in vitro and in vivo. bioRxiv https://doi.org/10.1101/2021.05.28.446234
17. Shola DTN, Yang C, Han C, et al (2021) Generation of mouse model (KI and CKO) via Easi-CRISPR BT. In: Singh SR, Hoffman RM, Singh A (eds) Mouse genetics: methods and protocols. Springer, New York, pp 1–27
18. Nakayama T, Grainger RM, Cha SW (2020) Simple embryo injection of long single-stranded donor templates with the CRISPR/Cas9 system leads to homology-directed repair in Xenopus tropicalis and Xenopus laevis. Genesis 58:e23366. https://doi.org/10.1002/dvg.23366
19. Prykhozhij SV, Rajan V, Ban K, Berman JN (2021) CRISPR knock-in designer: automatic oligonucleotide design software to introduce point mutations by gene editing methods. ReGEN Open 1:53–67. https://doi.org/10.1089/regen.2021.0025
20. Inoue YU, Morimoto Y, Yamada M et al (2021) An optimized preparation method for long ssDNA donors to facilitate quick knock-in mouse generation. Cell 10:1–15. https://doi.org/10.3390/cells10051076
21. Jao L, Wente SR, Chen W (2013) Efficient multiplex biallelic zebra fish genome editing using a CRISPR nuclease system. Proc Natl Acad Sci 110:1–6. https://doi.org/10.1073/pnas.1308335110
22. Prykhozhij S V, Cordeiro-Santanach A, Caceres L, Berman JN (2020) Genome editing in zebrafish using high-fidelity Cas9 nucleases: choosing the right nuclease for the task BT. In: Sioud M (ed) RNA interference and CRISPR technologies: technical advances and new therapeutic opportunities. Springer, New York, pp 385–405
23. Lin B, Sun J, Fraser IDC (2021) Single-tube genotyping for small insertion/deletion mutations: simultaneous identification of wild type, mutant and heterozygous alleles. Biol Methods Protoc 5:1–11. https://doi.org/10.1093/biomethods/bpaa007
24. Prykhozhij SV, Steele SL, Razaghi B, Berman JN (2017) A rapid and effective method for screening, sequencing and reporter verification of engineered frameshift mutations in zebrafish. Dis Model Mech 10:811–822. https://doi.org/10.1242/dmm.026765
25. Kakui H, Yamazaki M, Shimizu KK (2021) PRIMA: a rapid and cost-effective genotyping method to detect single-nucleotide differences using probe-induced heteroduplexes. Sci Rep 11:1–10. https://doi.org/10.1038/s41598-021-99641-x
26. Connelly JP, Pruett-Miller SM (2019) CRIS.Py: a versatile and high-throughput analysis program for CRISPR-based genome editing. Sci Rep 9:1–8. https://doi.org/10.1038/s41598-019-40896-w
27. Lindsay H, Burger A, Biyong B et al (2016) CrispRVariants charts the mutation spectrum of genome engineering experiments. Nat Biotechnol 34:701–702. https://doi.org/10.1038/nbt.3628
28. Boel A, Steyaert W, De Rocker N et al (2016) BATCH-GE: Batch analysis of next-generation sequencing data for genome editing assessment. Sci Rep 6:30330. https://doi.org/10.1038/srep30330
29. Touroutine D, Tanis JE (2020) A rapid, superselective method for detection of single nucleotide variants in caenorhabditis elegans. Genetics 216:343–352. https://doi.org/10.1534/genetics.120.303553

Chapter 20

Generation of Transgenic Fish Harboring CRISPR/Cas9-Mediated Somatic Mutations Via a tRNA-Based Multiplex sgRNA Expression

Tomoya Shiraki and Koichi Kawakami

Abstract

The controlled expression of Cas9 and/or sgRNA in transgenic zebrafish made it possible to knock out a gene in a spatially and/or temporally controlled manner. This transgenic approach can be more useful if multiple sgRNAs are efficiently expressed since we can improve the biallelic frame-shift mutation rate and circumvent the functional redundancy of genes and genetic compensation. We developed the tRNA-based system to express multiple functional sgRNAs from a single transcript in zebrafish and found that it is applicable to the transgenic expression of multiple sgRNAs. In this chapter, we describe a procedure for the generation of plasmids containing multiple sgRNAs flanked by tRNAs and a method to induce multiple CRISPR/Cas9-mediated genome modifications in transgenic zebrafish.

Key words CRISPR/Cas9, Genome editing, Somatic mutations, sgRNA, tRNA, Transgenic

1 Introduction

The clustered regularly interspaced short palindromic repeats (CRISPR)/Cas9 system is a bacterial adaptive immune system and has been applied for genome engineering in various organisms including zebrafish [1–7]. The CRISPR/Cas9 system has greatly accelerated genome engineering research owing to its simplicity, efficiency, and versatility. The CRISPR/Cas9 system consists of the endonuclease Cas9 and two short RNAs, crRNA and tracrRNA. In the synthetically reconstituted system, the two short RNAs can be fused into a single chimeric guide RNA (sgRNA) consisting of an ~20-nt sequence complementary to the target site and a scaffold sequence recognized by Cas9. Cas9 is directed to the specific genomic site by sgRNA and introduces double-strand DNA breaks (DSBs). Co-delivery of Cas9 and sgRNA into cells enables genome editing via DSB repair by either nonhomologous end-joining (NHEJ) or homology-directed repair (HDR) mechanism.

James F. Amatruda et al. (eds.), *Zebrafish: Methods and Protocols*, Methods in Molecular Biology, vol. 2707,
https://doi.org/10.1007/978-1-0716-3401-1_20,

In zebrafish, the CRISPR/Cas9 technology enables rapid generation of knockout fish lines by simply injecting an sgRNA and the Cas9 mRNA into fertilized eggs [6, 7]. Furthermore, transgenic fish that express sgRNAs under the control of the U6 promoter and the Cas9 gene under the control of a temporally and/or spatially controlled enhancer/promoter were generated to analyze gene function in a specific cell type and/or later developmental stages [8–10]. A possible drawback of this transgenic method is the incomplete disruption of target genes due to the mosaicism of the introduced mutations. CRISPR/Cas9-mediated mutagenesis relies on the NHEJ repair system that usually causes small insertions or deletions (indels), and thus theoretically two-thirds of them result in frame-shift, and thereby only four-ninths of the target sites are expected to become biallelic frame-shift (null) mutations. On the other hand, substantial cases of functional redundancy [11] and genetic compensation [12] (or transcriptional adaptation) have been reported in zebrafish. To circumvent these problems, it is required to efficiently express multiple sgRNAs targeting multiple sites in a gene or multiple genes in zebrafish.

RNA polymerase III (Pol III)-dependent U6 promoter has been used to express an sgRNA with defined start and end sites. One simple way to express multiple sgRNAs is to use multiple U6 promoters [9, 10]. However, the insertion of multiple promoters at one locus in transgenic fish may cause unexpected complications in gene expression (i.e., silencing). The other way to express multiple sgRNAs is to process a single transcript containing multiple sgRNAs into functional sgRNAs by a posttranscriptional processing mechanism such as Csy4, ribozymes, and tRNA-based system. The tRNA-based system utilizing the endogenous tRNA processing machinery (Fig. 1) was first developed in rice [13], and we found that it also works in zebrafish [14]. The tRNA-based system has the advantages of higher efficiency and lower toxicity compared to other methods, ribozymes and Csy4, respectively, in zebrafish. In this chapter, we provide a detailed protocol for the construction of plasmids containing tRNA-sgRNA tandem repeats and show examples of transgenic fish that have somatic mutations in genes responsible for dark pigment formation.

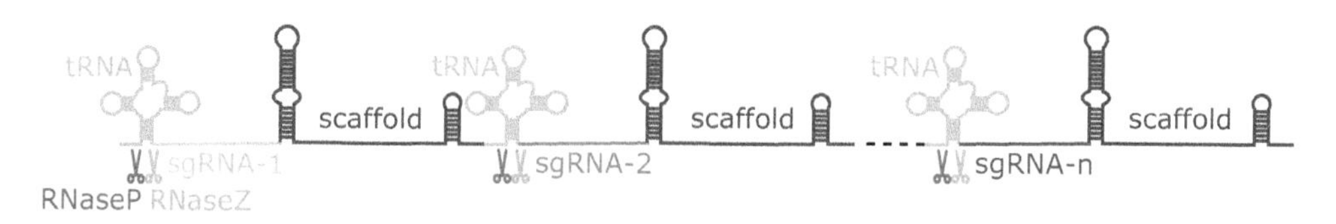

Fig. 1 The schematic diagram of the polycistronic tRNA-sgRNA system for multiplex genome editing. The polycistronic tRNA-sgRNA is cleaved by endogenous RNase P and RNase Z at specific sites, and each gRNA is eventually excised from the precursor transcript

2 Materials

2.1 Microinjection of sgRNAs, Cas9 mRNA, and Plasmids

1. Cas9 expression plasmid: pCS2 + hSpCas9 [15] (Addgene #51815).
2. *Tol2* expression plasmid: pCS-zT2TP.
3. *Not*I restriction enzyme.
4. mMessage mMachine SP6 kit (Thermo Fisher).
5. mini Quick Spin RNA Columns (Roche).
6. Forward primer to generate the DNA template for sgRNA synthesis: 5′-TAATACGACTCACTATA**GGNNNNNNNNNNNNNNNNNN**GTTTAAGAGCTATGC-3′. The nucleotides in bold letters represent the 20-nt protospacer sequence corresponding to target sites.
7. Reverse primer to generate the DNA template for sgRNA synthesis: 5′- AAAAGCACCGACTCGGTGCCACTTTTT-CAAGTTGATAACGGACTAGCCTTATTTAAACTTGCTATGCTGTTTCCAGCATAGCTCTTAAAC-3′. Universal for all gRNAs.
8. PrimeSTAR GXL DNA Polymerase (TaKaRa) or Kod One (TOYOBO).
9. QIAquick PCR purification kit (Qiagen).
10. MEGAshortscript T7 Transcription Kit (Thermo Fisher).
11. Nuclease-free water.
12. KCl.
13. Phenol Red.
14. Glass capillaries with filament GD-1 (Narishige).
15. Puller.
16. Microinjector.
17. Dissecting stereomicroscope.

2.2 Heteroduplex Mobility Assay (HMA)

1. Proteinase K (ProK).
2. Genome extraction buffer for PCR (10 mM Tris–HCl [pH 8.0], 0.2% Tween-20, 2 mM EDTA).
3. Heat block or hybridization oven set at 55 °C.
4. Ex Taq DNA polymerase (TaKaRa).
5. Gene-specific primers to amplify 100–200 bp fragments surrounding the targeted genomic locus by PCR.

2.2.1 Heteroduplex Mobility Assay with Polyacrylamide Gel Electrophoresis (PAGE)

1. Gel: SuperSep DNA, 15%, 17well (FUJIFILM Wako).
2. Electrophoresis device: EasySeparator (FUJIFILM Wako).
3. Running buffer (25 mM Tris–HCl, 192 mM glycine).
4. Ethidium bromide solution.
5. 100 bp DNA ladder marker.

2.2.2 Heteroduplex Mobility Assay with MultiNA (SHIMADZU)

1. MultiNA (SHIMADZU).
2. DNA-500 kit (SHIMADZU).
3. SYBR Gold (Thermo Fisher).
4. pUC19 DNA/MspI (HpaII) Marker (Thermo Fisher).

2.3 Construction of Plasmids Containing tRNA-sgRNA Tandem Repeats

1. pgRNA-Dr-tRNA Plasmids (Fig. 2a): Dr-tRNA^{Gly}(GCC), Dr-tRNA^{Asn}(GTT), Dr-tRNA^{Gln}(CTG), Dr-tRNA^{His}(GTG), Dr-tRNA^{Leu}(CAG), Dr-tRNA^{Met}(CAT), Dr-tRNA^{Ser}(GCT), Dr-tRNA^{Thr}(AGT), Dr-tRNA^{Lys}(CTT). Sequences of each tRNA are shown in Table 1.
2. Forward and reverse primers to amplify tRNA-flanked sgRNAs for construction. Primers are designed as shown in Table 2.
3. PrimeSTAR GXL DNA Polymerase (TaKaRa) or Kod One (TOYOBO).
4. In-Fusion HD Cloning Kit (TaKaRa).
5. pT2TS-ubb:Cas9;u6c:Dr-tRNAGly(GCC)-sgRNA-scaffold plasmid (Fig. 2b) or other all-in-one plasmids containing Cas9 and tRNA-sgRNA-scaffold driven by certain promoters.
6. *Dpn*I restriction enzyme.
7. Ampicillin.
8. LB medium (Bacto tryptone, Bacto yeast extract, and NaCl) and LB agar.
9. Primers for colony PCR and sanger sequencing. zU6c_seq Fw: 5′-TAAGCGTTTGCAGGTTTGCC-3′ and h-globin-pA_seq Fw: 5′-AAAGGGAATGTGGGAGGTCAGTG-3′ for the pT2TS-ubb:Cas9;u6c:Dr-tRNAGly(GCC)-sgRNA-scaffold plasmid.
10. QIAprep Spin Miniprep Kit (Qiagen).

3 Methods

3.1 Design and Preparation of sgRNAs

1. Find typically three sgRNAs with high activity for each target gene. Search for at least five sgRNA target sequences for each target locus using CRISPRscan (http://www.crisprscan.org/) [16]. Each target sequence consists of a 20-nt protospacer followed by the NGG, a 3-nt PAM sequence. Since T7 RNA

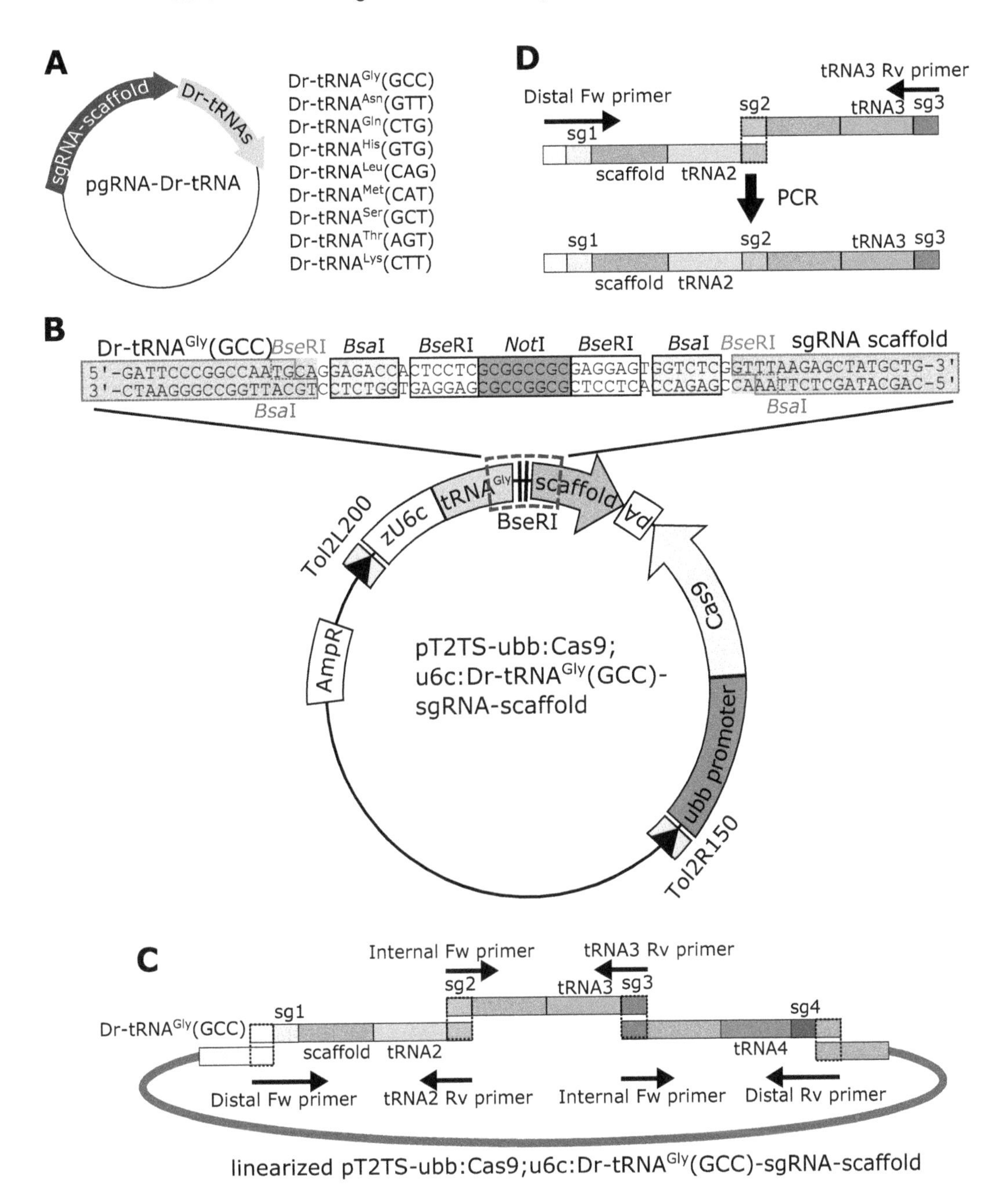

Fig. 2 A detailed method for the one-step cloning of multiple tRNA-sgRNAs into pT2TS-ubb:Cas9;u6c:Dr-tRNAGly(GCC)-sgRNA-scaffold. (**a**) The schematic illustration of pgRNA-Dr-tRNA plasmids used for the generation of plasmids containing tRNA-sgRNA tandem repeats. Zebrafish tRNA genes were artificially synthesized and inserted directly downstream of the sgRNA scaffold. (**b**) The schematic diagram of pT2TS-ubb:Cas9;u6c:Dr-tRNAGly(GCC)-sgRNA-scaffold. (**c**) The schematic diagram of one-step seamless cloning of four sgRNAs. (**d**) The schematic diagram of overlap extension PCR to fuse two fragments

Table 1
Sequences of tRNA genes

tRNA name	Sequence
Dr-tRNA^{Gly}(GCC)	gtgaGCATTGGTGGTTCAGTGGTAGAATTCTCGCCTGCCACGCGGGAGGCCCGGGTTCGATTCCCGGCCAATGCA
Dr-tRNA^{Lys}(CTT)	gttctcatcaGCCCGGCTAGCTCAGTCGGTAGAGCATGAGACTCTTAATCTCAGGGTCGTGGGTTCGAGCCCCACGTCGGGCG
Dr-tRNA^{Asn}(GTT)	gctatctGTCTCTGTGGCGCAATCGGTTAGCGCGTTCGGCTGTTAACCGAAAGGTTGGTGGTTCGAGCCCACCCAGGGACG
Dr-tRNA^{Met}(CAT)	gcctgaagGTTTCCGTAGTGTAGTGGTTATCACGTTCGCCTCATACGCGAAAGGTCCCCAGTTCGAAACTGGGCGGAAACA
Dr-tRNA^{Gln}(CTG)	gacttgaGGTTCCATGGTGTAATGGTTAGCACTCTGGACTCTGAATCCAGCGATCCGAGTTCAAATCTCGGTGGGACCA
Dr-tRNA^{Ser}(GCT)	ggaaaatGACGAGGTGGCCGAGTGGTTAAGGCGATGGACTGCTAATCCATTGTGCTTTGCACGCATGGGTTCGAATCCCATCCTCGTCG
Dr-tRNA^{Thr}(AGT)	gcagcGGCGCCGTGGCTTAGTTGGTTAAAGCGCCTGTCTAGTAAACAGGAGATCCTGGGTTCGAATCCCAGCGGTGCCT
Dr-tRNA^{His}(GTG)	gctcGCCGTGATCGTACAGTGGTTAGTACTCTGCGTTGTGGCCGCAGCAACCCCGGTTCGAATCCGGGTCACGGCA
Dr-tRNA^{Leu}(CAG)	gcatGTCAGGATGGCCGAGTGGTCTAAGGCGCTGCGTTCAGGTCGCAGTCTCCCCTGGAGGCGTGGGTTCGAATCCCACTTCTGACA

tRNA 5′ leader sequences are shown in lower case

Table 2
Primer sequences for amplification of sgRNA-tRNA fragments

Primer name	Sequence	Template vector
Distal Fw primer	GATTCCCGGCCAATGCANNNNNNNNNNNNNNNNNNNN**GTTTAAGAGCTATGCTGGAA**	
Internal Fw primer	NNNNNNNNNNNNNNNNNNNN**GTTTAAGAGCTATGCTGGAA**	
Asn Rv primer	NNNNNNNNNNNNNNNNNNNNCGTCCCTGGGTGGGCTCGAA	Dr-tRNA^{Asn}(GTT)
Met Rv primer	NNNNNNNNNNNNNNNNNNNNTGTTTCCGCCCAGTTTCG	Dr-tRNA^{Met}(CAT)
Ser Rv primer	NNNNNNNNNNNNNNNNNNNNCGACGAGGATGGGATTCGAAC	Dr-tRNA^{Ser}(GCT)
His Rv primer	NNNNNNNNNNNNNNNNNNNNTGCCGTGACCCGGATTCGAAC	Dr-tRNA^{His}(GTG)
Thr Rv primer	NNNNNNNNNNNNNNNNNNNNAGGCACCGCTGGGATTCGAAC	Dr-tRNA^{Thr}(AGT)
Gln Rv primer	NNNNNNNNNNNNNNNNNNNNTGGTCCCACCGAGATTTGAA	Dr-tRNA^{Gln}(CTG)
Leu Rv primer	NNNNNNNNNNNNNNNNNNNNTGTCAGAAGTGGGATTCGAAC	Dr-tRNA^{Leu}(CAG)
Lys Rv primer	NNNNNNNNNNNNNNNNNNNNCGCCCGACGTGGGGCTCGAA	Dr-tRNA^{Lys}(CTT)
Distal Rv primer	**CAGCATAGCTCTTAAAC**NNNNNNNNNNNNNNNNNNNNNNNNNNNNNNNNNNNNNNNN	Dr-tRNA X

The nucleotides in bold letters denote the sgRNA scaffold sequence and the underlined letters represent the tRNA sequences. 20-nt Ns represent sgRNA target sequence, and for Rv primers the sequences are converted into its reverse complement. The underlined 20-nt Ns of Distal Rv primer is the same as the underlined sequence of tRNA Rv primer for the corresponding template vector

polymerase requires the 5′-GG for efficient transcription, replace the 5′end of the target site with GG if the 5′end of the target site does not begin with GG (*see* **Note 1**).

2. Prepare DNA templates for sgRNA synthesis by fill-in PCR. Synthesize a 52-nt forward DNA oligo for each sgRNA with the sequence 5′- TAATACGACTCACTATA**GGNNNNNNNNNNNNNNNNNN**GTTTAAGAGCTATGC-3′. The forward oligo consists of the 17-nt T7 promoter, the 20-nt of the target sequence (shown in bold), and a constant 15-nt tail for annealing. Synthesize a 90-nt universal reverse primer with the sequence 5′- AAAAGCACCGACTCGGTGCCACTTTTTCAAGTTGATAACGGACTAGCCTTATTTAAACTTGCTATGCTGTTTCCAGCATAGCTCTTAAAC-3′. Here, we use a modified version of the sgRNA scaffold [17].
3. Generate the DNA template for sgRNA synthesis by fill-in PCR. PCR is performed with 5 μM of each primer concentration using PrimeSTAR GXL DNA Polymerase or Kod One. A 127 bp PCR product is generated using PrimeSTAR GXL DNA Polymerase following these parameters: 1 min at 98 °C, 30 cycles of 10 s at 98 °C, 30 s at 45 °C, and 15 s at 68 °C, and a final step at 68 °C for 7 min.
4. PCR products are purified using QIAquick PCR purification kit columns. From a 10-μL of PCR reaction mixture, 2–4 μg of DNA is collected. The length of the PCR products can be examined by agarose gel electrophoresis.
5. Synthesize sgRNAs from the 100–150 ng of DNA templates in the 10-μL reaction mixture by in vitro transcription using MEGAshortscript T7 Transcription Kit. Incubate the mixture at 37 °C for 2–4 h.
6. Add 1 μL of TURBO DNase I and incubate at 37 °C for 30 min to remove the DNA template. The synthesized sgRNAs are precipitated with ammonium acetate/ethanol and resuspended in 30 μL of nuclease-free water. Determine the concentration of sgRNA by a spectrophotometer (typically 300–1000 ng/μL) and dilute the sgRNA to 250 ng/μL (10× stock).

3.2 Microinjection of Cas9 mRNA and sgRNAs

1. Linearize 4 μg of pCS2 + hSpCas9 by *Not*I digestion. Examine the size of DNA fragments produced from the digestion by agarose gel electrophoresis.
2. Purify the linearized plasmid using a QIAquick PCR purification kit column.
3. Cas9 mRNA is transcribed from 1 μg of the linearized plasmid by using the mMESSAGE mMACHINE SP6 Transcription Kit according to the manufacturer's instruction.

4. Purify the Cas9 mRNA by using the mini Quick Spin RNA Columns and the following ethanol precipitation. Store at −80 °C as a 3× concentrated stock (600 ng/μL).
5. Prepare the injection needle using a puller and cut the tip of the needle with a razor.
6. Prepare 2–4 μL injection mixture: 25 ng/μL sgRNA, 200 ng/μL Cas9 mRNA, 0.2 M KCl, and 0.05% phenol red.
7. Inject 1–2 nL of the injection solution into 1-cell stage zebrafish embryos.
8. Incubate the embryos in E3 buffer at 28.5 °C.

3.3 Evaluation of the Efficiency of sgRNAs by the Heteroduplex Mobility Assay

We evaluated the efficiency of sgRNAs by the heteroduplex mobility assay, in which DNA heteroduplexes containing mismatches within a double-strand DNA migrate more slowly than homoduplexes.

1. For each target site, design a pair of primers for the amplification of the 100–200 bp genomic region containing the target site.
2. Genomic DNA is isolated from a mixture of three embryos at 24 h postfertilization (hpf) or one larva at 120 hpf. Add 100 μL of 50–100 μg/mL ProK in the genome extraction buffer for PCR.
3. Incubate at 55 °C for at least 2 h.
4. Incubate at 95 °C for 10 min to inactivate the ProK.
5. Amplify the DNA fragment surrounding the target site from 1 μL of the genomic DNA for 10 μL of the reaction mixture by PCR using ExTaq. The cycling condition is as follows: 94 °C for 2 min, followed by 35 or 40 cycles of 30 s at 94 °C, 30 s at 55 °C, and 30 s at 72 °C.
6. Add 2 μL of 6× loading dye to the PCR samples.
7. The resulting amplicons are analyzed by PAGE using SuperSep DNA gels (15% polyacrylamide gel). Alternatively, the PCR products are analyzed by MultiNA.

3.4 Construction of Plasmids Containing tRNA-sgRNA Tandem Repeats

1. For the construction of plasmids containing tRNA-sgRNA tandem repeats, we use nine plasmids each containing one of nine zebrafish endogenous tRNA genes (Dr-tRNA^{Gly}(GCC), Dr-tRNA^{Lys}(CTT), Dr-tRNA^{Asn}(GTT), Dr-tRNA^{Met}(CAT), Dr-tRNA^{Gln}(CTG), Dr-tRNA^{Ser}(GCT), Dr-tRNA^{Thr}(AGT), Dr-tRNA^{His}(GTG), and Dr-tRNA^{Leu}(CAG)) directly downstream of a modified sgRNA scaffold [17] (Fig. 2a). Sequences of each tRNA are shown in Table 1 (*see* **Note 2**).
2. Linearize the pT2TS-ubb:Cas9;u6c:Dr-tRNAGly(GCC)-sgRNA-scaffold vector (*see* **Note 3**) (Fig. 2b) by *Bse*RI. *Bse*RI

produces cohesive ends that allow seamless cloning of sgRNA sequences. Only *Bse*RI is available due to the existence of *Bsa*I sites within the *ubiquitin b* (*ubb*) promoter.

3. Prepare tRNA-sgRNA fragments by PCR amplification (Fig. 2c). Design primers according to Table 2. Internal primers contain ~20-nt sgRNA sequences used as homology arm for seamless ligation, while distal primers include 15-nt overlap with *Bse*RI-treated pT2TS-ubb:Cas9;u6c:Dr-tRNAGly (GCC)-sgRNA-scaffold. Each sgRNA fragment is amplified from the corresponding pgRNA-drtRNA vector by PCR using PrimeSTAR GXL DNA Polymerase or Kod One. PCR is performed with the concentration of 1 ng/μL plasmid template using PrimeSTAR GXL following these parameters: 1 min at 98 °C, 20 cycles of 10 s at 98 °C, 30 s at 55 °C, and 20 s at 68 °C.
4. Run the PCR products on 2% agarose gel. The expected sizes are ~200 bp.
5. Purify the PCR products using a QIAquick PCR purification kit column.
6. (Optional) For cloning more than four sgRNAs (three fragments), we fuse fragments by overlap extension PCR (Fig. 2d) to improve the efficiency of following seamless cloning. Run PCR with the concentration of 5 ng/μL each fragment without primers using PrimeSTAR GXL following these parameters: 1 min at 98 °C, 15 cycles of 10 s at 98 °C, 30 s at 55 °C, and 20 s at 68 °C. Add primer pair (final concentration of 0.3 μM each) and run PCR following these parameters: 20 cycles of 10 s at 98 °C, 30 s at 55 °C, and 30 s at 68 °C. Examine the size of PCR products on 2% agarose gel. Purify the PCR products using a QIAquick PCR purification kit column.
7. Digest the PCR products with *Dpn*I to remove the template plasmid DNA.
8. The multiple tRNA-sgRNA fragments are cloned into the linearized pT2TS-ubb:Cas9;u6c:Dr-tRNAGly(GCC)-sgRNA-scaffold by seamless cloning using In-Fusion (*see* **Notes 4** and **5**). An example of the case of four sgRNAs is shown in Fig. 2c. Prepare the In-Fusion mixture: 2 ng each of ~200-bp tRNA-sgRNA fragments, 4 ng each of ~400-bp tRNA-sgRNA fragments, and ~ 50 ng of vector backbone in 2.5 μL reaction volume.
9. Incubate at 50 °C for 15 min.
10. Transform the mixture into chemically competent cells and plate cells on Ampicillin plates.

11. (Optional) Select clones with correct inserts by colony PCR using ExTaq. Perform colony PCR using the primers below: zU6c_seq Fw: 5′-TAAGCGTTTGCAGGTTTGCC-3′, h-globin-pA_seq Fw: 5′-AAAGGGAATGTGGGAGGTCA GTG-3′
12. Purify the recombinant plasmid DNA from bacterial cultures using the QIAprep Spin Miniprep Kit according to the manufacturer's instruction.
13. Verify the recombinant colonies by Sanger sequencing with zU6c_seq Fw and h-globin-pA_seq primers.

3.5 Generation of Transgenic Fish and Detection of Somatic Mutations

1. Linearize 2 μg of pCS-zT2TP by *Not*I digestion. Examine the size of DNA fragments produced from the digestion by agarose gel electrophoresis.
2. Purify the linearized plasmid using a QIAquick PCR purification kit column.
3. *Tol2* mRNA is transcribed from 1 μg of the linearized plasmid by using the mMESSAGE mMACHINE SP6 Transcription Kit according to the manufacturer's instruction.
4. Purify the *Tol2* mRNA by using the mini Quick Spin RNA Columns and the following ethanol precipitation. Store at −80 °C as a 10x concentrated stock (250 ng/μL).
5. Prepare the injection needle using a puller and cut the tip of the needle with a razor.
6. Prepare 2–4 μL injection mixture: 25 ng/μL plasmid, 25 ng/μL *Tol2* mRNA, 0.2 M KCl, and 0.05% phenol red.
7. Inject 1–2 nL of the injection solution into 1-cell stage zebrafish embryos.
8. Raise F0 founders to adult fish.
9. Cross the potential F0 founder with wild-type fish.
10. Genomic DNA is isolated from a mixture of ~10 embryos at 1 day postfertilization (dpf). Add 100 μL of 50–100 μg/mL ProK in the genome extraction buffer for PCR. Prepare genomic DNA for PCR as described above.
11. Design primer pairs to amplify the DNA fragment specifically derived from the transgene such as Cas9 (Fig. 3a). Amplify the DNA fragment by PCR using ExTaq. The cycling condition is as follows: 94 °C for 2 min, followed by 35 or 40 cycles of 30 s at 94 °C, 30 s at 55 °C, and 30 s at 72 °C.
12. Examine the somatic mutation by HMA as described above (Fig. 3b) and analyze the phenotype of the transgenic line (Fig. 3c).

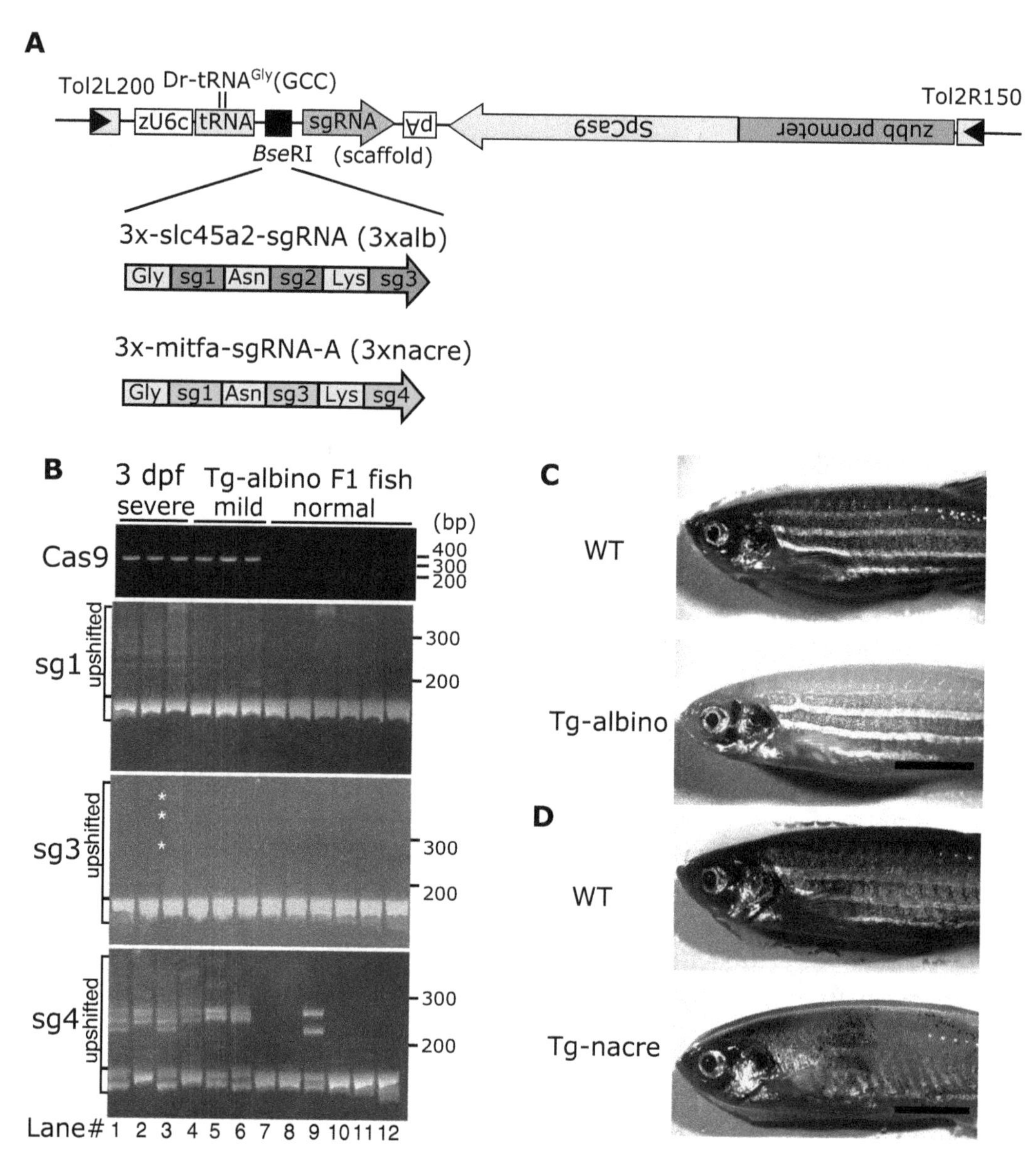

Fig. 3 Generation of transgenic *albino* and *nacre* zebrafish by using tRNA-based multiplexed sgRNA expression. (**a**) Structure of DNA construct used for the generation of transgenic *albino* and *nacre* zebrafish. Three distinct sgRNAs targeting the *albino* (*slc45a2*) locus or the *nacre* (*mitfa*) locus were cloned at the *Bse*RI site of an all-in-one Tol2 vector, pT2TS-ubb:Cas9;u6c:Dr-tRNAGly(GCC)-sgRNA-scaffold, and are transcribed as a single polycistronic RNA under the zebrafish u6c promoter. Cas9 is driven by the ubb promoter and its direction is opposite to that of the u6c:sgRNAs. (**b**) Detection of mutations at multiple genomic sites of the slc45a2 locus in transgenic-albino larvae. (Top) Genotyping of transgenic fish was performed by PCR using primers on SpCas9 genes. (Bottom three panels) Detection of mutations at multiple genomic sites of the slc45a2 locus by heteroduplex mobility assay (HMA). Mutations were introduced in all the three targeting sites in a transgenic larva (lane#3) that showed severe albino phenotype. White asterisks on the HMA gel image of sg3 denote faint heteroduplex bands. (Adapted from [14]) (**c**) Representative images of the lateral views of adult wild-type (WT; upper) and transgenic-albino (Tg-albino; lower) fish. (Adapted from [14]) (**d**) Representative images of the lateral views of adult wild-type (WT; upper) and transgenic-nacre (Tg-nacre; lower) fish. Scale bars: 5 mm

4 Notes

1. Wild-type Cas9 can tolerate single base mismatches between the genome and the 5′end of the sgRNA [18, 19]. SP6 RNA polymerase efficiently transcribes the sgRNA starting with 5′-GG or 5′-GA [19]. If SP6 RNA polymerase is used for transcription, replace the T7 promoter sequence (5′--TAATACGACTCACTATA-3′) of the forward primer to produce the DNA template with the SP6 promoter sequence (5′-ATTTAGGTGACACTATA-3′). Forward primer for SP6 is 5′-ATTTAGGTGACACTATA**GRNNNNNNNNNNNNN NNNNNN**GTTTAAGAGCTATGC-3′. The nucleotides in bold letters represent the 20-nt protospacer sequence corresponding to target sites. Note that some high-fidelity variants of Cas9 require a perfect match even in the 5′end of the sgRNA for high editing efficiency.
2. We use zebrafish tRNA genes that have neither the *Bsa*I nor *Bse*RI site, both used for sgRNA cloning, and that do not possess TTTT, which is the termination signal for RNA polymerase III (Pol III).
3. The pT2TS-ubb:Cas9;u6c:Dr-tRNAGly(GCC)-sgRNA-scaffold vector contains the ubb promoter and the U6 promoter to drive the ubiquitous expression of Cas9 and tRNA-sgRNA, respectively. The ubb and U6 promoters can be changed to appropriate Pol-II promoters for your research purposes. Note that tRNA itself has Pol-III promoter activity and the tRNA-sgRNA fragment is constitutively expressed even without promoters [20]. Using hammerhead (HH) ribozyme that has self-cleaving activity [21], upstream of the first gRNA may help to circumvent the endogenous promoter activity of tRNAs.
4. Other seamless cloning methods can also be used according to the manufacturer's instructions. Primers should be designed with the appropriate homology length for each method.
5. Alternatively, the tRNA-gRNA fragment can be artificially synthesized according to the sequences in Tables 1 and 2.

Acknowledgments

This work was supported by grants from the Ministry of Education, Culture, Sports, Science and Technology (MEXT) in Japan: KAKENHI Grant Number 19K16196 and 21H02463.

References

1. Jinek M, Chylinski K, Fonfara I, Hauer M, Doudna JA, Charpentier E (2012) A programmable dual-RNA-guided DNA endonuclease in adaptive bacterial immunity. Science 337: 816–821
2. Cong L, Ran F, Cox D, Lin S, Barretto R, Habib N, Hsu P, Wu X, Jiang W, Marraffini L et al (2013) Multiplex genome engineering using CRISPR/Cas systems. Science 339: 819–822
3. Mali P, Yang L, Esvelt KM, Aach J, Guell M, DiCarlo JE, Norville JE, Church GM (2013) RNA-guided human genome engineering via Cas9. Science 339:823–826
4. Hsu PD, Lander ES, Zhang F (2014) Development and applications of CRISPR-Cas9 for genome engineering. Cell 157:1262–1278
5. Doudna JA, Charpentier E (2014) The new frontier of genome engineering with CRISPR-Cas9. Science 346:1258096
6. Chang N, Sun C, Gao L, Zhu D, Xu X, Zhu X, Xiong JW, Xi JJ (2013) Genome editing with RNA-guided Cas9 nuclease in Zebrafish embryos. Cell Res 23:465–472
7. Hwang WY, Fu Y, Reyon D, Maeder ML, Tsai SQ, Sander JD, Peterson RT, Yeh JRJ, Joung JK (2013) Efficient genome editing in zebrafish using a CRISPR-Cas system. Nat Biotechnol 31:227–229
8. Ablain J, Durand EM, Yang S, Zhou Y, Zon LI (2015) A CRISPR/Cas9 vector system for tissue-specific gene disruption in zebrafish. Dev Cell 32:756–764
9. Yin L, Maddison LA, Li M, Kara N, Lafave MC, Varshney GK, Burgess SM, Patton JG, Chen W (2015) Multiplex conditional mutagenesis using transgenic expression of Cas9 and sgRNAs. Genetics 200:431–441
10. Di Donato V, De Santis F, Auer TO, Testa N, Sánchez-Iranzo H, Mercader N, Concordet JP, Del BF (2016) 2C-Cas9: a versatile tool for clonal analysis of gene function. Genome Res 26:681–692
11. Howe K, Clark MD, Torroja CF, Torrance J, Berthelot C, Muffato M, Collins JJEJE, Humphray S, McLaren K, Matthews L et al (2013) The zebrafish reference genome sequence and its relationship to the human genome. Nature 496:498–503
12. Rossi A, Kontarakis Z, Gerri C, Nolte H, Hölper S, Krüger M, Stainier DYR (2015) Genetic compensation induced by deleterious mutations but not gene knockdowns. Nature 524:230–233
13. Xie K, Minkenberg B, Yang Y (2015) Boosting CRISPR/Cas9 multiplex editing capability with the endogenous tRNA-processing system. Proc Natl Acad Sci U S A 112:3570–3575
14. Shiraki T, Kawakami K (2018) A tRNA-based multiplex sgRNA expression system in zebrafish and its application to generation of transgenic albino fish. Sci Rep 8:1–14
15. Ansai S, Kinoshita M (2014) Targeted mutagenesis using CRISPR/Cas system in medaka. Biol Open 3:362–371
16. Moreno-Mateos MA, Vejnar CE, Beaudoin JD, Fernandez JP, Mis EK, Khokha MK, Giraldez AJ (2015) CRISPRscan: designing highly efficient sgRNAs for CRISPR-Cas9 targeting in vivo. Nat Methods 12(10):982–988
17. Chen B, Gilbert LA, Cimini BA, Schnitzbauer J, Zhang W, Li GW, Park J, Blackburn EH, Weissman JS, Qi LS et al (2013) Dynamic imaging of genomic loci in living human cells by an optimized CRISPR/Cas system. Cell 155:1479–1491
18. Hwang WY, Fu Y, Reyon D, Maeder ML, Kaini P, Sander JD, Joung JK, Peterson RT, Yeh JRJ (2013) Heritable and precise zebrafish genome editing using a CRISPR-Cas system. PLoS One 8:1–9
19. Gagnon JA, Valen E, Thyme SB, Huang P, Ahkmetova L, Pauli A, Montague TG, Zimmerman S, Richter C, Schier AF (2014) Efficient mutagenesis by Cas9 protein-mediated oligonucleotide insertion and large-scale assessment of single-guide RNAs. PLoS One 9:5–12
20. Knapp DJHF, Michaels YS, Jamilly M, Ferry QRV, Barbosa H, Milne TA, Fulga TA (2019) Decoupling tRNA promoter and processing activities enables specific Pol-II Cas9 guide RNA expression. Nat Commun 10:1490
21. Lee RTH, Ng ASM, Ingham PW (2016) Ribozyme mediated gRNA Generation for in vitro and in vivo CRISPR/Cas9 mutagenesis. PLoS One 11:1–12

Chapter 21

Scalable CRISPR Screens in Zebrafish Using MIC-Drop

Saba Parvez, Tejia Zhang, and Randall T. Peterson

Abstract

CRISPR-Cas9 is a powerful tool to interrogate gene function in a targeted and systematic manner. Although the technology has been scaled up for large-scale genetic screens in cell culture, similar scale screens in vivo have been extremely challenging due to the cost, labor, and time required to generate and keep track of thousands of mutant animals. We reported the development of Multiplexed Intermixed CRISPR Droplets (MIC-Drop), a platform that makes large-scale reverse genetic screens possible in zebrafish. In this chapter, we provide a detailed protocol to conduct large-scale genetic screens using this novel platform.

Key words CRISPR-Cas9, Zebrafish, Genetic screen, Reverse genetics, High-throughput, F0 CRISPR screen, In vivo

1 Introduction

The zebrafish has proven to be a powerful model system for large-scale genetic screens [1–9]. However, most large-scale genetic screens in zebrafish are done using forward genetic methods such as ENU-based mutagenesis [1, 3] or insertional mutagenesis [5, 6]. Although high-throughput, the mutations generated through forward genetics approaches are random and therefore require huge efforts to identify the causative mutation(s) behind an observed phenotype. The advent of reverse genetic techniques, such as TALENs [10, 11], ZFNs [12], and CRISPR-Cas9 [13], has enabled targeted genetic perturbation but thus far has been limited in throughput. Traditionally, genetic screens using reverse genetic techniques have been done by brute-force scaling up of single gene knockout approaches, where one gene is targeted from a single injection needle and the F0 mosaic animal are genotyped and crossed for several generations to make the homozygous loss-of-function mutants [7, 8, 14–16]. This multistep process makes large-scale reverse genetic screens in zebrafish cumbersome and expensive. Recent developments using CRISPR-Cas9 has shown

James F. Amatruda et al. (eds.), *Zebrafish: Methods and Protocols*, Methods in Molecular Biology, vol. 2707,
https://doi.org/10.1007/978-1-0716-3401-1_21,

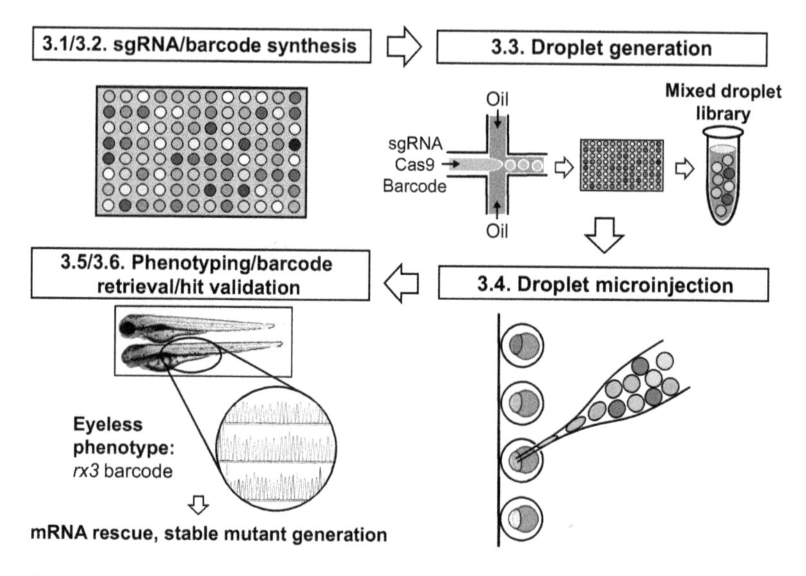

Fig. 1 MIC-Drop workflow

promise in enabling large-scale targeted genetic screens in zebrafish. Several labs have shown that injecting ribonucleoprotein (RNP) containing Cas9 and multiplexed sgRNAs in zebrafish embryos generates biallelic mutation, and the resulting F0 mutants recapitulate known homozygous loss-of-function mutant phenotypes, thus opening the doors to scalable F0 CRISPR screens [9, 17–21]. However, despite this advancement, large-scale screens still need to be performed by targeting one gene from a single injection needle, thus providing limited economy of scale.

We have developed Multiplexed Intermixed CRISPR Droplets (MIC-Drop), one of the first scalable platforms to enable large-scale and targeted genetic perturbation screens in whole animals (Fig. 1) [22]. MIC-Drop combines multiplexed sgRNAs, microfluidics, and DNA barcoding to obviate the issues faced by traditional reverse genetic screens. Multiplexed sgRNAs ensure efficient biallelic mutation in the injected embryos, allowing phenotyping in F0 fish. The use of microfluidics enables generation of microdroplets containing RNP complexes targeting many genes. Droplets targeting tens to hundreds of different genes are intermixed and injected into embryos from a single microinjection needle, with one droplet injected per embryo. Finally, the use of a unique DNA barcode associated with each gene facilitates raising the injected embryos en masse. Only embryos that display a desired phenotype are isolated, the DNA barcode from them recovered and sequenced to rapidly identify the causative genetic perturbation. Here, we provide a step-by-step protocol for conducting large-scale CRISPR screens using the MIC-Drop platform.

2 Materials

2.1 Zebrafish Husbandry and Maintenance

1. Wild-type adult zebrafish of desired genetic background. We have used Tu/TuAB strains for our screens; however, other strains can also be used. The GRCz10/11 reference genome sequences are based on the Tu strain. Please be aware that polymorphisms across different strains may affect gene editing efficiency.
2. Mating tanks.
3. E3 medium (5 mM NaCl, 0.15 mM KCl, 0.33 mM $CaCl_2$, and 0.33 mM $MgSO_4$) with methylene blue (0.5 mg/L).
4. Tea strainer.
5. Petri dish.
6. Disposable transfer pipettes.
7. 28.5 °C incubator.

2.2 MIC-Drop Library Generation

1. Gene-specific oligos targeting gene(s)-of-interest (Integrated DNA Technologies (IDT)).
2. Constant region oligo with sequence 5′-AAAAGCACCGACTCGGTGCCACTTTTTCAAGTTGATAACGGACTAGCCTTATTTTAACTTGCTATTTCTAGCTCTAAAAC-3′ (IDT).
3. 96-well PCR plates and adhesive PCR sealing plates (ThermoFisher Scientific).
4. PCR strip tubes and caps (ThermoFisher Scientific).
5. Multichannel pipettes and tips (Rainin).
6. Phusion HS Flex DNA polymerase and 5x HF buffer (New England Biolabs).
7. 100 mM dNTP set (Invitrogen).
8. DMSO (Fisher Scientific).
9. Nuclease- free water (Ambion).
10. Thermocycler (BioRad).
11. GeneJet PCR purification kit (ThermoFisher Scientific).
12. ZR-96 DNA Clean and Concentrator-5 (Zymo Research).
13. Agarose (Fisher Scientific).
14. Ethidium bromide, 10 mg/mL (Invitrogen).
15. 1x TAE buffer (40 mM Tris base, 20 mM acetic acid, 1 mM EDTA).
16. Standard gel electrophoresis and imaging system.
17. Nanodrop spectrophotometer (ThermoFisher Scientific).
18. MegaScript SP6 in vitro transcription kit (Invitrogen).

19. Ribolock RNAse inhibitor (40 U/μL) (ThermoFisher Scientific).
20. RNase Away (ThermoFisher Scientific).
21. RNA Clean and Concentrator-5 (Zymo Research).
22. ZR-96 RNA Clean and Concentrator (Zymo Research).
23. 0.5% Sterile Phenol Red solution (Sigma-Aldrich).
24. EnGen Spy Cas9 NLS and 10× NEBuffer r3.1 (New England Biolabs).
25. Barcode forward primer (/5BiosG/CGTAATACGACTCAC TATAGGGCTTCAGCCAAGGAAGCTACATTTAGGTGAC ACTATAGn) (IDT), where /5BiosG/ refers to 5′ biotin modification of the oligo.
26. Barcode reverse primer (/5BiosG/GCTAGTTATTGCTCAG CGGGTCTTGTTTCTCGGTGTGCTTGCTATTTCTAGC TCTAAAAC) (IDT), where /5BiosG/ refers to 5′ biotin modification of the oligo.
27. Novec HFE-7500 Engineered Fluid (3 M).
28. 008-Surfactant-1 g (Ran Biotechnologies).
29. QX200 Droplet generator and cartridge holder (BioRad).
30. DG8 cartridges and gaskets (BioRad).

2.3 MIC-Drop Library Injection, Phenotyping, and Validation

1. Glass Capillaries (World Precision Instruments).
2. Flaming/Brown Micropipette puller (Sutter Instrument).
3. Microloader tips (Eppendorf).
4. Picoliter injector (Harvard Apparatus) with micromanipulator (Narishige). (Other injector may also be used; however, the pressure settings may vary from what is specified in this manuscript).
5. Microinjection molds for arranging and injecting embryos.
6. Stereo microscope for injection and phenotyping.
7. Dumont Tweezer, style 5 (Electron Microscopy Sciences).
8. Proteinase K, recombinant PCR grade (Roche).
9. 2x PCR lysis buffer (20 mM Tris (pH 8), 4 mM EDTA, 0.4% Triton X-100).
10. 100 mM dUTP solution (ThermoFisher Scientific).
11. Uracil DNA glycosylase (UDG) (New England Biolabs).
12. GoTaq DNA polymerase and 5× buffer (Promega).
13. Exonuclease I (New England Biolabs).
14. Shrimp alkaline phosphatase (New England Biolabs).

15. Barcode amplification forward primer (5′-GTGTAAAACGAC GGCCAGTATGGCACCAACTCGATGACGTAATACGACT CACTATAGGGC-3′) (IDT).
16. Barcode amplification reverse primer (5′- CAGGAAACAGCT ATGACATAGTCCTGCTGTACCAGGCGTCTGCTAGTTA TTGCTCAGCGG-3′) (IDT).

3 Methods

3.1 sgRNA Design and Synthesis

Several online tools are available for designing sgRNAs for CRISPR-Cas9-mediated gene knockout in zebrafish [23–25]. We have primarily used CHOPCHOP, for which we explain the sgRNA generation steps below.

1. Access the CHOPCHOP website at https://chopchop.cbu.uib.no/. Enter the Refseq/ENSEMBL/gene ID or genomic coordinates of the gene "Target". Search for targets "In" Danio rerio (GRCz10/GRCz11 reference database), "Using" CRISPR/Cas9, "For" gene knockout.
2. Under "Options" → "Cas9", change the "5′ requirements for sgRNA" to GN or NG. sgRNAs that start with GN are best for transcription with SP6 polymerase. Avoid adding supernumerary "GN" to sgRNA as they reduce target editing efficiency. Keep remaining settings at default. Click "Find Target Sites!". Depending on how busy the server is, it may take a few seconds to several minutes to get the sgRNA target sequences.
3. Use the following criteria to pick four sgRNAs. If the gene-of-interest has ohnologs, pick two sgRNAs per ohnolog:
   (a) Non-overlapping sgRNAs of length 19 or 20 bases that start with GN.
   (b) sgRNAs that target exons coding for the first 50% of the gene. Targeting only the early exons may still result in functional proteins from alternative translation start sites. Alternatively, one can also pick sgRNAs that target known functional domains.
   (c) sgRNAs with high efficiency (>40).
   (d) sgRNAs with GC content between 45 and 75%.
   (e) sgRNAs with zero off-targets. If sgRNA with zero off-targets cannot be designed, pick sgRNAs with three mismatches ("3 MM") with the off-target site of which at least one mismatch is in the seed region (10 nucleotides 5′ to the PAM site).
   (f) sgRNAs with low self-complementarity (<3).
   (g) sgRNAs without poly Ts in the sequence.

4. Copy the sgRNA spacer sequences (without PAM). Add an SP6 polymerase sequence (ATTTAGGTGACACTATA) to the 5′ of the sgRNA sequences and 20 nucleotides of the sgRNA backbone sequence to the 3′ end to generate the gene-specific oligos. The oligos should have the following sequence: ATTTAGGTGACACTATA $GN_{18/19}$ GTTTTA-GAGCTAGAAATAGC, where $GN_{18/19}$ is the spacer sequence.
5. Design gene-specific oligos for the desired genes. Order the gene-specific oligos and the constant region oligo. When targeting only a few genes, we order the gene-specific oligos at 25 nmol scale from IDT. For large-scale orders, we order the oligos at 500 pmol scale in 96-well plates. The HPLC-purified constant region oligo is ordered in bulk.
6. Once received, resuspend the oligos in nuclease-free water to 100 μM. Oligos at 500 pmol scale are dissolved to 10 μM by adding 50 μL nuclease-free water to the wells and pipetting to resuspend.
7. Generate the DNA templates individually (if DNA templates will be used for barcode generation) or in a pooled format. We use a fill-in PCR using the gene-specific oligos and the constant region oligo to generate the DNA templates [26].
   (a) If doing individual synthesis, make a 50 μL PCR reaction mix containing the following: 2 μM gene-specific oligo, 2 μM constant region oligo, 1× HF buffer, 200 μM dNTPs, 3% DMSO (v/v), and 1 U of Phusion HS Flex DNA polymerase.
   (b) If doing pooled synthesis, make a 50 μL PCR reaction mix containing the following: 0.5 μM of each of the four gene-specific oligos, 2 μM constant region oligo, 1× HF buffer, 200 μM dNTPs, 3% DMSO (v/v), and 1 U of Phusion HS Flex DNA polymerase.
8. Perform a fill-in PCR in a thermocycler at the following settings: 98 °C for 2 min, 50 °C for 10 min, 72 °C for 10 min.
9. PCR cleanup the samples using a PCR cleanup kit as per manufacturer's instructions. When handling large number of samples, we use a ZR-96 DNA Clean and Concentrator-5 kit. Elute the PCR products in 15–20 μL of nuclease-free water. Typical DNA template yield is 100–200 ng/μL.
10. If DNA templates were generated individually, pool template DNA targeting the same gene/ohnologs in an equimolar ratio in a separate PCR tube/plate. Since the concentrations are typically similar after DNA synthesis, equal volumes of each template can be combined.

11. Set up in vitro transcription reaction to generate sgRNAs from the DNA templates. A typical 10 μL reaction contains the following: 5 mM ATP, 5 mM CTP, 5 mM GTP, 5 mM UTP, 1× reaction buffer, 1 U of RNAse inhibitor, 1× SP6 enzyme mix, and 20 ng/μL pooled DNA templates. Mix and incubate the reaction at 37 °C overnight (~16–18 h).
12. Add 10 μL nuclease-free water to bring the sample volume to 20 μL. Add 1 μL Turbo DnaseI to the samples. Mix and incubate for 15 min at 37 °C.
13. Clean up sgRNAs using RNA cleanup kits. For large number of genes, we use a ZR-96 RNA Clean and Concentrator-5 kit. Follow manufacturer's instructions. Elute sgRNAs in 10–15 μL nuclease-free water.
14. Measure sgRNA concentration using Nanodrop. To get a more accurate reading, we dilute 1 μL of the sgRNAs 10 folds prior to concentration measurement. Typical sgRNA yield is 1000–3000 ng/μL.
15. Check integrity of a few randomly sampled sgRNAs by gel electrophoresis. To eliminate secondary RNA structure and to ensure a single band, heat ~500–800 ng of the sgRNAs at 70 °C for 5 min. Rapidly cool the sample on ice. Resolve the sample on a 2% agarose gel.

3.2 Barcode Generation

The spacer sequences in the DNA template used for in vitro transcription are unique to the gene being targeted and can therefore be used as barcodes. We generate the DNA barcode by extending any one of the DNA templates (generated in **step 9** of Subheading 3.1) and by adding universal primer sequences for amplification and unambiguous sequencing (Fig. 2). Once generated, the same set of barcodes can be used several times.

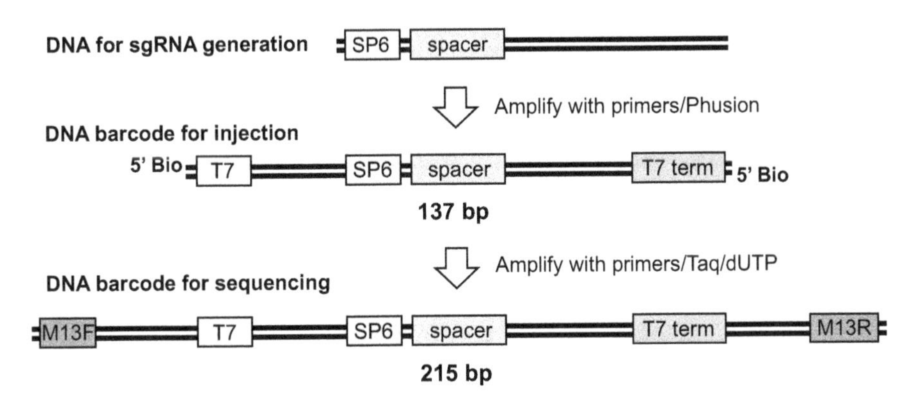

Fig. 2 DNA barcode generation. DNA barcode for injection was PCR-amplified from one of the gene-specific DNA templates used in sgRNA synthesis; addition of 5′biotinylation allows barcode enrichment in case of degradation. For sequencing, an additional PCR amplification was performed on the injection barcode to attach M13F and M13R sequencing primer sites

1. Set up a 50 μL PCR reaction containing the following: 0.5 μM Barcode forward primer, 0.5 μM Barcode reverse primer, 1x HF buffer, 200 μM dNTPs, 3% (v/v) DMSO, 1 U Phusion HS Flex DNA polymerase, and 1 ng/μL template DNA.
2. Amplify the barcode in a thermal cycler at the following settings: 98 °C for 30 s, [98 °C for 10 s, 53 °C for 30 s, 72 °C for 10 s] × 30 cycles, 72 °C for 2 min.
3. Clean up barcodes using a PCR cleanup kit. For a large number of samples, we use a ZR-96 DNA Clean and Concentrator-5 kit. Elute barcodes in 25–30 μL nuclease-free water. Measure DNA concentration using a Nanodrop and dilute to 100 ng/μL, if necessary.

3.3 Droplet Generation

1. Make a 3% Fluorosurfactant-HFE (3% FS-HFE hereafter) solution by dissolving 1 g of 008-fluorosurfactant in 33 mL of Novec HFE-7500.
2. Make the RNP mix in a total volume of 19 μL by combining 5000 ng of pooled sgRNAs, 4.2 μL of 20 μM EnGen Spy Cas9 NLS, and 2.5 μL of 10× r3.1 buffer. Mix by pipetting and incubate the samples at room temperature for 5–10 min.
3. Add 2.5 μL of the 100 ng/μL DNA barcode and 3.5 μL of 0.5% Phenol Red dye. The total volume of the RNP mix should be 25 μL.
4. Place a cartridge in the cartridge holder (Fig. 3a). Mix and transfer 20 μL of the RNP mix to the wells labeled "Sample." If making droplets for less than eight samples, fill the remaining wells with 1× NEBuffer r3.1.
5. Transfer 70 μL 3% HFE to each of the wells labeled "Oil." Seal the cartridge with a gasket. Place the cartridge holder in a QX200 droplet generator. The droplet generation will commence once the lid is closed.
6. Once droplet generation is over, use a Rainin 200 μL multichannel pipette set at 35 μL to transfer the droplets from the cartridge to tubes containing 30–50 μL of 3% FS-HFE. Check the integrity of the droplets by visual inspection (Fig. 3b). Droplets should be uniform in size (Fig. 3c). Store droplets at 4 °C. Droplets stored at 4 °C are stable for months, however, the Cas9 enzyme will lose activity upon prolonged storage. For best results, we recommend injecting the droplets within 1–2 weeks of generation.
7. Repeat **steps 4–6** above if generating droplets for more than eight samples.
8. If intermixing droplets targeting several genes, add 50 μL of 3% FS-HFE in a separate PCR tube. Add 2 μL of droplets targeting each gene. Because of the high density of 3% FS-HFE, the

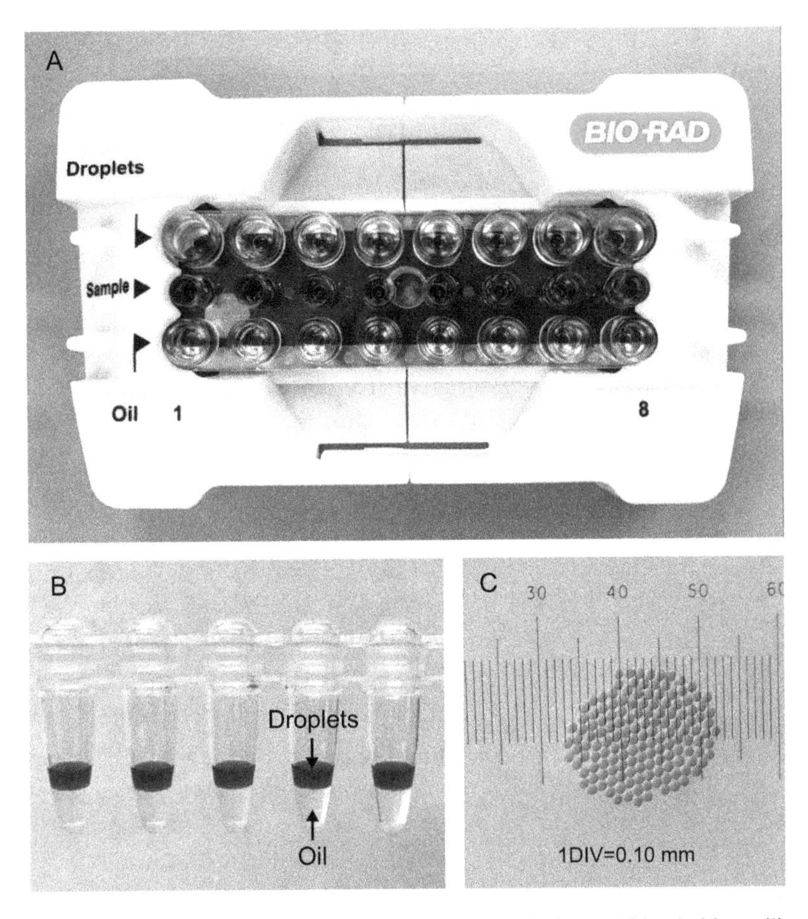

Fig. 3 Droplet preparation. (**a**) The QX200 cartridge in the cartridge holder, with samples and 3% FS-HFE added to the designated locations in the cartridge. (**b**) Droplets following transfer into microcentrifuge tubes containing 3% FS-HFE. Droplets (with Phenol Red) float on top of the oil due to lower density. (**c**) Droplets are uniform in size

droplets float on the oil surface (Fig. 3b). Make sure to pipette droplets from the top to ensure even representation of droplets targeting each gene. Gently mix the droplets using a P200 pipette. Check the integrity of the droplets by visual inspection. Avoid dropping the tubes containing the droplets or vigorous mixing of the droplets. Transfer and gentle mixing should not cause droplets to coalesce. Store intermixed droplets at 4 °C and inject within 1–2 weeks of droplet generation.

3.4 Droplet Microinjection

1. Pull microinjection needles prior to injection. Use the following setting on a P1000 Sutter Instrument: Heat: 565, Pull: 64, Velocity: 77, Time: 80, and Pressure: 500. Use freshly pulled needles for injection. Dust and debris in the needles can cause clogging or droplets to break apart.

2. Set up mating crosses using established protocols the evening before the injection day.
3. On the day of injection, pull dividers. Allow the fish to lay eggs for 5–10 min. Collect embryos in a tea strainer. Wash with E3 to get rid of fish waste. Transfer embryos to a petri dish.
4. Use a microloader tip to transfer 3–4 μL of the intermixed droplets along with a small volume of the oil into a microinjection needle (Fig. 4a). Droplets should make up 70–80% of the total volume transferred. There should be approximately 500–700 droplets in the injection needle. Gently flick the needle to remove any air bubbles.
5. Connect the needle to the injection setup. Trim the needle using a pair of fine tweezers to create an opening of size ~10–20 μm.
6. Due to density difference between the aqueous droplets and 3% FS-HFE, the droplets will collect at the top of the injection needle. Use the "Clear" setting on the injector to clear out most of the oil at the tip.
7. Transfer 50–100 embryos to the injection mold. Tuck them in the grooves of the injection mold for injection.
8. Once most of the 3% FS-HFE is cleared out from the microinjection needle, embryo injection can start. The typical injector settings for the Harvard Apparatus injectors are as follows: 40–70 ms for injection time and 6–9 psi for injection pressure.
9. The droplets will be separated from each other by a small section of oil as they move toward the needle tip. Push the foot pedal on the injector until a droplet reaches the needle tip and a part of it is ejected in the injection mold (visible via Phenol Red dye) (Fig. 4b, **step 1**). Now, insert the needle in a zebrafish embryo. Push the foot pedal to inject the remaining droplet in the embryo. It should take two to three pushes of the foot pedal to inject most of the droplet (Fig. 4b, **step 2**).
10. Pull out the needle from the embryo just before the oil separating the current and next droplet reaches the needle tip. If needed, adjust the time and pressure on the injection apparatus: one only needs to inject 50–75% of the droplet to get efficient gene knockout efficiency. Injecting only part of the droplets is recommended as this ensures reduced toxicity due to excess RNP and DNA barcode injection.
11. Push out the oil between two adjacent droplets into the injection mold until a small volume of the next droplet is ejected (Fig. 4b, **step 3**). Insert the needle into the subsequent embryo to inject the droplet (Fig. 4b, **step 4**).
12. Inject droplets in the desired number of embryos. We regularly inject 300–600 droplets in one morning (2–3 h).

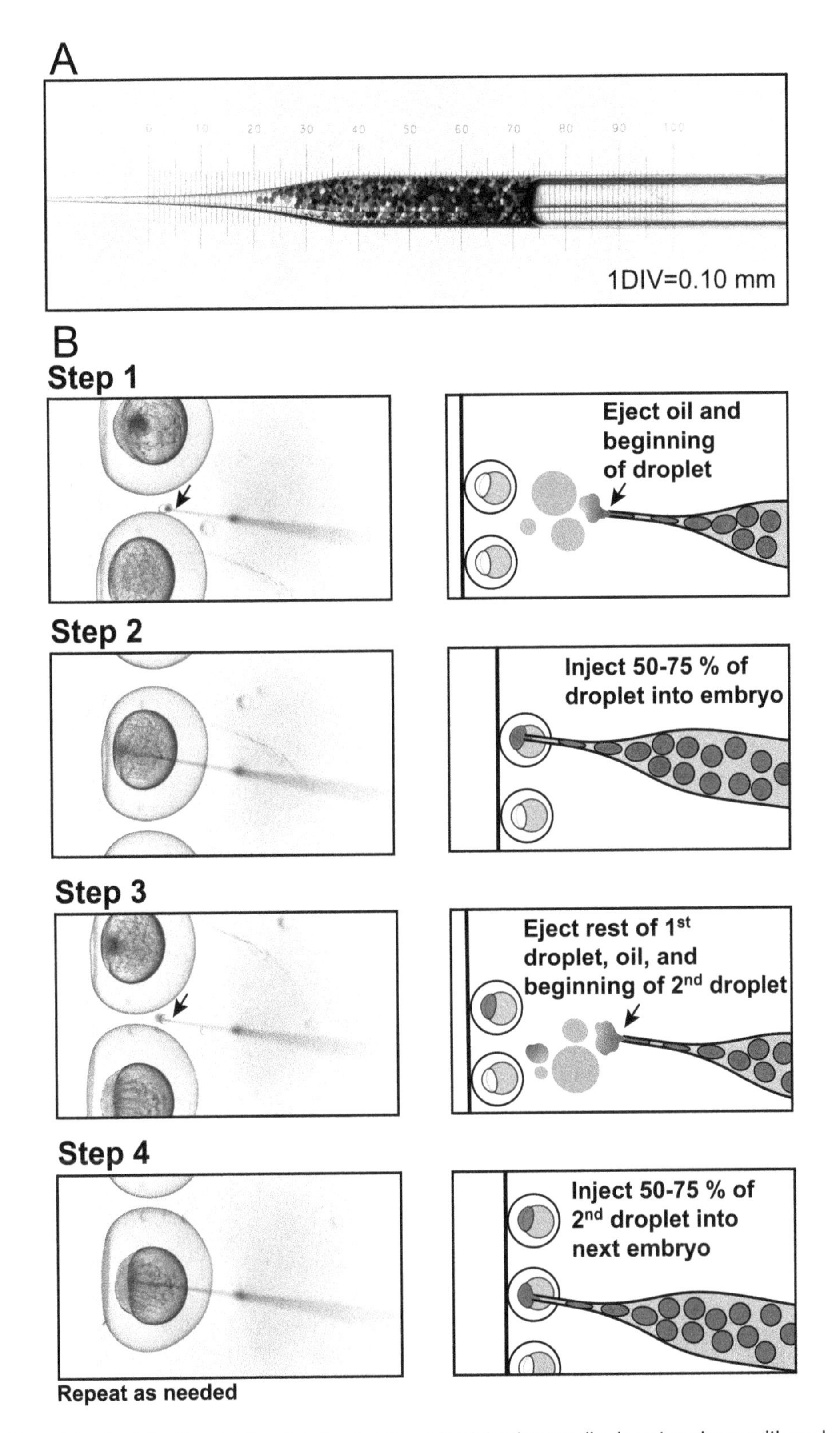

Fig. 4 Droplet microinjection. (**a**) Droplets loaded in a microinjection needle do not coalesce with each other. Food coloring has been added to the droplets to facilitate visualization. (**b**) Stepwise illustration of droplet delivery into 1-cell zebrafish embryos. Illustrated schematics are provided next to images acquired during microinjection. Arrows point to tip of microinjection needle

13. Transfer the injected embryos to a petri dish. Wash once with E3 to get rid of any 3% FS-HFE and residual RNP and barcodes. Transfer and split embryos to new petri dish so that there are only 50–60 embryos per petri dish.
14. Grow embryos in an incubator set at 28.5 °C until phenotyping at the desired time point.

3.5 Phenotyping and Barcode Retrieval

1. Count and remove any dead embryos at 12–24 h postfertilization (hpf).
2. Count and remove any embryos that show gross morphological defects resulting from nucleic acid toxicity.
3. Screen embryos for phenotype-of-interest at the desired time point.
4. Isolate embryos with the desired phenotype. If unhatched, dechorionate the embryos. Wash embryos once with E3.
5. Transfer embryos to a separate petri dish. Wash three times with E3, each time aspirating most of the E3 and refilling with fresh E3. Extensive washing is recommended to remove any residual DNA barcodes.
6. Prepare 2× PCR lysis buffer (Subheading 2.3, **step 9**). Add Proteinase K to a final concentration of 0.2 mg/mL.
7. Prepare a tube/PCR plate containing 10 μL 2× PCR lysis buffer containing Proteinase K.
8. Transfer the embryos with phenotype along with 10 μL E3 to separate tubes/wells of a PCR plate. Euthanasia may be necessary once embryos reach a specific stage, as per institute-specific animal care requirements.
9. Close the lid/seal the PCR plates. Incubate the samples at 50–55 °C overnight (14–16 h) to fully lyse the embryos.
10. Heat inactivate the Proteinase K at 95 °C for 10 min. The sample containing the DNA barcode can be stored at −20 °C for >1 month or can be used right away for DNA amplification.
11. To amplify the DNA barcode, make a 15 μL PCR mix containing 0.5 μM each of barcode amplification forward and reverse primers, 200 μM dNTPs (70:30 of dTTP:dUTP), 0.375 U UDG, 0.375 U GoTaq polymerase, and 2 μL of the embryo lysate containing the DNA barcode. Prior to adding the lysate to the PCR mix, pipette the lysate to resuspend. Centrifuge at 3000–5000 × *g* for 5 min to pellet the debris. Transfer 2–3 μL of the supernatant to the PCR reaction mix. If handling many samples, make a master mix containing all the components except the lysate containing the DNA barcode. UDG and 70:30 of dTTP:dUTP is used in the PCR reaction to degrade any possible carryover contamination product prior to amplification.

12. Amplify the DNA barcode at the following settings: 37 °C for 10 min, 95 °C for 10 min, [95 °C for 30 s, 55 °C for 30 s, 72 °C for 15 s] × 34 cycles, 72 °C for 5 min. The initial incubation at 37 °C is performed to allow degradation of carryover contamination by the UDG enzyme, which is subsequently heat inactivated.
13. PCR cleanup can be performed either using column cleanup or enzymatically. Although column cleanup may provide cleaner amplicon and therefore better sequencing results, for a large number of samples, enzymatic cleanup is more efficient and cost-effective. To perform enzymatic cleanup, add 1 μL of Exonuclease I and 0.5 μL of shrimp alkaline phosphatase to 5 μL of the PCR product. Pipette to mix the samples and then incubate at 37 °C for 15 min. Heat inactivate the samples at 95 °C for 15 min.
14. Dilute the samples 3–5×. The samples are now ready for Sanger sequencing. Sequence the samples with standard M13F or M13R primers.

3.6 Identification of Candidate "Hits" and Validation

1. Tally the number of barcodes for each gene and list the observed phenotypes.
2. For validation and follow-up, prioritize the genes with a high frequency of barcodes and those that yield consistent phenotypes.
3. Perform secondary validation of the candidate "hit" by injecting just the Cas9 and sgRNAs into 50–100 embryos. Count the embryos with the expected phenotype and also measure the phenotypic penetrance.
4. The candidate "hit" can also be validated by checking if the phenotype is rescued by coinjection of the mRNA expressing the targeted gene along with the sgRNAs.
5. Germline-stable mutants can then be generated of the validated "hits" using established protocol for further characterization and mechanistic investigation.

4 Notes

1. For genes with multiple splice isoforms, the default CHOPCHOP output is sgRNAs that target all isoforms. In rare cases when no common sgRNAs targeting all isoforms can be designed (e.g., when splice isoforms use distinct exons), change the default setting of "Intersection" to "Union" on CHOPCHOP under Options → General → Isoform consensus determined by. The default setting on CHOPCHOP is

"Intersection," which searches for sgRNAs only in regions present in *all* isoforms of the gene. "Union" searches for sgRNAs in regions present in *any* isoform. Pick four sgRNAs targeting different isoforms of the gene.

2. The optimal number of sgRNAs that should be used for F0 CRISPR screens has not been tested. Higher number of sgRNAs increase the chance of generating on-target biallelic loss-of-function mutations, but may also increase off-target gene editing. If validated sgRNAs with high on-target efficiencies are known, the number of sgRNAs targeting each gene can be reduced.
3. The injection pressure settings will vary depending on the injector system being used. The typical setting for the Harvard Apparatus injector is specified in Subheading 3.4, **step 8**. Occasionally, during droplet injection, we encounter high negative pressure that causes the droplets to not eject/move back up the injection needle after each push of the foot pedal. This is generally due to excess oil in the needle and can be resolved by using the "Clear" button on the injector to push out the excess oil. Additionally, the $P_{balance}$ can be adjusted to a less negative value.
4. The droplets that are injected in the beginning will have more oil separating them and therefore will require multiple pushes of the foot pedal to clear out the oil between two adjacent droplets. Later droplets will be closer together. Reduce the pressure as needed such that it still requires two to three pushes to inject the droplets and one to two pushes of the foot pedal to clear out the oil in between two droplets.
5. In all screens, we recommend including the following controls: (a) Injection control: droplets targeting a gene that results in an easily measurable phenotype. *Rx3*, *tyr*, *tbx16*, etc. can all serve as excellent injection controls; (b) Positive control: droplets targeting a gene that results in the phenotype-of-interest; and (c) Uninjected control: uninjected embryos to compare general viability and health of the clutches.
6. In our lab, an experienced user regularly injects 300–500 embryos in 2–3 h. Therefore, in effect, 300–500 genes can be targeted in a single injection session. However, to account for random distribution of droplets, incomplete phenotypic penetrance, and to ensure redundancy in phenotypes in the injected embryos, we target up to 50 genes in one session. This allows us to have a coverage of 6–10 embryos injected for each gene targeted. Phenotypes observed across multiple embryos are more likely to be true "hits" rather than random injection artifacts.

7. Thorough washing of embryos prior to barcode retrieval is crucial. Carryover contamination will obfuscate sequencing results, making it difficult to identify "hits."
8. We define candidate "hits" as genes whose barcodes are identified at a higher frequency that what would be expected from random sampling. We use a binomial probability of less than 0.05 as the cutoff to determine the candidate "hits." However, this cutoff may need to be adjusted if the screened phenotype-of-interest overlaps with common injection-related phenotypes. Similarly, if the screened phenotype-of-interest is unlikely to be spurious, "hits" that do not meet the above-mentioned cutoff can still be considered for secondary validation and follow-up.

References

1. Driever W, Solnica-Krezel L, Schier AF et al (1996) A genetic screen for mutations affecting embryogenesis in zebrafish. Development 123: 37–46
2. Mullins MC, Acedo JN, Priya R, Solnica-Krezel L, Wilson SW (2021) The zebrafish issue: 25 years on. Development 148: dev200343
3. Haffter P, Granato M, Brand M et al (1996) The identification of genes with unique and essential functions in the development of the zebrafish, Danio rerio. Development 123:1–36
4. Kimble J, Nüsslein-Volhard C (2022) The great small organisms of developmental genetics: caenorhabditis elegans and drosophila melanogaster. Dev Biol 485:93–122
5. Amsterdam A, Burgess S, Golling G et al (1999) A large-scale insertional mutagenesis screen in zebrafish. Genes Dev 13:2713–2724
6. Golling G, Amsterdam A, Sun Z et al (2002) Insertional mutagenesis in zebrafish rapidly identifies genes essential for early vertebrate development. Nat Genet 31:135–140
7. Varshney GK, Pei W, LaFave MC et al (2015) High-throughput gene targeting and phenotyping in zebrafish using CRISPR/Cas9. Genome Res 25:1030–1042
8. Pei W, Xu L, Huang SC et al (2018) Guided genetic screen to identify genes essential in the regeneration of hair cells and other tissues. Npj. Regen Med 3:11
9. Shah AN, Davey CF, Whitebirch AC, Miller AC, Moens CB (2015) Rapid reverse genetic screening using CRISPR in zebrafish. Nat Methods 12:535–540
10. Sander JD, Cade L, Khayter C et al (2011) Targeted gene disruption in somatic zebrafish cells using engineered TALENs. Nat Biotechnol 29:697–698
11. Cade L, Reyon D, Hwang WY et al (2012) Highly efficient generation of heritable zebrafish gene mutations using homo- and heterodimeric TALENs. Nucleic Acids Res 40:8001–8010
12. Foley JE, Yeh JR, Maeder ML et al (2009) Rapid mutation of endogenous zebrafish genes using zinc finger nucleases made by Oligomerized Pool ENgineering (OPEN). PLoS One 4:e4348
13. Hwang WY, Fu Y, Reyon D et al (2013) Efficient genome editing in zebrafish using a CRISPR-Cas system. Nat Biotechnol 31:227–229
14. Shin U, Nakhro K, Oh C-K et al (2021) Large-scale generation and phenotypic characterization of zebrafish CRISPR mutants of DNA repair genes. DNA Repair 107:103173
15. Thyme SB, Pieper LM, Li EH et al (2019) Phenotypic landscape of schizophrenia-associated genes defines candidates and their shared functions. Cell 177:478–491.e420
16. Sun Y, Zhang B, Luo L et al (2019) Systematic genome editing of the genes on zebrafish chromosome 1 by CRISPR/Cas9. Genome Res 30: 118–126
17. Burger A, Lindsay H, Felker A et al (2016) Maximizing mutagenesis with solubilized CRISPR-Cas9 ribonucleoprotein complexes. Development 143:2025–2037
18. Wu RS, Lam II, Clay H et al (2018) A rapid method for directed gene knockout for screening in G0 zebrafish. Dev Cell 46:112–125.e114

19. Hoshijima K, Jurynec MJ, Klatt Shaw D et al (2019) Highly efficient CRISPR-Cas9-based methods for generating deletion mutations and F0 embryos that lack gene function in zebrafish. Dev Cell 51:645–657.e644
20. Kroll F, Powell GT, Ghosh M et al (2021) A simple and effective F0 knockout method for rapid screening of behaviour and other complex phenotypes. elife 10:e59683
21. Quick RE, Buck LD, Parab S, Tolbert ZR, Matsuoka RL (2021) Highly efficient synthetic CRISPR RNA/Cas9-based mutagenesis for rapid cardiovascular phenotypic screening in F0 zebrafish. Front Cell Dev Biol 9. https://doi.org/10.3389/fcell.2021.735598
22. Parvez S, Herdman C, Beerens M et al (2021) MIC-drop: a platform for large-scale in vivo CRISPR screens. Science 373:1146–1151
23. Labun K, Montague TG, Krause M et al (2019) CHOPCHOP v3: expanding the CRISPR web toolbox beyond genome editing. Nucleic Acids Res 47:W171–W174
24. Concordet J-P, Haeussler M (2018) CRISPOR: intuitive guide selection for CRISPR/Cas9 genome editing experiments and screens. Nucleic Acids Res 46:W242–W245
25. Varshney GK, Carrington B, Pei W et al (2016) A high-throughput functional genomics workflow based on CRISPR/Cas9-mediated targeted mutagenesis in zebrafish. Nat Protoc 11:2357–2375
26. Gagnon JA, Valen E, Thyme SB et al (2014) Efficient mutagenesis by Cas9 protein-mediated oligonucleotide insertion and large-scale assessment of single-guide RNAs. PLoS One 9:e98186

Chapter 22

The Goldfish Genome and Its Utility for Understanding Gene Regulation and Vertebrate Body Morphology

Yoshihiro Omori and Shawn M. Burgess

Abstract

Goldfish, widely viewed as an ornamental fish, is a member of *Cyprinidae* family and has a very long history in research for both genetics and physiology studies. Among *Cyprinidae*, the chromosomal locations of orthologs and the amino acid sequences are usually highly conserved. Adult goldfish are 1000 times larger than adult zebrafish (who are in the same family of fishes), which can make it easier to perform several types of experiments compared to their zebrafish cousins. Comparing mutant phenotypes in orthologous genes between goldfish and zebrafish can often be very informative and provide a deeper insight into the gene function than studying the gene in either species alone. Comparative genomics and phenotypic comparisons between goldfish and zebrafish will provide new opportunities for understanding the development and evolution of body forms in the vertebrate lineage.

Key words Phenotypic variety, Domestication, GWAS, Whole-genome duplication

1 Introduction

Strains of goldfish with high morphological diversity have been generated by artificial selection over the past millennium. Goldfish breeders, mainly in East Asia, have developed many genetic strains with a wide variety of colorations and different body, fin, eye, hood, and scale morphologies (Figs. 1 and 2). From the tenth to seventeenth centuries A.D. (Song and Ming dynasties in China), goldfish underwent domestication and several representative phenotypes, including twin tails, long fins, and short bodies, were established. The first records of humans breeding goldfish for aesthetic purposes were in China over 1000 years ago. Breeding efforts spread to Japan around the sixteenth century and thereafter also in Europe and North America, which generated many novel phenotypes and combinations. Currently, at least 200 variants and 80 genetically established strains are produced and maintained [1].

James F. Amatruda et al. (eds.), *Zebrafish: Methods and Protocols*, Methods in Molecular Biology, vol. 2707,
https://doi.org/10.1007/978-1-0716-3401-1_22,

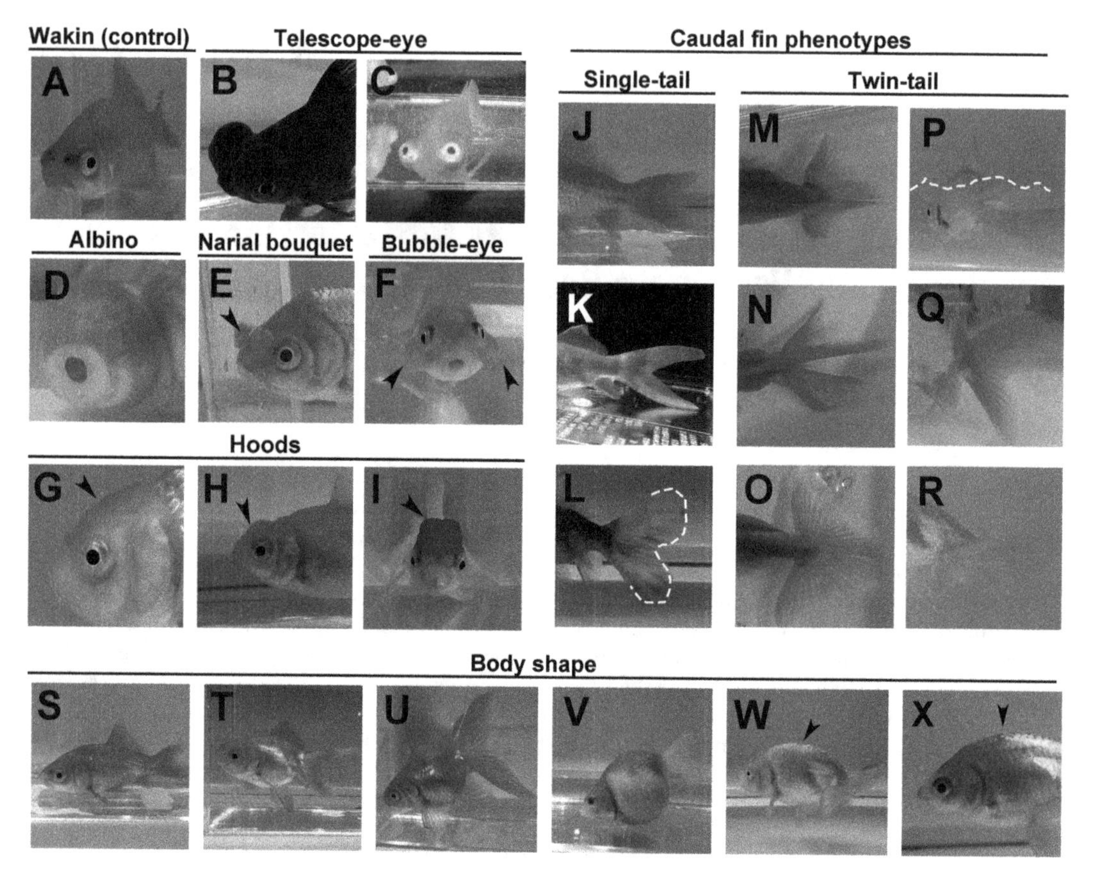

Fig. 1 The variety of morphological phenotypes in goldfish strains. The common goldfish (single-tailed Red Wakin strain) shows a morphologically "normal" appearance (**a**). Enlarged and protuberant eyeballs are observed in the Black Telescope eye (**b**) and Albino Celestial (**c**) strains. The eyes of Albino Celestial goldfish are upwardly directed (**c**). The pink pupils in Albino Telescope eye (**d**). Hypertrophy of the nasal septum forms "narial bouquets," which develop in the Pompon strain (**e**). In bubble-eye goldfish, large vesicles arise from the lower ocular orbit (arrowhead in **f**). A thickened epidermis (hood) on the head (arrowheads in **g**–**i**) develops in the Ranchu (**g**), Hama-nishiki (**h**), and Redcap Oranda strains (**i**).(**j**–**r**) The various caudal fin morphologies in goldfish strains. The normal single tail in the common goldfish (**j**), long single tail in the Comet (**k**), and heart-shaped tail in the Bristol Shubunkin strain (**l**). The normal twin tails in Ranchu (**m**), long twin tails in Oranda (**n**), butterfly tail that spreads horizontally in the Butterfly-tail strain (**o**), peacock tail in the Jikin strain (**q**), and nonelongated twin tails in the Osaka Ranchu strain (**r**). In the Tosakin strain, the lower tail lobes extend laterally and curl forward (dotted line in **p**). (**s**–**x**) A variety of body shapes in goldfish strains, namely, the normal wild-type body of the common goldfish (**s**), shortened and globular body of Oranda (**t**) and Ranchu goldfish (**w**), severely shortened body and increased body depth of Ryukin goldfish (**u**), and severely shortened and round body of Pearlscale (**v**). The dorsal-fin-loss phenotype is observed in Ranchu (**w**) and Osaka Ranchu (**x**). (Reprinted with permission from Kon et al. [6])

Goldfish have several advantages as a model system for understanding the molecular basis of the morphological diversity of vertebrates. Goldfish and zebrafish belong to the same broad family, *Cyprinidae*. Therefore, it is relatively straightforward to find orthologous genes between zebrafish and goldfish, and phenotypic

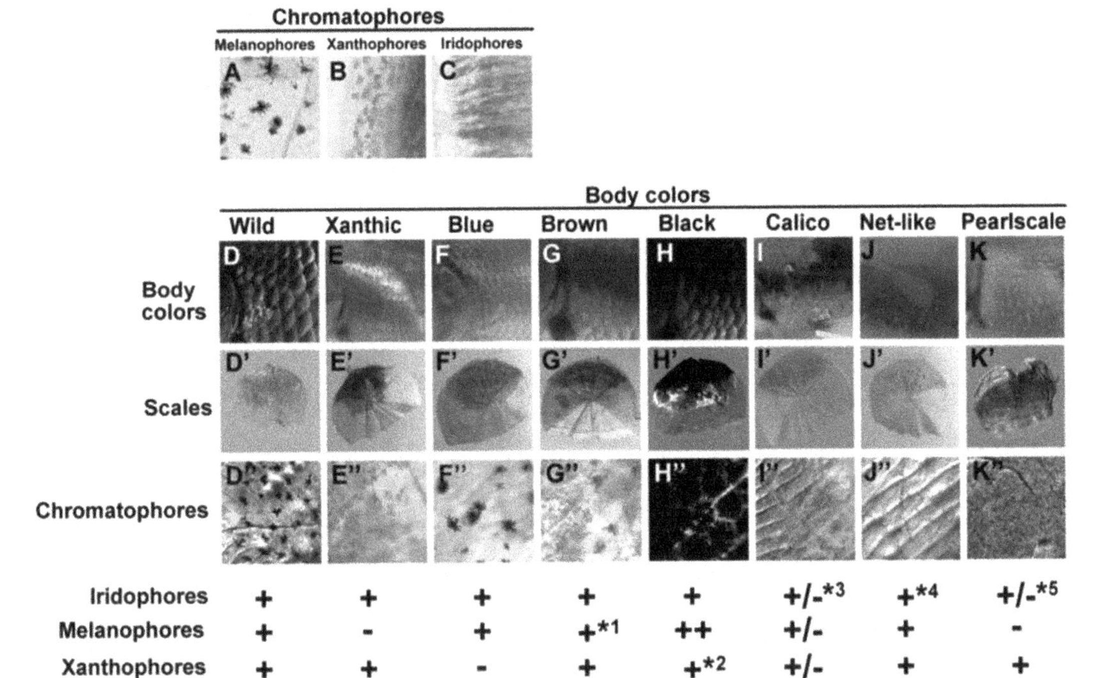

Fig. 2 The variety of body colors, scales, and chromatophores of goldfish strains. Goldfish have three types of chromatophore: melanophores (**a**), xanthophores (**b**), and iridophores (**c**). The goldfish body surface (**d**–**k**), scales (**d′**–**k′**), and chromatophores on the scales (**d″**–**k″**) are indicated. The presence of the three chromatophore types is indicated below the microscopic images. The images are from wild goldfish (**d**), xanthic in Oranda (**e**), metallic blue in Seibun (**f**), brown in Chocolate Oranda (**g**), black in Black Telescope eye (**h**), calico in Shubunkin (**i**), net-like transparent scales in Momiji Wakin (**j**), and the Pearlscale phenotype, also called the domed scale, in the Pearlscale strain (**k**). (Reprinted with permission from Omori and Kon [52], Kon et al. [6]). *1: There are brown-colored melanocytes on the scale surface. *2: Xanthophores are difficult to observe due to the enrichment of melanophores. *3: Several different scale types exist on the body surface of an individual. *4: Partial existence of iridophores on the scale. *5: Both types of scales are present on the body surface of an individual

comparison because the two species is facilitated through their similar development and body structure. Domesticated goldfish strains have biologically interesting phenotypes that have not been identified in mutant zebrafish, including the formation of hoods (a thickened epidermis on the head), narial bouquets, variations in body proportion, fin shapes, scale morphologies, and multiple variations in body coloration (Figs. 1 and 2). Zebrafish mutant screening has focused mainly on mutations arising during development using mutagens such as ethylnitrosourea (ENU) [2], retroviral integration [3], or radiation [4]. Phenotypes observed in adult zebrafish were also identified [5], but the analysis of such mutations has traditionally been laborious and time-consuming, and has not been done so extensively.

In the case of goldfish, for hundreds of years, breeders in East Asia have used ponds to artificially select for phenotypes that appear in adult goldfish. Hence, the mutants in goldfish strains are, by definition, nonlethal during development and are always fertile. In addition, the common carp and goldfish share a whole-genome duplication (WGD) that occurred approximately 14 million years ago that has resulted in a duplication of most genes. The potential redundancy of genes generated by the additional WGD may allow for the avoidance of lethality caused by mutations in early developmental stages. Thus, it is expected that novel mutations or phenotypes may be identified in goldfish that have not been found in zebrafish mutant screening. Genome-wide analyses of different goldfish strains is revealing clues of the molecular mechanisms underlying the phenotypes of ornamental varieties [6, 7].

Zebrafish and medaka may have disadvantages associated with their small sizes. The adult Wakin goldfish reach approximately 35 cm in length, with a body mass of 1 kg, which is more than 1000 times that of adult zebrafish (4–4.5 cm, 0.6–0.8 g) [8]. The goldfish egg, embryos, and larvae are also larger than those of zebrafish [9]. Therefore, goldfish can be advantageous for collecting sufficient biomaterials needed to identify and characterize rare cell types using techniques such as single-cell RNA-seq analysis. In addition, goldfish may be useful for experiments involving microsurgery or micromanipulation [10]. In this chapter, we will discuss the utility of the goldfish genome both for establishing it as a modern fish model as well as the use of the goldfish genome for comparative genomic studies involving zebrafish. We will also discuss goldfish mutants with phenotypes similar and/or related to those of zebrafish mutants (Table 1) and the opportunity goldfish presents for identifying nonlethal mutations that impact body form or function.

A number of characteristic phenotypes have been found in goldfish strains that were not identified by screening zebrafish or medaka. Why do goldfish strains have such a variety of phenotypes? The recent WGD that occurred in the ancestors of goldfish and the generation of excess gene duplications in the genome appear to have impacted the diversification of this species (Fig. 7). The gene duplications appear to have facilitated the emergence of new sub- or neo-functionalizations of the duplicated genes since the original function can be retained by the other gene copy [11–13]. Alterations in function or dosage reductions of genes in one or both of the two subgenomes have likely contributed to the generation of diverse goldfish phenotypes that could be selected by breeders.

1. The goldfish chromosomes can be partitioned into L- and S-subgenomes based comparative genomic analysis of the interchromosomal relatedness.

Table 1
Variety of goldfish phenotypes, the representative strains, and the reported genes/genetic loci

Categories	Phenotypes	Feature	Representative strains in goldfish	Reported mutations, loci, and genetic inheritance in goldfish	Similar phenotypes and causative mutations in zebrafish	Related human disease
Eyes	Telescope-eye	Enlarged and protuberant eyes	Telescope-eye, Black Moor, Globe-eye, Celestial	D/d, Recessive (Matsui [29]), three types of mutations in *lrp2aL* (Kon et al. [6])	Mutations in *lrp2a* (Veth et al. [40])	Donnai-Barrow syndrome (OMIM #222448)
Fins	Long-fins	Elongation of caudal fins are remarkable. Other parts of fins also longer than the common goldfish	Comet, Shubunkin, Ryukin, Oranda	A missense mutation in *kcnk5bS* (Kon et al. [6])	*another longfin (alf)* mutant, a missense mutation in *kcnk5b* (Perathoner et al. [30])	
	Twin-tail	Double caudal fins	Ryukin, Oranda, Ranchu	A mutation in *chordinA (chdS)*, Recessive (Abe et al. [36])	*dino* mutant, a missense mutation in *chordin* (Fisher and Halpern [33]; Hammerschmidt et al. [34]; Schulte-Merker et al. [35])	
	Heart-shaped caudal fin	Heart-shaped caudal fin	Bristol Shubunkin	A region including *rpzS* (Kon et al. [6])	A mutation in *rpz* gene (Green et al. [53])	
	Dorsal-fin-loss	No dorsal fin	Ranchu	A region including *lrp6S* (Kon et al. [6])	Loss of function of *lrp6* (Kon et al. [6])	Tooth agenesis, selective, 7 (OMIM #616724)

2. In the S-subgenome, more gene loss, gene disruption, and sequence divergence were identified than in the L-subgenome, suggesting the L versions of the genes are the "dominant" ones for preserving function on average.
3. In addition, the frequency of single-nucleotide variants (SNVs) and indels was greater in the S-subgenome than in the L-subgenome among goldfish strains [6, 14]. Thus, the two goldfish subgenomes have evolved asymmetrically.
4. So far, six loci associated with goldfish phenotypes (twin tails, long tail, telescope eye, dorsal fin loss, albinism, and heart-shaped tail) have been reported. In all of these loci, the function of at least one paralog of the S-subgenome is affected. The mutations of four genes, *chordin*, *kcnk5b*, *lrp6*, and *rpz*, are on the S-subgenome, while *oca2* has mutations in both the S- and the L- subgenomes. In goldfish strains with telescope-eyes, *lrp2a* has mutations in the L-subgenome, but its paralog on the S-subgenome appears to have already been lost and it is a singleton in the goldfish genome.
5. These observations provide a hypothesis that the phenotypes caused by mutations in the S-subgenome are more commonly isolated by breeders selecting goldfish mutants during the history of goldfish domestication.

2 Materials

2.1 Fish

Adult goldfish strains were obtained from a commercial supplier (Meito Suien Co., Ltd., Aichi, Japan). Wakin goldfish was obtained from a provider in Yatomi (Aichi, Japan) and bred in Nagahama Institute of Bioscience and Technology. Albino Celestials and Albino Ranchu were obtained from the Aichi Fisheries Research Institute.

2.2 Anesthetization of Goldfish

1. Tricaine (Nacalai, 100 mg/L).
2. Plastic tank (1 L).
3. Fish net (12 cm × 10 cm).
4. Forceps (10 cm).

2.3 Genomic DNA Purification from Goldfish Tissues

1. TissueLyser II (QIAGEN).
2. Blood & Cell Culture DNA Maxi Kit (QIAGEN).
3. Pippin pulse electroporation system (NIPPON Genetics).
4. Agarose gel electrophoresis system (Mupid).
5. Agarose (Nacalai).
6. Ethidium bromide (Wako).

7. M220 Focused-ultrasonicator (Covaris).
8. TruSeq DNA PCR-Free Library Prep Kit (Illumina).
9. Agencourt AMPure XP (Beckman Coulter).
10. Agilent 2100 Bioanalyzer (Agilent Technologies).

2.4 NGS Sequencers

1. PacBio RS II sequencer (Pacific Biosciences).
2. HiSeq 2500 sequencer (Illumina).

2.5 Voltage Clamp Recording

1. ND96 (+) solution (93.5 mM NaCl, 2 mM KCl, 1.8 mM $CaCl_2$, 2 mM $MgCl_2$, 5 mM HEPES [pH 7.5, adjusted with NaOH]).
2. iTEV90 multielectrode clamp amplifier (HEKA).

2.6 Identification of the 13-kbp Insertion in the Telescope-Eye Goldfish Genome

1. KOD polymerase (TOYOBO).
2. Big Dye terminator sequencing kit (Thermo Fisher Scientific).
3. ABI3700 DNA sequencer (Thermo Fisher Scientific).

2.7 CRISPR/Cas9-Mediated Gene Disruption of lrp6 in Zebrafish

1. MEGAshortscript T7 Transcription Kit (Thermo Fisher Scientific).
2. Cas9 3NSL enzyme (Integrated Data Technologies).
3. FemtoJet microinjector (Eppendorf).
4. Femtotip II microcapillaries (Eppendorf).
5. 1.5% agarose molds.

3 Methods

3.1 Introduction to Comparative Genomics

There had been no fewer than three efforts to sequence the *Carassius auratus* genome [15–17], demonstrating the broad interest the scientific community had in obtaining a high-quality assembly for the goldfish. There are a number of reasons why the goldfish genome is of interest to developmental biologists and evolutionary biologists as well as its potential impact on commercial husbandry of ornamental goldfish. There are a number of closely related carp species of which the grass carp (*Ctenopharyngodon idella*), the common carp (*Cyprinus carpio*), and the goldfish are the most broadly studied. The grass carp has 25 chromosomes and evolutionarily split from common carp and goldfish, approximately 14.4 million years ago [15]. After that split, the ancestor of common carp and goldfish went through a whole-genome duplication event. The most likely cause was a rare nondisjunction event after two different species of carp bred, resulting in a viable tetraploid offspring known as an "allotetraploid." After this genome duplication occurred, the common carp and goldfish speciated, roughly

11 million years ago. This "recent" speciation between common carp and goldfish provides an interesting opportunity for comparative genomics during the slow "rediplodization" process, i.e., when the function of the two duplicated genes diverges enough that they are no longer fully redundant [18].

3.2 Comparative Genomics Using the Goldfish and Zebrafish Genomes

We have generated a UCSC browser-enabled data hub for visualizing the goldfish genome (https://research.nhgri.nih.gov/goldfish/) as well as partnering with Ensembl to display the genome in their genome browser (https://ensembl.org/Carassius_auratus/). One of the early forms of comparative genomics was to look for regions of sequence conservation that were not associated with protein coding. These regions went by many names [19–21], such as "conserved noncoding element" or CNE, and are thought to represent gene regulatory regions, i.e., "enhancers" [22]. The number of conserved regions varies based on the evolutionary "distance" between species, e.g., a human to primate comparison generates very large regions of conservation and comparing humans to fish reveals very few sequences conserved outside of the coding regions. For many years, zebrafish did not have another available fish genome that was close enough to identify CNEs with a good signal-to-noise ratio. Recent addition of the grass carp [23], the common carp [24], and now the goldfish [15] gave us three genomes, all with approximately 60 million years of separation from the zebrafish genome, which is roughly similar to the human-to-rodent comparisons of Prabhakar et al. [25].

1. In order to optimize the efficiency of enhancer detection, it is beneficial to choose genomic comparisons close enough to not lose critical information, but far enough to filter out less informative sequence conservation [25]. We performed a four-way whole genome comparison between the fish and generated visualization tracks for the zebrafish and goldfish genome UCSC web browsers.
2. The UCSC custom track is available through the public hubs of the UCSC browser. To load the track, click the "track search" button just above the various pull-down menus that activate tracks. Click the "include public hubs" option and search for "Conservation and CRISPR targets." Toggle visibility to "show" and then click on "return to browser."
3. There are two tracks that represent the same four-way comparison in slightly different ways (Fig. 3). The first track generates a histogram that shows base by base how well conserved the zebrafish genomic sequence is across the three carp species (Fig. 3a). The second track (Fig. 3b) identifies the boundaries of sequences that meet the definition of conserved sequences (>200 bp with >85% identity).

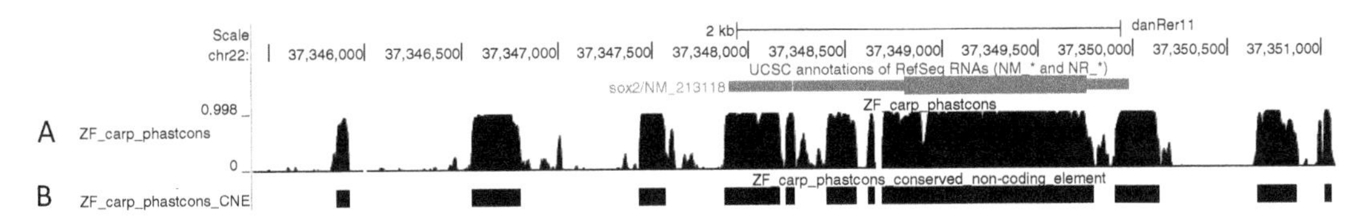

Fig. 3 The chromosomal region of the *sox2* locus (in blue) in zebrafish on the UCSC genome browser. (**a**) The histogram representing regions of homology across the four fish species: zebrafish, goldfish, common carp, and grass carp. (**b**) The genomic regions that reach the definition of a conserved element (≥200 bp; ≥85% identity) are marked with black bars

4. In the example provided, the zebrafish *sox2* gene is represented in blue while putative regulatory regions can be both upstream and downstream of the mRNA. While CNEs can be a robust method for identifying candidate regulatory sequences, it is essential to combine such analysis with other genetic or genomic methods to establish functionality. Deletion by CRISPR-Cas9, cloning enhancers upstream of expression reporters, or co-localization of CNEs with ATAC-seq "peaks" are three examples of how CNEs can be tested for roles in gene regulation.

3.3 Comparisons Within the Goldfish Genome

Because the whole-genome duplication shared between the common carp and goldfish is fairly recent, it provides opportunities for genomic comparisons between in essence, four (sometimes 8!) different copies of every gene. Susumu Ohno proposed that large jumps in evolution are facilitated by gene duplication, which allows one copy of the gene to mutate into a new function [18]. Whole-genome duplications have occurred multiple times in vertebrate evolution, and each time, after the duplication event, the organisms begin the slow return to diploid status through the three processes: gene inactivation, subfunctionalization (functions are split between the two genes), and neofunctionalization (one of the genes obtains a new function). With sufficient gene expression data across goldfish tissues and/or developmental timepoints, you can compare expression for each "ohnolog" (homologous loci matching across the genome duplication) with the existence or absence of conserved sequences in each chromosomal loci.

1. One of the goldfish genome assemblies is available for viewing through a UCSC-enabled genome browser at https://research.nhgri.nih.gov/goldfish/.
2. Gene expression from seven tissues can be activated by toggling the "expression RPM" track, gene models by toggling the "maker_gene" track, and the four-way conservation alignments by toggling the "phastcons" track. Genes of interest are then searched by name in the search window at the top of the page. Most genes will give two different hits in the genome.

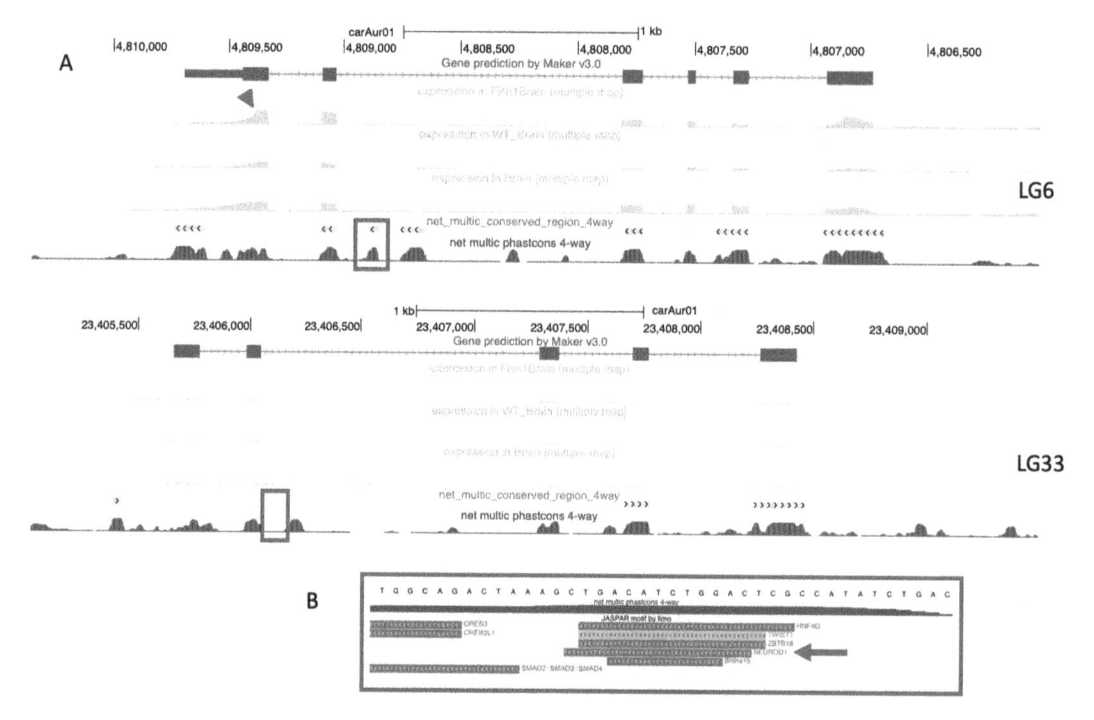

Fig. 4 The chromosomal regions of two copies of the goldfish *fkbp11* gene on linkage group 6 (top) and on linkage group 33. (**a**) RNA-seq data from three brain samples are shown in yellow. On LG6, active transcription is marked by a red arrowhead. No expression is seen from the *fkbp11* gene on LG33. Conserved sequences are shown by blue histograms. A position of CNE identified in LG6, that is missing in LG33, is marked with a red box. (**b**) A magnified view of CNE on LG6 with JASPAR transcription factor predictions. A predicted site for NeuroD binding, a transcription factor known to drive expression in neuronal tissue, is marked with a red arrow

For an example, see Fig. 4a. The *fkbp11* gene shows different expression between the copy on chromosome 6 that is expressed in the brain and the copy on chromosome 33 that is not. Most of the CNE are present in both copies of the gene. The CNE on chr6 that is marked with a red box is not present in the copy on chr33. Figure 4b shows a magnification of the region on chr6 with JASPAR-predicted transcription factor binding sites [26]. The red arrow shows a strongly predicted NEUROD1 binding site within this CNE, a transcription factor with a known role in driving neuronal cell expression [27].

3. Differential analyses such as these can be done on an individual gene basis or globally using computational methods. It is potentially possible to compare expression in goldfish with the homologous genes in the common carp to understand gene expression changes that brought about speciation or compare the expression of each ohnolog with that of the zebrafish's single copy to see if we can better understand the process of species radiation and functional evolution.

3.4 Identifying Loci Associated with Goldfish Phenotypes

There are many historical examples of human's "intervening" with natural selection in order to generate agricultural species, both plants and animals, with traits that make them more useful. A smaller subset of animals, usually raised as pets, were bred in part or purely for aesthetics. Two classic examples of this are dogs and goldfish. Obtaining a genomic sequence for a species that has undergone human preferential breeding provides new opportunities to link observable phenotypes to the underlying genotypic variation. This has been done quite successfully in dogs where completion of the genome has opened up a wide variety of new avenues of research ranging from morphology to disease studies [28]. Goldfish represent a similar opportunity as there are roughly 200 recognized varieties of goldfish. Several key attributes define the various varieties: number of fins and fin shape, body shape, normal or protruding eyes, scale color, and presence or absence of a "hood" on the head.

A strategy for identifying mutations in goldfish strains is to leverage the existing varieties directly by analyzing the genomes from individuals of many different strains. In this case, it is necessary to analyze dozens of different strains, some of which have the target phenotype, while others do not [6].

Genome-wide association studies (GWAS) have been used to find the associations between SNVs detected by whole-genome sequencing and phenotypes in various goldfish strains [6]. The general principle of GWAS using whole-genome resequencing data is shown in Fig. 5, in which the dorsal fin loss is used as an example. Whole-genome sequencing is used to find all genetic polymorphisms in the genome of all individuals. Most of the SNVs do not correspond to the presence or absence of the phenotype because they can segregate independently from the causative mutation (e.g., SNV1). In an unlinked region like this genomic locus, the score ($-\log P$) on the Manhattan plot is low. On the other hand, in very rare cases, the presence or absence of the phenotype and the SNVs in the genome occurs together more often than by chance (e.g., SNV2). In these loci, the score on the Manhattan plot is high. Since haplotype blocks (regions where recombination between two or more commonly occurring SNVs are rare) are often linked to the causative variant, the SNVs with high scores on the Manhattan plot will cluster together at loci located close to the causative variant, forming a peak on the Manhattan plot. There is a high probability that the variants responsible for the observed phenotypes are located within such peaks, although it is not guaranteed. Candidate genes within the linked region are identified and sequence variation that might impact the candidate genes are identified.

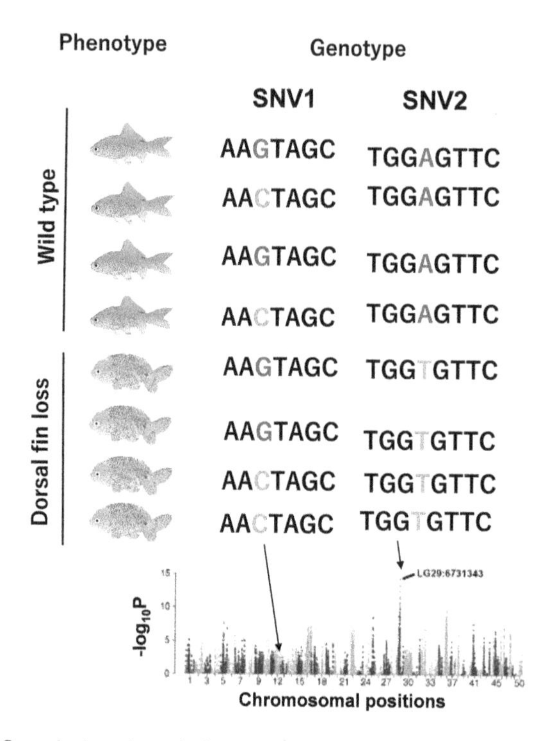

Fig. 5 GWAS analysis using whole-genome resequencing data. In regions where the genetic polymorphisms do not correspond to the presence or absence of a phenotype (e.g., SNV1), the LOD score (−log *P*) on the Manhattan plot is low. In rare cases, the presence or absence of the phenotype is in linkage disequilibrium with the genotype (e.g., SNV2). In this genomic region, the score on the Manhattan plot is high across several SNVs. There is a high probability that the mutation of the responsible gene is located within this peak

3.4.1 Genomic DNA Purification

1. The goldfish were anesthetized using Tricaine (100 mg/L) in a tank (1 L). The caudal fins of goldfish were dissected using forceps (10 cm). High molecular weight genomic DNA was purified using a TissueLyser II (QIAGEN) and the Blood & Cell Culture DNA Maxi Kit (QIAGEN). For genome resequencing, 20-kb bands of purified genomic DNA are identified by agarose gel electrophoresis. The genomic DNA was sheared to an average size of 600 bp using an M220 Focused-ultrasonicator (Covaris). Paired-end libraries were prepared using a TruSeq DNA PCR-Free Library Prep Kit (Illumina). They were purified using Agencourt AMPure XP (Beckman Coulter). The DNA concentration of the libraries was measured using an Agilent 2100 Bioanalyzer (Agilent Technologies). HiSeq 2500 sequencers (Illumina) were used for 250-bp paired-end sequencing reads.
2. To perform a GWAS, the following steps are conducted: generation of FASTQ raw sequence data by whole-genome sequencing using NGS (approximately 10× coverage per individual), followed by mapping of these data to the goldfish

genome using the BWA-mem (v 0.7.16a) program using default settings.

3. Next GATK (v4.0.4.0) is used to perform variant calling to determine the genotype across the entire genome of the analyzed population. All calls from individuals were combined into one dataset. Following the removal of variants with a read depth < 3, variants with a call rate $< 100\%$ and minor allele frequency $< 1\%$ were filtered out.
4. Finally, GWAS analysis is used to identify associations between phenotypes and genotypes prepared using the PLINK (v1.90b4.5) program using default settings. Manhattan plots are created using the R (v3.4.3) and qqman packages.

3.4.2 GWAS Analysis 1: The Long-Tail Phenotype

Many goldfish strains except the Wakin and Ranchu groups possess the longfin phenotype, in which all five fins, especially caudal fins, are longer than those of the wild-type fish (Fig. 1k, n). The longfin phenotype is observed in both single- and twin-tailed goldfish strains. A previous study suggested that the long-tail phenotype in goldfish is probably dominant [29].

1. GWAS analysis identified a notable association between a locus on LG45 and the long-tail phenotype (Fig. 6a) [6]. In the vicinity of this genomic location, *kcnk5b* was identified as a strong candidate gene. Mutations in its zebrafish ortholog (*another longfin*, *alf*) were reported to cause a phenotype of a longer caudal fin [30]. KCNK5 is a two-pore domain potassium channel that produces background K^+ currents over a large range of membrane potentials [31, 32], and the zebrafish *alf* mutations increase the K^+ conductance of the channel to cause the hyperpolarization of cells.
2. The mutation V165E was identified in the long-tail goldfish strains, but not in normal finned strains. It is located in the vicinity of the mutated amino acid W169L (Fig. 6b, c), which was identified in the zebrafish *alf* mutant.
3. To test whether mutations in goldfish *kcnk5b* cause a change in K^+ conductance, we performed voltage-clamp recordings using oocytes injected with mRNA of wild-type and mutant goldfish *kcnk5b*. Following overnight incubation at 18 °C, the transmembrane current was measured in ND96 (+) solution using an iTEV90 multielectrode clamp amplifier (HEKA). The cells were initially clamped at −80 mV and then subjected to 500 ms voltage steps from −100 mV to +60 mV in 20 mV increments. K^+ conductance in the *Xenopus* oocytes injected with mutant *kcnk5b* (W169L) mRNA significantly increased, suggesting that the goldfish mutant *kcnk5b* causes significant hyperpolarization of cells. This evidence supports the hypothesis that the *kcnk5b* mutation is causative of the longfin phenotype in goldfish strains.

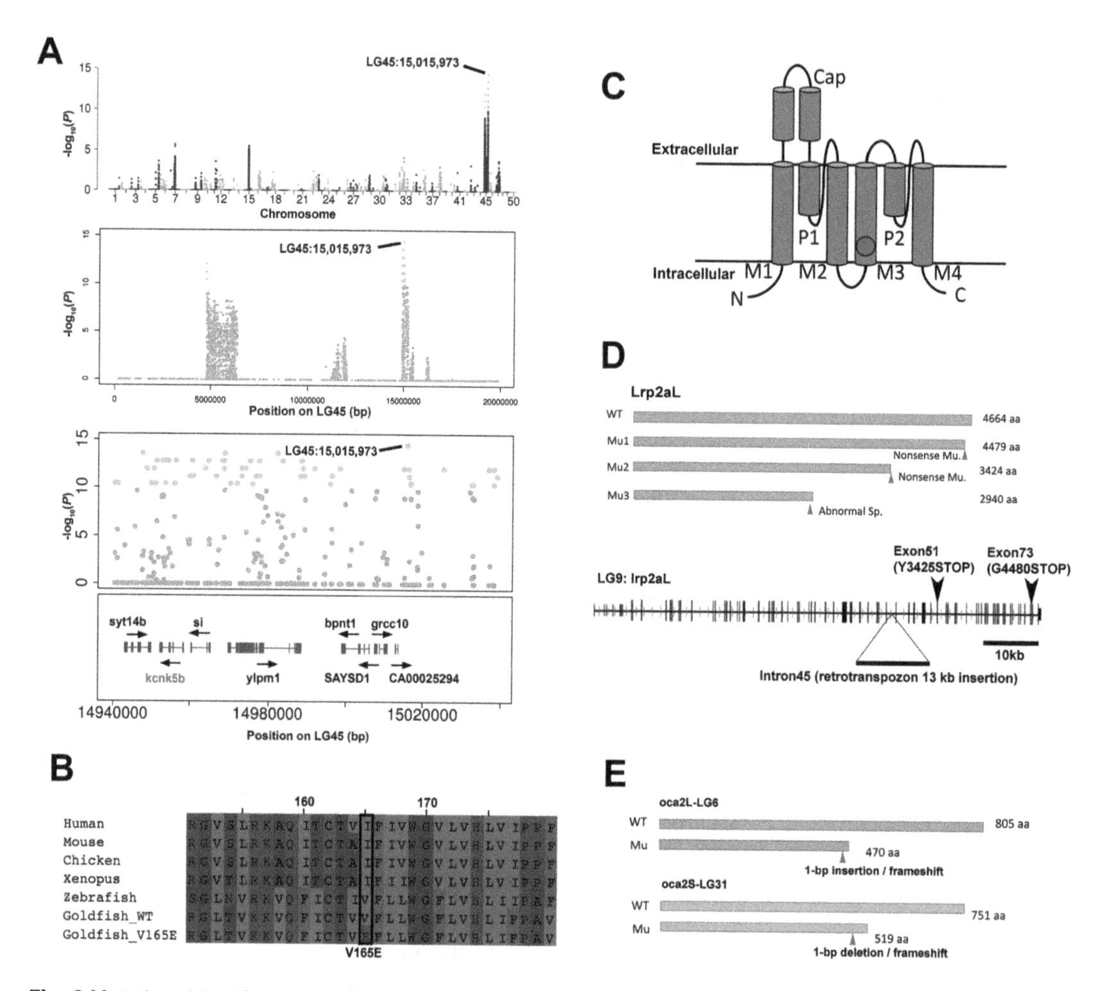

Fig. 6 Mutations identified in goldfish strains. (**a**) Manhattan plot showing the results of GWAS for the long-tail phenotype. Green dots represent variants with $P < 1.0 \times 10^{-10}$. The highest peak was found on LG45. *kcnk5bS* is located 59 kb upstream from the variant with the lowest *P*-value. (**b**) Protein sequence alignment of *kcnk5b* orthologs around V165. The hydrophobic amino acids are conserved at the V165 site of goldfish (black line box), among all species (**c**). The membrane topology of Kcnk5b, a member of the two-pore-domain potassium channel containing four transmembrane helices (M1, M2, M3, and M4), two channel pore helices (P1 and P2), and an extracellular cap (Cap). (**d**) Residue V165 is located at the M3 helix (red circle). Three types of mutations were identified in goldfish strains with the telescope-eye phenotype in lrp2a. The 13-kb retrotransposon insertion was identified in intron 45 of *lrp2a* L in strains with the telescope-eye phenotype. (**e**) Frameshift mutations in *oca2* in both L- and S-subgenomes in the albino strains. (Reprinted with permission from Kon et al. [6])

3.4.3 GWAS Analysis 2: The Twin-Tail Phenotype

Wild-type goldfish as well as most teleost fish have a single caudal fin. However, many goldfish strains possess duplicated caudal fins and have a laterally bifurcated caudal axial skeletal system, which is known as the twin-tail phenotype (Fig. 1m–r). This phenotype is one of the most remarkable features observed in goldfish strains, but not in other ornamental fish species such as zebrafish and

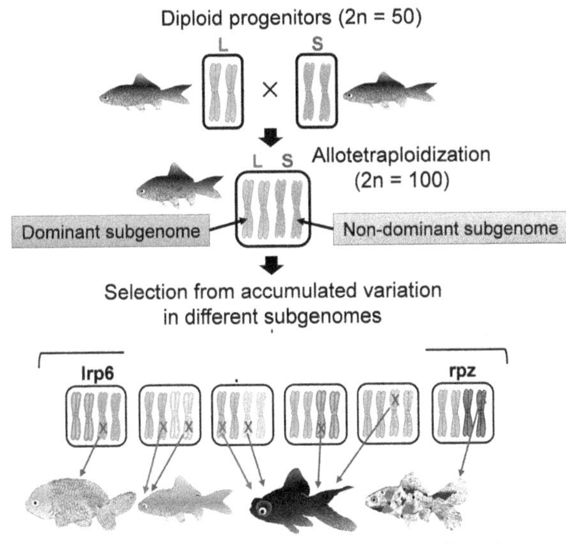

Fig. 7 Why do goldfish strains have such variety of phenotypes? Ancient WGD in the teleost and recent WGD in the ancestor of goldfish are depicted. The goldfish chromosomes are partitioned into L- and S-subgenomes, which is a basis of asymmetric subgenome evolution. More gene loss, gene disruption, and sequence divergence occur in the S-subgenome. In all six loci associated with goldfish phenotypes (twin tails, long tail, telescope eye, dorsal fin loss, albinism, and heart-shaped tail), the function of at least one paralog on the S-subgenome is affected. The diversified mutations accumulated in ohnologs in the asymmetrically evolved subgenomes may explain why the nonlethal, diverse morphological and body-color phenotypes could be selected by goldfish breeders. (Reprinted with permission from Kon et al. [6])

medaka. Mutations in the zebrafish *chordin* gene, which regulates axial skeleton patterning, were reported to produce a phenotype similar to that in the twin-tail goldfish [33–35].

1. Previously, using a candidate gene approach, a common nonsense mutation in the goldfish *chordinA* (*chdS*) gene was identified in several twin-tail goldfish strains [36].
2. GWAS of the twin-tail phenotype using a 27-strain dataset also identified the strongest association between the genomic variants of LG40 in the vicinity of the *chdS* gene [6]. As is the case with most goldfish genes, there are two copies of chordin: *chdS* and *chdL*. No significant association was found between *chdL*, on LG15 and the twin-tail phenotype, therefore *chdL* does not appear to contribute to the twin-tail phenotype, at least in the strains analyzed in that study [6].

3.4.4 GWAS Analysis 3: The Telescope-Eye Phenotype

Goldfish with a "telescope-eye" possess enlarged protuberant eyeballs (Fig. 1b, c). Such goldfish show an increased diameter of the eyeballs, but the diameter of the lens is relatively unaffected [37]. Therefore, the telescope-eyes are myopic [38]. The surface area of the retina in telescope-eyes is increased, making the retina thinner than wild-type [37]. As the enlargement of telescope-eyes is probably associated with increased intraocular pressure [39], similar to that observed in zebrafish *lrp2a* mutants (*bugeye*) [40], goldfish with telescope-eyes could be used as a model for studying the mechanism of intraocular pressure regulation in human glaucoma.

1. To identify the genomic locus associated with the telescope-eye phenotype in goldfish, GWAS was conducted using whole-genome sequence data [6].
2. A mild association between genomic variants on chromosome 9 and the telescope-eye phenotype was identified. A strong candidate gene for the telescope-eye phenotype, *lrp2a*, was identified close to this locus. *lrp2* mutations have been shown to cause an enlarged eye phenotype in adult zebrafish, mice, and humans [40, 41] [42]. Based on an analysis of *Lrp2*-null mice, *Lrp2* was shown to act as a sonic hedgehog (shh) clearance receptor and to regulate normal formation of the retinal margin [41].
3. Detailed sequence analysis of *lrp2a* variants in telescope-eye goldfish using the IGV (v2.4.0) program identified two nonsense mutations in *lrp2* genes (Fig. 6d) and an insertion in intron 45 in several goldfish individuals with telescope-eyes (Fig. 6d).
4. We designed the primers to amplify the insertion, cloned 13-kbp PCR fragment from genomic DNA of telescope-eye goldfish using KOD polymerase (TOYOBO), and sequenced the PCR fragment using the Big Dye terminator sequencing kit (Thermo Fisher Scientific) and an ABI3700 DNA sequencer (Thermo Fisher Scientific). We found a 13-kbp retrotransposon was inserted in the intron, which caused the loss of normal *lrp2a* expression. Overall, three independent loss-of-function mutations in the *lrp2a* gene were identified in strains with the telescope-eye phenotype.
5. Interestingly, the *lrp2a* gene was found only in the L-subgenome, and *lrp2a* ortholog on the S-subgenome has been lost during the evolution of the goldfish genome, even in the wild-type common goldfish. Based on these results, loss-of-function mutations in *lrp2a* on the L-subgenome are most likely the cause of the telescope-eye phenotype in goldfish.

3.4.5 GWAS Analysis 4: The Dorsal Fin Loss Phenotype

Strains of the Ranchu group show the phenotype of dorsal fin loss (Fig. 1w, x), which is most likely recessive [29].

1. To identify the loci associated with this phenotype, a GWAS was conducted using eight strains showing the dorsal-fin-loss phenotype being treated as cases and the other 20 strains being treated as controls.
2. This analysis identified the highest association between a genomic variant on LG29 and the dorsal-fin-loss phenotype. A 26-kb homozygous haplotype containing three genes in all eight goldfish strains with the dorsal-fin-loss phenotype was identified at this chromosomal location.
3. One of the three genes is the Wnt coreceptor-encoding *lrp6*, which is thought to be a candidate gene, since Wnt signaling is essential for fin formation in teleost fish [43–45]. In *Xenopus laevis*, it was reported that partial knockdown of *lrp6* affects the formation of the fin fold at the larval stage [46].
4. No obvious mutation in the *lrp6*-coding region of the dorsal-fin-loss goldfish was found. However, the expression level of *lrp6* was significantly decreased, suggesting that a mutation of the regulatory element causes reduced production of the *lrp6* mRNA at the embryonic stage.
5. Since no goldfish genome editing method has yet been established, to observe the effect of loss of function of *lrp6*, CRISPR/Cas9-mediated gene disruption of *lrp6* in zebrafish was conducted. gRNA was synthesized using a MEGAshortscript T7 Transcription Kit (Thermo Fisher Scientific), with PCR products containing the target-specific sequence.
6. Cas9 3NSL enzyme (Integrated Data Technologies) and gRNA were injected into zebrafish embryos at the single-cell stage using a FemtoJet microinjector (Eppendorf) and Femtotip II microcapillaries (Eppendorf).
7. Images of 4 dpf zebrafish larvae were taken using the M165FC stereomicroscopy system (Leica). The ratio of the unaffected dorsal fin fold length to body length was measured using ImageJ.
8. Injection with *lrp6* gRNA and Cas9 caused a partial defect in the formation of the dorsal fin fold in 4 dpf zebrafish larvae, supporting the hypothesis that *lrp6* is essential for formation of the dorsal part of the fin in *Cyprinidae*. Thus, zebrafish is useful model for confirming genotype/phenotype correlations in goldfish mutants.

3.4.6 Strain-Specific Variant (SSV) Analysis: Albinism

Goldfish strains with orange or red bodies lack melanophores in the skin and scales, but they usually retain black pigment in the retina. The pigment cells in the skin derive from neural crest cells, whereas retinal pigmented epithelia originate from optic lobe neuroepithelial cells [47]. The mutation that causes the xanthic body-color phenotype appears to affect only the pigment cells derived from neural crest cells. In many cases, xanthic goldfish strains possess black pupils that reflect the black retinal pigmented epithelia. In contrast, albino goldfish strains have pink pupils due to depigmentation of the retinal pigmented epithelia (Fig. 1d), since the melanin loss occurs in both the body surface and the retina. Albino goldfish also show the phenotype of a lack of dark body coloration in juveniles. A traditional genetic study reported that albinism in goldfish is double recessive for two independently assorting autosomal loci, namely, *m/m* and *s/s* [48].

1. We analyzed homozygous strain-specific variants (SSVs) as follows. SSVs were identified using SnpSift (v4.2) program. If >2 SSVs were identified within a 50-kb region, they were recognized as a cluster. A SSV-enriched region (SSVR) was defined as a cluster >100 kb that contained >5 SSVs. In SSV analysis, 172 genes with SSVs in Albino Celestial goldfish were identified. In vertebrates, 143 body-color-related genes have been reported. It was found that only *oca2* overlapped between the two groups. *Oca2* encodes an anion transporter regulating the pH of melanosomes. In several vertebrate species such as humans, mice, and zebrafish, loss-of-function mutations of *oca2* are known to lead to albinism [49–51].
2. The critical mutations identified in both *oca2* ohnologs (*oca2L* and *oca2S*) of Albino Celestial using the IGV program were frameshift indels, resulting in the production of truncated proteins comprising 470 and 519 amino acids, respectively (Fig. 6e). The same homozygous mutations of both *oca2* ohnologs were found in five albino strains similar to Albino Celestial. Thus, these frameshift mutations in both *oca2* ohnologs on both the L- and S-subgenomes are most likely to be the common causative mutations of albinism in goldfish strains.

4 Conclusions

What advantages does the genomics study of goldfish provide for comparative genomics research? Comparing goldfish to zebrafish is roughly the same as comparing mice to humans. Aligning human chromosomes to the mouse ones reveals long stretches of homology, but there are also significant regions of rearrangement, possibly explaining the very different body forms between the two species. In contrast, goldfish and zebrafish are members of Cyprinidae, and the chromosomal locations of orthologs and the amino acid sequences are highly conserved, making it easier to form comparisons. Not surprisingly, the basic body form of wild-type goldfish and wild-type zebrafish are more similar than the human to mouse comparison. Similar to how mouse has become the most popular mammalian model organism, zebrafish has had an enormous rise in popularity over the last few decades, because it is amenable to genomic manipulations and its generation time is relatively short. However, adult goldfish are 1000 times larger than adult zebrafish, making it easier to perform microsurgeries, physiological readings, or other organismal manipulations in goldfish. These features of goldfish may be very useful for challenging studies such as single-cell RNA-seq analysis of rare cell types or electrophysiological analyses in the brain. In these respects, the two species can compensate for each other's weaknesses. Comparison of mutant phenotypes in orthologous genes between goldfish and zebrafish can be very informative and provide a deeper insight into the gene function than single mutants might yield. Goldfish mutants will often have a milder phenotype due to the WGD compared to those of zebrafish, and hypomorphic mutations are often more informative than full gene knockouts. The comparative genomics and phenotypic comparisons between goldfish and zebrafish will provide new and exciting opportunities for understanding the relationships between genotype, phenotype, and the evolution of body forms in the vertebrate lineage.

References

1. Nasu M, Ohuchi Y (2016) Nishikigoi and Goldfish. Tokyo, Seibundo Shinkosha
2. Grunwald DJ, Streisinger G (1992) Induction of recessive lethal and specific locus mutations in the zebrafish with ethyl nitrosourea. Genet Res 59(2):103–116
3. Golling G et al (2002) Insertional mutagenesis in zebrafish rapidly identifies genes essential for early vertebrate development. Nat Genet 31(2):135–140
4. Walker C, Streisinger G (1983) Induction of mutations by gamma-rays in pregonial germ cells of zebrafish embryos. Genetics 103(1): 125–136
5. Henke K et al (2017) Genetic screen for post-embryonic development in the zebrafish (Danio rerio): dominant mutations affecting adult form. Genetics 207(2):609–623
6. Kon T et al (2020) The genetic basis of morphological diversity in domesticated goldfish. Curr Biol 30(12):2260–2274 e6
7. Braasch I (2020) Genome evolution: domestication of the allopolyploid goldfish. Curr Biol 30(14):R812–R815
8. Balon EK (2004) About the oldest domesticates among fishes. J Fish Biol 65(s1):1–27
9. Tsai HY et al (2013) Embryonic development of goldfish (Carassius auratus): a model for the study of evolutionary change in developmental mechanisms by artificial selection. Dev Dyn 242(11):1262–1283
10. Portavella M, Torres B, Salas C (2004) Avoidance response in goldfish: emotional and temporal involvement of medial and lateral telencephalic pallium. J Neurosci 24(9): 2335–2342
11. Van de Peer Y, Maere S, Meyer A (2009) The evolutionary significance of ancient genome duplications. Nat Rev Genet 10(10):725–732
12. Brawand D et al (2014) The genomic substrate for adaptive radiation in African cichlid fish. Nature 513(7518):375–381
13. Paape T et al (2018) Patterns of polymorphism and selection in the subgenomes of the allopolyploid Arabidopsis kamchatica. Nat Commun 9(1):3909
14. Luo J et al (2020) From asymmetrical to balanced genomic diversification during rediploidization: subgenomic evolution in allotetraploid fish. Sci Adv 6(22):eaaz7677
15. Chen Z et al (2019) De novo assembly of the goldfish (Carassius auratus) genome and the

evolution of genes after whole-genome duplication. Sci Adv 5(6):eaav0547
16. Chen D et al (2020) The evolutionary origin and domestication history of goldfish (Carassius auratus). Proc Natl Acad Sci U S A 117(47):29775–29785
17. Li JT et al (2021) Parallel subgenome structure and divergent expression evolution of allotetraploid common carp and goldfish. Nat Genet 53(10):1493–1503
18. Ohno S, Wolf U, Atkin NB (1968) Evolution from fish to mammals by gene duplication. Hereditas 59(1):169–187
19. Bejerano G et al (2004) Ultraconserved elements in the human genome. Science 304(5675):1321–1325
20. Margulies EH et al (2003) Identification and characterization of multi-species conserved sequences. Genome Res 13(12):2507–2518
21. Woolfe A et al (2005) Highly conserved non-coding sequences are associated with vertebrate development. PLoS Biol 3(1):e7
22. Pennacchio LA et al (2006) In vivo enhancer analysis of human conserved non-coding sequences. Nature 444(7118):499–502
23. Wang Y et al (2015) The draft genome of the grass carp (Ctenopharyngodon idellus) provides insights into its evolution and vegetarian adaptation. Nat Genet 47(6):625–631
24. Xu P et al (2014) Genome sequence and genetic diversity of the common carp, Cyprinus carpio. Nat Genet 46(11):1212–1219
25. Prabhakar S et al (2006) Close sequence comparisons are sufficient to identify human cis-regulatory elements. Genome Res 16(7): 855–863
26. Fornes O et al (2020) JASPAR 2020: update of the open-access database of transcription factor binding profiles. Nucleic Acids Res 48(D1): D87–D92
27. Pang ZP et al (2011) Induction of human neuronal cells by defined transcription factors. Nature 476(7359):220–223
28. Wayne RK, Ostrander EA (2007) Lessons learned from the dog genome. Trends Genet 23(11):557–567
29. Matsui Y (1935) Kagaku to Shumi Kara Mita Kingyo no Kenkyuu. Tokyo, Seizando Syoten
30. Perathoner S et al (2014) Bioelectric signaling regulates size in zebrafish fins. PLoS Genet 10(1):e1004080
31. Enyedi P, Czirjak G (2010) Molecular background of leak K+ currents: two-pore domain potassium channels. Physiol Rev 90(2): 559–605
32. Lesage F, Barhanin J (2011) Molecular physiology of pH-sensitive background K (2P) channels. Physiology (Bethesda) 26(6): 424–437
33. Fisher S, Halpern ME (1999) Patterning the zebrafish axial skeleton requires early chordin function. Nat Genet 23(4):442–446
34. Hammerschmidt M et al (1996) dino and mercedes, two genes regulating dorsal development in the zebrafish embryo. Development 123:95–102
35. Schulte-Merker S et al (1997) The zebrafish organizer requires chordino. Nature 387(6636):862–863
36. Abe G et al (2014) The origin of the bifurcated axial skeletal system in the twin-tail goldfish. Nat Commun 5:3360
37. Raymond PA, Hitchcock PF, Palopoli MF (1988) Neuronal cell proliferation and ocular enlargement in Black Moor goldfish. J Comp Neurol 276(2):231–238
38. Easter SS Jr, Hitchcock PF (1986) The myopic eye of the Black Moor goldfish. Vis Res 26(11): 1831–1833
39. Raymond P et al (1984) The telescopic eyes of Black Moor goldfish: elevated intraocular pressure and altered aqueous outflow pathways. Invest Ophthalmol Vis Sci 25:282
40. Veth KN et al (2011) Mutations in zebrafish lrp2 result in adult-onset ocular pathogenesis that models myopia and other risk factors for glaucoma. PLoS Genet 7(2):e1001310
41. Christ A et al (2015) LRP2 acts as SHH clearance receptor to protect the retinal margin from Mitogenic Stimuli. Dev Cell 35(1):36–48
42. Pober BR, Longoni M, Noonan KM (2009) A review of Donnai-Barrow and facio-oculo-acoustico-renal (DB/FOAR) syndrome: clinical features and differential diagnosis. Birth Defects Res A Clin Mol Teratol 85(1):76–81
43. Kawakami Y et al (2006) Wnt/beta-catenin signaling regulates vertebrate limb regeneration. Genes Dev 20(23):3232–3237
44. Nagayoshi S et al (2008) Insertional mutagenesis by the Tol2 transposon-mediated enhancer trap approach generated mutations in two developmental genes: tcf7 and synembryn-like. Development 135(1):159–169
45. Tatsumi Y et al (2014) TALEN-mediated mutagenesis in zebrafish reveals a role for r-spondin 2 in fin ray and vertebral development. FEBS Lett 588(24):4543–4550
46. Hassler C et al (2007) Kremen is required for neural crest induction in Xenopus and

promotes LRP6-mediated Wnt signaling. Development 134(23):4255–4263

47. Bharti K et al (2006) The other pigment cell: specification and development of the pigmented epithelium of the vertebrate eye. Pigment Cell Res 19(5):380–394
48. Yamamoto T-O (1973) Inheritance of albinism in the goldfish, Carassius auratus. Jpn J Genet 48(1):53–64
49. Lee ST et al (1994) Diverse mutations of the P gene among African-Americans with type II (tyrosinase-positive) oculocutaneous albinism (OCA2). Hum Mol Genet 3(11):2047–2051
50. Brilliant MH (2001) The mouse p (pink-eyed dilution) and human P genes, oculocutaneous albinism type 2 (OCA2), and melanosomal pH. Pigment Cell Res 14(2):86–93
51. Beirl AJ et al (2014) oca2 Regulation of chromatophore differentiation and number is cell type specific in zebrafish. Pigment Cell Melanoma Res 27(2):178–189
52. Omori Y, Kon T (2019) Goldfish: an old and new model system to study vertebrate development, evolution and human disease. J Biochem 165(3):209–218
53. Green J et al (2009) A gain of function mutation causing skeletal overgrowth in the *rapunzel* mutant. Dev Biol 334(1):224–234

INDEX

A

B

C

D

E

James F. Amatruda et al. (eds.), *Zebrafish: Methods and Protocols*, Methods in Molecular Biology, vol. 2707, https://doi.org/10.1007/978-1-0716-3401-1,

O

P

R

S

T

U

V

W

Z

Printed in the USA
CPSIA information can be obtained
at www.ICGtesting.com
LVHW080719111023
760793LV00006B/10

9 781071 634004